MW0071OO23

ฅตผาใผทดน

ServSafe
Coursebook

National Restaurant Association
EDUCATIONAL FOUNDATION
250 South Wacker Drive, Suite 1400, Chicago, IL 60606

National Restaurant Association
EDUCATIONAL FOUNDATION

DISCLAIMER

The information presented in this book has been compiled from sources and documents believed to be reliable and represents the best professional judgment of the National Restaurant Association Educational Foundation. However, the accuracy of the information presented is not guaranteed, nor is any responsibility assumed or implied, by the National Restaurant Association Educational Foundation for any damage or loss resulting from inaccuracies or omissions.

Laws may vary greatly by city, county, or state. This book is not intended to provide legal advice or establish standards of reasonable behavior. Operators who develop food safety related policies and procedures as part of their commitment to employee and customer safety are urged to use the advice and guidance of legal counsel.

Table of Contents

UNIT I THE SANITATION CHALLENGE

UNIT II THE FLOW OF FOOD THROUGH THE OPERATION

UNIT III CLEAN AND SANITARY FACILITIES AND EQUIPMENT

UNIT IV SANITATION MANAGEMENT

APPENDIXES

GLOSSARY

INDEX

International Food Safety Council

A Strategic Initiative of the
National Restaurant Association
EDUCATIONAL FOUNDATION

In 1993, the National Restaurant Association Educational Foundation recognized the need for food safety awareness and created the International Food Safety Council (Council). The Council's mission is to heighten the awareness of the importance of food safety education throughout the restaurant and foodservice industry.

The Council envisions a future where foodborne illness no longer exists. Through its educational programs, publications and awareness campaigns the Council reaches foodservice instructors, restaurant operators, suppliers, manufacturers, distributors, as well as organizations such as healthcare facilities, colleges and universities, supermarkets, associations and other stakeholders in the industry.

Founding Sponsors

American Egg Board
847.296.7043
www.aeb.org

Campbell Soup Company
800.879.7687
www.campbellsoup.com

ECOLAB, Inc.
800.352.5326
www.ecolab.com

FoodHandler Inc.
800.338.4433
www.foodhandler.com

Heinz North America
800.547.8924
www.heinz.com

KatchAll/San Jamar
800.533.6900
www.katchall.com
www.sanjamar.com

National Cattlemen's Beef Association
303.694.0305
www.beef.org
www.beeffoodservice.org

SYSCO Corporation
281.584.1390
www.sysco.com

Tyson Foods, Inc.
800.424.4253
www.tyson.com

Campaign Sponsors

Activ USA, Inc.
877.228.4879
www.activusa.net

Alaska Seafood Marketing Institute
800.806.2497
www.alaskaseafood.org

Atkins Temptec
800.284.2842
www.atkinstemptec.com

Bunzl Distribution
888.997.4515
www.bunzldistribution.com

Cargill Foods
800.CARGILL
www.cargillfoods.com

Colgate-Palmolive Company
888.276.0783
www.colpalipd.com

Cooper Instrument Corporation
800.835.5011
www.cooperinstrument.com

Daydots International
800.321.3687
www.daydots.com

Farquharson Enterprises, Ltd.
800.773.1455

International Dairy-Deli-Bakery Association™
608.238.7908
www.iddba.com

International Foodservice Manufacturers Association (IFMA)
312.540.4400
www.foodserviceworld.com/ifma

Johnson Wax Professional
800.558.2332
www.jwp.com

Jones Dairy Farm
800.635.6637
www.jonessausage.com

Lipton Foodservice
800.884.4841
www.liptonfs.com

Monsanto Company
314.694.1000
www.monsanto.com

Orkin Exterminating Company, Inc.
800.ORKIN.NOW
www.orkin.com

Procter & Gamble Company
800.817.6710
www.pg.com

Produce Marketing Association
302.738.7100
www.pma.com

Reckitt Benckiser, Inc.
800.677.9218
www.reckittprofessional.com

U.S. Foodservice, Inc.
410.312.7100
www.usfoodservice.com

For more information on the National Restaurant Association Educational Foundation's International Food Safety Council, visit our Web site at www.foodsafetycouncil.org or contact us at:
International Food Safety Council
250 South Wacker Drive, Suite 1400
Chicago, IL 60606
800.456.0111
312.715.1010 In Chicagoland

Acknowledgements

The development of the ServSafe Coursebook would not have been possible without the expertise of our many advisors and manuscript reviewers. The National Restaurant Association Educational Foundation is pleased to thank the following people for the time and effort they dedicated to this project.

Marie-Luise Baehr, Sodexho Marriott Services

Cheryl Barsness, Alaska Seafood Marketing Institute

Jeff Carletti, KatchAll Industries

Elaine Cash, Daydots International

Larry Clark, Travel Centers of America

Gary DuBois, Taco Bell Corporation

Deborah Fitzgerald, Cooper Instrument Corporation

Kristen Forrestal, Darden Restaurants, Inc.

Peter Good, Peter Good Seminars, Inc.

David Goronkin, Buffets, Inc.

Joe Grosdidier, Ecolab, Inc.

Steven F. Grover, R.E.H.S., National Restaurant Association

Margaret Hardin, Ph.D, National Pork Producers Council

Harold Harlan, Ph.D., National Pest Control Association

Jean Hayden, Ohio Department of Health

Alice M. Heinze, RD, MBA, American Egg Board

Michael P. Hiza, RD, Manchester Community Technical College

Jane Lindeman, National Cattlemen's Beef Association

Mark McFarlane, Briazz

Kathy Means, Produce Marketing Association

Bruce Moilan, Chester-Jensen Co., Inc.

Nevin B. Montgomery, National Frozen Food Association, Inc.

Mauro Mordini, Lipton Foodservice

Robert Munnis, CSFP, Atkins Technical, Inc.

Alain Porte, MBA, SmithKline Beecham Pharmaceuticals

David J. Poulter, Buffets, Inc.

Mary Sandford, Burger King Corporation

Jill A. Snowdon, American Egg Board

J.R. (Jim) Starnes, FMP, Tyson Foods, Inc.

Lisa Wright, Foodmaker, Inc./Jack in the Box Restaurants

A Message From

THE NATIONAL RESTAURANT ASSOCIATION EDUCATIONAL FOUNDATION

Food safety is non-negotiable. Serving safe food is not an option. It's our obligation as restaurant and food service professionals. Proper training is one of the best ways to create a culture of food safety within our establishments.

By opening this book you have made a significant commitment to promoting food safety. We applaud you for that commitment.

The ServSafe® program has become the industry standard in food-safety training and is accepted in almost all United States jurisdictions that require employee certification. The ServSafe program provides accurate, up-to-date information for all levels of employees on all aspects of handling food, from receiving and storing to preparing and serving. You will learn science-based information on how to run a safe establishment—information all of your employees need to have in order to be a part of the food-safety team.

Your food-safety education does not end once you are certified in the ServSafe program. You have the responsibility to take your knowledge back to the unit and make your coworkers part of the food-safety culture. You will qualify to participate in the International Food Safety Council, which will help you share your knowledge. The Council, a coalition of restaurant and foodservice professionals created by the National Restaurant Association Educational Foundation in 1993, is a valuable resource that promotes the importance of food-safety training.

Whether you are new to the restaurant industry or continuing your career, your participation in ServSafe training will make you more qualified to serve safe food and to spread that knowledge throughout the industry.

Thank you for making the commitment to food-safety training and becoming an active part of the food-safety culture within the rapidly growing restaurant, foodservice, and hospitality industry.

Features of the ServSafe Coursebook

We have designed the ServSafe Coursebook to enhance your ability to learn and retain comprehensive food-safety knowledge. Here are the key features you can expect to find in each chapter of this book.

TEST YOUR FOOD-SAFETY KNOWLEDGE:

Each chapter begins with five True or False questions, which are designed to test your prior knowledge of some of the concepts presented in the chapter. To find the answer to each question, and for further explanation, go to the section and page number indicated.

TABLE OF CONTENTS:

The chapter content is organized under the major headings identified in this section.

LEARNING OBJECTIVES:

The learning objectives identify what you should be able to do after completing the chapter. These objectives are linked to the tasks required to keep your establishment safe.

KEY TERMS:

Terms that are important for a thorough understanding of the chapter content are identified in this section and appear in the order in which they occur in the chapter text. Throughout the chapter, these terms are highlighted in green. Each key term has also been defined in the Glossary.

EXHIBITS:

Throughout each chapter exhibits have been placed which visually reinforce or review key concepts presented in the text. The Exhibits, which are referenced in the chapter text by chapter number and a letter, include charts, photographs, illustrations, and tables.

Key Point

Throughout each chapter, icons will appear in the margins of the page. These icons emphasize concepts presented in the text that are important to your understanding of food safety.

ICONS:

Throughout each chapter, icons will appear in the margins of the page. These icons emphasize concepts presented in the text that are important to your understanding of food safety.

Key Point icons are the most common type of icon to appear in the chapters. However, concept icons related to HACCP, personal hygiene, health, cross-contamination, and time-temperature abuse appear throughout. Look for these icons as you read through the chapter.

A CASE IN POINT:

These real-world scenarios give you the opportunity to apply concepts that you have learned in the chapter.

TRAINING TIPS FOR THE CLASSROOM:

These tips appear at the end of the chapter and are designed to give you training tools that you can use to teach others the food-safety concepts presented in the chapter. Each training tip begins with an objective, which tells you what your learner should be able to do after completing the activity. These tips are best suited for teaching food-safety in a classroom environment.

TRAINING TIPS ON THE JOB:

These tips appear at the end of the chapter and are designed to help you implement food-safety concepts back at your establishment with your personnel. Specifically, these are designed to aid you in sharing the food-safety knowledge you have acquired. Each training tip begins with a statement that describes the purpose of the activity.

DISCUSSION QUESTIONS:

These open-ended questions are designed to make you think about some of the more important food-safety concepts presented in the chapter.

MULTIPLE-CHOICE STUDY QUESTIONS:

These questions are designed to test your knowledge of the food-safety concepts presented in the chapter. These questions are similar to the ones that you will find on the certification exam. If you have difficulty answering these questions, review the content further.

ADDITIONAL RESOURCES:

In this section you will find resources such as books, articles, and Web sites which will enable you to further explore the food-safety concepts presented in each chapter.

ANSWERS:

The answers for the following are found in the Instructor's Guide: Test Your Food-Safety Knowledge, A Case in Point, Discussion Questions, and Multiple-Choice Study Questions.

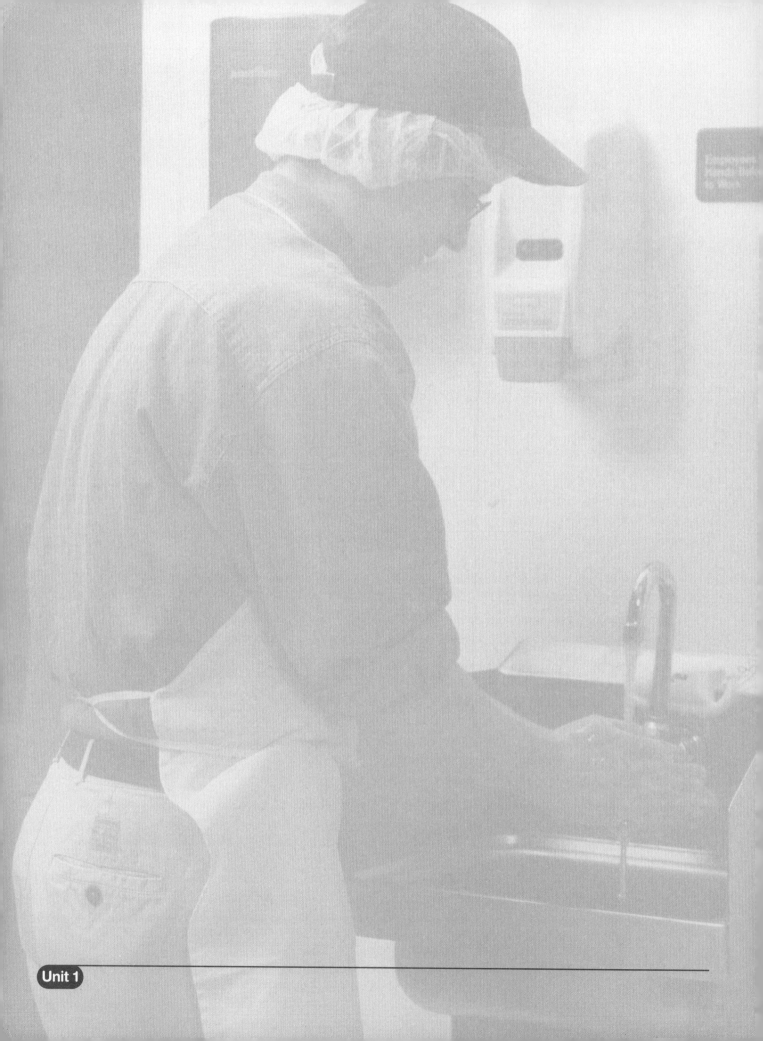

UNIT 1

THE SANITATION CHALLENGE

Food safety is an integral element of our organizational culture, which incorporates mutual respect and trust among guests, employees and vendors. Our guests expect not only a high quality dining experience, but a safe dining experience as well. The ServSafe Food Safety Training Program is the foundation we use to build awareness at all levels in the organization. Every manager in our system has been ServSafe certified. Commitment to food safety has the same organizational relevance as great tasting food, excellent service, and a clean environment.

David Goronkin
Executive Vice President, Operations
Buffets, Inc.

Chapter 1
Providing Safe Food

Knowledge

TEST YOUR FOOD-SAFETY KNOWLEDGE

1. **True or False:** Improperly washed hands can cause a foodborne illness. *(See Poor Personal Hygiene, page 1-9.)*

2. **True or False:** Young children may be more likely than adults to become ill from contaminated food. *(See Populations at High Risk for Foodborne Illness, page 1-5.)*

3. **True or False:** Food has been time-temperature abused any time it has been allowed to remain at temperatures favorable to the growth of microorganisms. *(See Time-Temperature Abuse, page 1-8.)*

4. **True or False:** A foodhandler's dirty clothing can cross-contaminate food. *(See Practicing Good Personal Hygiene, page 1-9.)*

5. **True or False:** Potentially hazardous foods are generally dry, low in protein, and highly acidic. *(See Foods Most Likely to Become Unsafe, page 1-6.)*

Key Terms

Foodborne illness
Outbreak of foodborne illness
Warranty of sale
Reasonable care defense
Hazard Analysis Critical Control Point (HACCP)
Flow of food
Model Food Code
Immune system
Potentially hazardous foods

Heat-treated
Ready-to-eat foods
Contamination
Biological, chemical, and physical hazards
Cross-contamination
Food-contact surface
Personal hygiene
Clean
Sanitary

Learning Objectives

After completing this chapter, you should be able to:

○ Explain the dangers of foodborne illness.

○ Identify high-risk populations for foodborne illness and explain why they are at risk.

○ Identify the characteristics of potentially hazardous foods.

○ Identify three types of contamination associated with food.

○ Describe how foodborne illness occurs.

○ Describe four key practices that can help ensure food safety.

When diners eat out, they expect safe food, clean surroundings, and well-groomed workers. Overall, the restaurant and foodservice industry does a good job of meeting these demands, but there is still room for improvement.

The risk of foodborne illness impacts the industry. Several factors account for this. They likely include the following.

○ The emergence of new foodborne pathogens (disease-causing organisms)

○ The importation of food from countries where food-safety practices may not be well developed

○ Changes in the composition of foods, which may leave fewer natural barriers to the growth of microorganisms

○ Increases in the purchase of take-out and home meal replacement (HMR) foods

○ Changing demographics, with an increased number of individuals at high risk for contracting foodborne illness

○ Employee turnover rates that make it difficult to manage an effective food-safety system

In the face of these challenges, all establishments must take the necessary steps to help ensure that the food they serve is safe. The first step is to develop a food-safety system that includes effective and ongoing employee training.

THE DANGERS OF FOODBORNE ILLNESS

The greatest dangers to food safety are foodborne illnesses. A **foodborne illness** is a disease that is carried or transmitted to people by food. The Centers for Disease Control and Prevention (CDC) defines an **outbreak of foodborne illness** as an incident in which two or more people experience the same illness after eating the same food. A foodborne illness is confirmed when laboratory analysis shows that a specific food is the source of the illness.

Each year, millions of people become ill from foodborne illness, although the majority of cases are not reported and do not occur at restaurants or other foodservice establishments. However, the cases that are reported and investigated help us understand some of the causes of illness, and what we, as restaurant and foodservice professionals, can do to control these causes in each of our establishments. The most commonly reported causes of foodborne illnesses are failure to properly cool foods, failure to cook and hold foods at the proper temperature, and poor personal hygiene.

Key Point

An outbreak of foodborne illness is an incident in which two or more people experience the same illness after eating the same food.

Fortunately, every restaurant and establishment, no matter how large or small, can take steps to ensure the safety of the food it prepares and serves to its customers.

The Costs of Foodborne Illness

National Restaurant Association figures show that a foodborne-illness outbreak can cost an establishment thousands of dollars. It can even cause an establishment to close.

If your establishment is implicated in an outbreak of foodborne illness, your costs may include increased insurance premiums and lawyer and court fees. You may have to pay for testing food supplies and employees, and may spend time and money retraining employees and cleaning and sanitizing the establishment. Food supplies that may or may not be contaminated will have to be discarded. Other risks could include lowered employee morale and absenteeism, embarrassment and bad publicity, loss of customers and sales, and loss of prestige and reputation *(see Exhibit 1a)*.

Today, customers are very willing to sue to obtain compensation for injuries they feel they have suffered as a result of the food they were served.

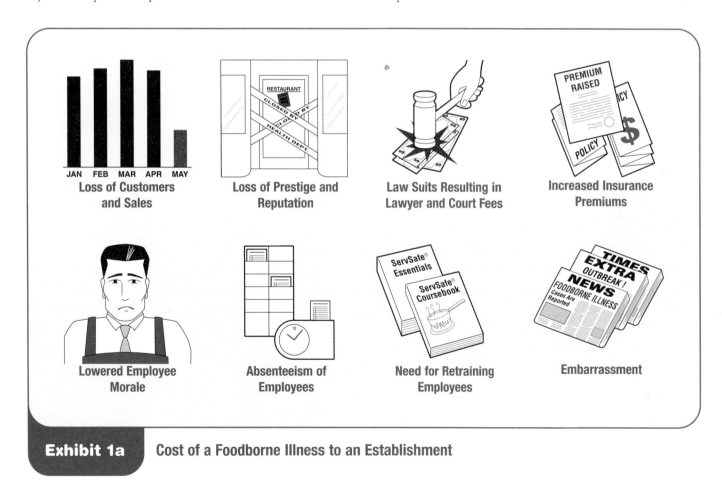

Exhibit 1a   Cost of a Foodborne Illness to an Establishment

Under the federal Uniform Commercial Code, a plaintiff bringing about a lawsuit must prove all of the following.

○ The food was unfit to be served.

○ The food caused the plaintiff harm.

○ In serving the food, the establishment operator violated the **warranty of sale,** that is, the rules stating how the food must be handled.

If the plaintiff wins the lawsuit, he or she can be awarded two types of damages. Compensatory damages are awarded for lost work, lost wages, and medical bills. Punitive damages are awarded in addition to normal compensation, to punish the defendant for wanton and willful neglect.

If you have a quality food-safety program in place, however, you can use a **reasonable care defense** against a food-related lawsuit. A reasonable care defense requires proof that your establishment did everything that could be reasonably expected to ensure that the food served was safe. Evidence of written standards, training practices, procedures such as a HACCP plan and its documentation, and positive inspection results are the keys to this defense.

The Benefits of a Food-Safety System

Serving safe food is vital to your establishment's success. A well-designed food-safety program can help protect your establishment's employees, customers, and reputation. Repeat business from customers and increased job satisfaction among employees can lead to higher profits and better service. Your establishment may benefit directly from reduced or minimized insurance costs, and will most likely benefit by reducing health-code violations and becoming less open to lawsuits claiming injury and negligence.

An added benefit to a food-safety program is that by handling food safely, you also preserve its quality. Safe foodhandling will help maintain the appearance, flavor, texture, consistency, and nutritional value of food. Food that is stored, prepared, and served properly is more likely to provide the quality that your customers deserve and demand. Safe foodhandling can also lead to lower food costs due to less waste.

PREVENTING FOODBORNE ILLNESS

There are many challenges to preventing foodborne illness. These include high employee turnover rates, service to an increasing number of high-risk customers, and the service of potentially hazardous foods. Establishing a comprehensive food-safety management program, however, greatly reduces the likelihood of causing foodborne illness.

Training Employees in Food Safety

One of the challenges that managers typically face is a high employee turnover rate. Preparing and serving safe food in public establishments is a serious responsibility. All restaurant and foodservice employees need to be trained in the procedures that can protect the public and themselves from foodborne illnesses.

Food-Safety Programs

An establishment should have an effective, proactive program that is based on preventing food-safety hazards before they occur. A reactive program that corrects a problem after it has occurred is not an effective system. The **Hazard Analysis Critical Control Point (HACCP)** program, which is discussed throughout this text, is a proactive, comprehensive, science-based food-safety system that allows operators to continuously monitor their establishments and reduce the risk of foodborne illness.

The key to the HACCP system is the emphasis on how food flows through the operation. This **flow of food** is the path food takes from receiving and storage through preparation and cooking, holding, serving, cooling and reheating. An establishment's HACCP plan identifies the points in the operation where contamination or growth of microorganisms can occur. Control procedures can then be implemented based on the hazards identified at those points. HACCP will be covered in more detail in Chapter 9.

The National Restaurant Association and the Food and Drug Administration's (FDA) Model Food Code encourage an establishment to develop and use a HACCP-based food-safety system to prevent foodborne illness. The **Model Food Code** is a science-based reference for retail restaurants and establishments on how to prevent foodborne illness. Local, state, and federal regulators often use the Food Code as a model to help develop or update their own food-safety regulations, and to ensure consistency with national regulatory policy.

HACCP Principle

A HACCP system is designed to prevent food-safety hazards from occurring.

Populations at High Risk for Foodborne Illness

The demographics of our population show there is an increase in the percentage of people at high risk of contracting a foodborne illness *(see Exhibit 1b on the next page)*. These individuals include the following groups of people.

○ Infants and young children

○ Pregnant women

○ Elderly people

○ People taking certain medications, such as antibiotics and immunosuppressants

○ People with weakened immune systems (those who have recently had major surgery, are organ-transplant recipients, or who have pre-existing or chronic illnesses)

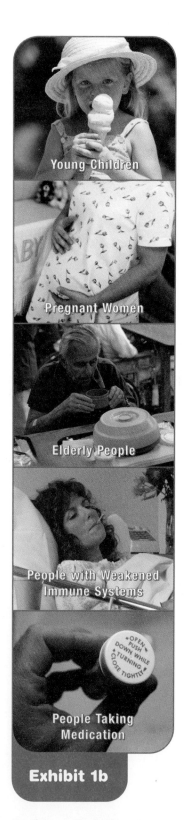

Young Children

Pregnant Women

Elderly People

People with Weakened
Immune Systems

People Taking
Medication

Exhibit 1b

**People at High Risk for
Foodborne Illness**

Young children are more at risk for contracting foodborne illnesses because they have not yet built up adequate **immune systems** (the body's defense system against illness) to deal with some diseases. This knowledge is especially important for quick-service restaurants since, according to a recent survey, 58 percent of families with children eat at such restaurants at least once a week.

Elderly people are more at risk because their immune systems and resistance may have weakened with age. In addition, as people age, their senses of smell and taste are diminished, so they may be less likely to detect "off" odors or tastes which indicate that food may be spoiled.

Foods Most Likely to Become Unsafe

Although any food can become contaminated, most foodborne illnesses are transmitted through foods in which microorganisms are able to grow rapidly. Such foods are classified as **potentially hazardous foods.** These foods typically have a history of being involved in foodborne-illness outbreaks, have a natural potential for contamination due to methods used to produce and process them, and are often moist, high in protein, and have a neutral or slightly acidic pH.

The FDA Model Food Code identifies potentially hazardous foods *(see Exhibit 1c)* as any food that consists in whole, or in part, of the following.

○ Milk or milk products

○ Shell eggs

○ Meats, poultry, and fish

○ Shellfish and edible crustacea (such as shrimp, lobster, crab)

○ Baked or boiled potatoes

○ Tofu or other soy-protein foods

○ Garlic-and-oil mixtures

○ Plant foods that have been **heat-treated** (cooked, partially cooked, or warmed)

○ Raw seeds and sprouts

○ Sliced melons

○ Synthetic ingredients (such as textured soy protein in hamburger supplement)

Care must be taken when handling **ready-to-eat foods,** which may also be considered unsafe because they are intended to be eaten without further washing or cooking. Proper cooking reduces the number of microorganisms on food to safe levels. Foods that have been properly cooked, as well as washed whole or cut fruits and vegetables, are considered ready-to-eat foods.

HACCP

Potentially hazardous foods are often moist, high in protein, and have a neutral or slightly acidic pH.

Hazard
Analizes
Creditcall
Control
Point

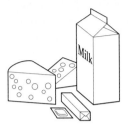

Milk and Milk Products

Sliced Melons

Garlic-and-Oil Mixtures

Minced Garlic In Oil

Poultry

Meat: Beef, Pork, Lamb

Shellfish and Crustacea

Fish

Sprouts and Raw Seeds

Baked or Boiled Potatoes

Shell Eggs

HAMBURGER **Mix**

Tofu

Soy-Protein Foods

Cooked Rice, Beans, or Other Heat-Treated Plant Foods

Exhibit 1c | **Potentially Hazardous Foods**

Disease-causing microorganisms are responsible for the majority of foodborne illness outbreaks.

A well-designed food-safety system will establish controls to prevent time-temperature abuse, cross-contamination, and poor personal hygiene.

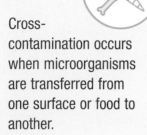

Cross-contamination occurs when microorganisms are transferred from one surface or food to another.

Potential Hazards to Food Safety

Unsafe food usually results from **contamination,** which is the presence of harmful substances not originally present in the food. Some food-safety hazards are introduced by humans or by the environment, and some occur naturally.

Food-safety hazards are divided into three categories: **biological hazards, chemical hazards,** and **physical hazards.**

- **Biological hazards** include certain bacteria, viruses, parasites, and fungi, as well as certain plants, mushrooms, and fish that carry harmful toxins.
- **Chemical hazards** include pesticides, food additives, preservatives, cleaning supplies, and toxic metals that leach from cookware and equipment.
- **Physical hazards** consist of foreign objects that accidentally get into the food, such as hair, dirt, metal staples, and broken glass.

By far, biological hazards pose the greatest threat to food safety. Disease-causing microorganisms are responsible for the majority of foodborne illness outbreaks.

HOW FOOD BECOMES UNSAFE

Foodborne illness is caused by several factors, which can be placed into one of three categories: time-temperature abuse, cross-contamination, and poor personal hygiene. Reported cases of foodborne illness usually involve more than one factor in each of these categories. A well-designed food-safety system will control these factors.

Time-Temperature Abuse: Food has been time-temperature abused any time it has been allowed to remain for too long at temperatures favorable to the growth of microorganisms. Common factors that have resulted in foodborne illness include the following.

- Failing to hold or store food at required temperatures
- Failing to cook or reheat foods to temperatures that kill microorganisms
- Failing to properly cool foods
- Preparing foods a day or more before they are served

Cross-Contamination: Cross-contamination occurs when microorganisms are transferred from one surface or food to another. Common factors that have resulted in foodborne illness include the following.

- Adding raw, contaminated ingredients to foods that receive no further cooking
- **Food-contact surfaces** (such as equipment or utensils) that are not cleaned and sanitized before touching cooked or ready-to-eat foods
- Allowing raw food to touch or drip fluids onto cooked or ready-to-eat food

○ Hands that touch contaminated (usually raw) food and then touch cooked or ready-to-eat food

○ Contaminated cleaning cloths that are not cleaned and sanitized before being used on other food-contact surfaces

Poor **Personal Hygiene:** Individuals with unacceptable personal hygiene can offend customers, contaminate food or food-contact surfaces, and cause illnesses. Common factors that have resulted in foodborne illness include the following.

○ Employees who fail to properly wash their hands after using the restroom or whenever necessary

○ Employees who cough or sneeze on food

○ Employees who touch or scratch sores, cuts, or boils and then touch food they are preparing or serving

KEY PRACTICES FOR ENSURING FOOD SAFETY

The key to food safety lies in controlling time and temperature throughout the flow of food, practicing good personal hygiene, preventing cross-contamination, and purchasing food supplies from approved suppliers. It is important to establish standard operating procedures that focus on these areas.

Controlling Time and Temperature

Microorganisms pose the largest threat to food safety. Like all living organisms, they cannot survive or reproduce outside certain temperature limits. As outlined in *Exhibit 1d,* time and temperature must be controlled throughout the flow of food. Each of these steps will be discussed in further detail in later chapters.

Practicing Good Personal Hygiene

Training employees in good personal hygiene is the responsibility of every manager. Features of a good personal hygiene program include the following.

○ **Proper handwashing.** Hands and fingernails should be washed and cleaned thoroughly before handling food, between each task, and before using food-preparation equipment.

○ **Strictly enforced rules regarding eating, drinking, and smoking.** These activities should be prohibited while preparing or serving food, or while in areas used for washing equipment and utensils.

○ **Preventing employees who are ill from working with food.** Cuts, burns, and sores must be properly cleaned and covered.

○ **General cleanliness.** Insist on daily bathing, clean hair, and clean clothing.

Receiving: Receive and store food quickly.

Storage: Store foods at their recommended temperatures.

Preparation: Minimize time spent in the temperature danger zone of 41°F (5°C) to 140°F (60°C).

Cooking: Cook food to its minimum safe internal temperature for the appropriate amount of time.

Holding: Hold hot foods at 140°F (60°C) or higher and cold foods at 41°F (5°C) or lower.

Cooling: Cool cooked food to 70°F (21°C) within two hours and from 70°F (21°C) to 41°F (5°C) or below in an additional four hours.

Reheating: Reheat foods to an internal temperature of 165°F (74°C) for fifteen seconds within two hours.

Exhibit 1d

Controlling Time and Temperature Throughout the Flow of Food

Preventing Cross-Contamination

Employees must be carefully trained to recognize and prevent cross-contamination of microorganisms between foods and food-contact surfaces. Some ways to prevent cross-contamination include the following.

○ Require employees to wash their hands frequently when working with raw foods. They should never touch raw foods and then touch ready-to-eat food without washing their hands.

○ Do not allow raw or contaminated food to touch or drip fluids onto cooked or ready-to-eat foods.

○ Clean and sanitize food-contact surfaces (such as equipment or utensils) that touch contaminated food before they come in contact with cooked or ready-to-eat foods.

○ Clean and sanitize cleaning cloths between each use.

Food-contact surfaces may be direct or indirect. A direct food-contact surface includes any surface of equipment or utensils that food normally touches, such as tableware, cutting boards, knives and other utensils used to prepare foods, and counters where food is prepared. An indirect food-contact surface is a surface that food may drain, drip, or splash onto during preparation, such as a backsplash of a counter. Food that splashes on an indirect food-contact surface may drain down onto a direct food-contact surface and contaminate it.

Food-contact surfaces, cleaning cloths, and sponges must be cleaned and sanitized to prevent cross-contamination. **Clean** simply means free of visible soil and refers only to outward appearance. **Sanitary,** on the other hand, means that the object is free from harmful levels of disease-causing organisms and other harmful contaminants. Hands must also be washed regularly to prevent cross-contamination.

Purchasing from Approved Suppliers

Use reputable and reliable suppliers to help avoid receiving contaminated foods. Suppliers' products and practices should meet your specifications as well as local, state, and federal regulations. Work closely with suppliers to set up effective receiving procedures, and make sure you address these in your purchase specification agreements. Reputable suppliers and distributors will address the following.

○ Deliver food at proper temperatures.

○ Use clean, and where appropriate, refrigerated trucks.

○ Train their employees in food-safety practices.

○ Use protective, leak-proof, durable packaging.

In addition, they should agree to adjust delivery schedules to meet your establishment's needs, so your deliveries do not arrive during busy work times; cooperate with your employees who inspect products when they are delivered; allow you to inspect their delivery vehicles and production facilities; and make available their inspection reports if asked.

THE FOOD-SAFETY RESPONSIBILITIES OF A MANAGER

The manager's basic food-safety responsibilities are to serve safe food to customers and to train employees in safe foodhandling practices. The manager's positive and supportive attitude toward food safety is critical. This attitude should be based on up-to-date knowledge of the regulations that affect the restaurant and foodservice industry.

Key Point

It is critically important for management to have a positive and supportive attitude toward food safety.

Meeting Food-Safety Regulations

To stay in operation, your establishment must comply with city, county, and state sanitation codes. The regulatory agency evaluating your establishment shares your commitment to serving safe food. Therefore, it is in your best interest to work with local authorities. During evaluations, the regulatory agency may identify deficiencies that require your attention, and may have the authority to assess fines and close an establishment that serves unsafe food.

The FDA recommends that local and state health departments hold the person in charge of a restaurant or establishment responsible for knowing and demonstrating the following information.

○ The diseases that are carried or transmitted by food and the symptoms of these diseases

○ Points in the flow of food where hazards can be prevented, eliminated, or reduced, and how procedures meet the requirements of the local code

○ The relationship between personal hygiene and the spread of disease, especially concerning cross-contamination, hand contact with ready-to-eat foods, and handwashing

○ How to keep injured or ill employees from contaminating food or food-contact surfaces

○ The need to control the length of time that potentially hazardous foods remain at temperatures where disease-causing microorganisms can grow

○ The hazards involved in the consumption of raw or undercooked meat, poultry, eggs, and fish

○ Safe cooking temperatures and times for potentially hazardous foods, such as meat, poultry, eggs, and fish

○ Safe temperatures and times for the refrigerated storage, hot holding, cooling, and reheating of potentially hazardous foods

○ Correct procedures for cleaning and sanitizing utensils and food-contact surfaces of equipment

○ The types of poisonous and toxic materials used in the operation, and how to safely store, dispense, use, and dispose of them

○ The need for equipment that is sufficient in number and capacity, and is properly designed, constructed, located, installed, operated, maintained, and cleaned

○ The sources of the water supply and the importance of keeping it clean and safe

○ How the operation complies with the principles of a HACCP-based food-safety system, as some local health departments require a HACCP-based system

○ The rights, responsibilities, and authorities the local code assigns to employees, managers, and the local health department

Marketing Food Safety

Make it clear to your employees and customers that your operation takes food safety very seriously.

Show your employees through actions, not just words, that top management is involved in and supports food-safety policies, and that food-safety training for managers and all employees is a high priority. Offer training courses, and update and evaluate them regularly. Discuss food-safety expectations. Document foodhandling procedures, use them in regular inspections, and update them as necessary. Show employees that safe foodhandling is appreciated—consider awarding certificates for training and giving out small rewards for good food-safety records. Set a good example by following all food-safety rules yourself.

Show your customers that your employees know and follow safety rules. Make sure that their appearance reflects your concern for food safety. Consider having employees wear food-safety pins or buttons, or use place mats and posters to get your message across. Be sure your employees can answer simple food-safety questions from customers.

When you are ServSafe certified, you can receive a free seal from the International Food Safety Council to post prominently in your establishment. For more information, visit the Council web site at www.foodsafetycouncil.org or call (800) 456-0111. Communicating your food-safety efforts will help assure your customers that you are committed to serving safe food.

RESPONDING TO AN OUTBREAK OF FOODBORNE ILLNESS

The primary responsibility of restaurant and foodservice employees is to ensure that customers are served quality food that has been prepared safely. By following standard operating procedures and implementing an effective food-safety program, all establishments should be able to meet this responsibility. *Appendix F* lists some guidelines to follow in the event of a complaint of foodborne illness.

SUMMARY

Foodborne illness can affect anyone. All restaurants and establishments must continually take steps to prevent it from occurring. The critical factors involved in food safety generally fall into three categories: food, employees, and facilities.

Food can become unsafe at any step in the flow of food. Control time and temperature and avoid contamination at each step.

You must work with reputable suppliers and implement strict receiving procedures to help ensure the delivery of safe food. Once the food arrives, it must be stored, prepared, cooked, held, served, cooled, and reheated using methods that maintain its safety.

People pose a major risk to safe food, especially foodhandlers who do not practice personal hygiene. You must carefully train, monitor, and reinforce food-safety principles in your establishment. Establishing a well-designed food-safety system can help protect your customers by preventing outbreaks of foodborne illness and can help the establishment avoid the potentially high costs associated with them. Regulatory agencies share your commitment to serving safe food.

A CASE IN POINT

Case Study

A pizza restaurant manager receives a call from a customer who purchased a take-out pizza the previous night. The customer claims that her children, ages three and five, are suffering from abdominal cramps, diarrhea, and fever. She suspects that the pizza was contaminated. The manager then asks whether anyone else in the house ate the pizza and whether they have the same symptoms. The customer says that both she and her husband ate the pizza, but they feel fine. The manager then asks what kind of pizza it was, and the customer tells him it was a mushroom and black olive pizza, with extra cheese. The manager then tells her that it couldn't have been the pizza, because it contained no meat products, and she and her husband are healthy. He suggests that her children have the stomach flu, wishes them well, and hangs up.

Did the manager handle this situation correctly? What did he do right? What did he do or assume that was wrong?

TRAINING TIPS

Training Tips for the Classroom

1. Food Safety on Trial: "But your honor, I thought *they were washing their hands.*"

Objective: *After completing this activity, class participants will be able to identify the legal liability faced by establishments.*

Directions: Present your class with a case study of an outbreak of foodborne illness. Use one of the Case In Point examples from this text or create a detailed fictitious case. Have the class assume that the outbreak has resulted in several law suits.

Break the class into the following teams: a plaintiff team (representing the persons who claim they became ill), a defense team (representing the establishment that served the food), and a jury. Allow about ten minutes for the plaintiff and defense to plan their presentations, then another ten minutes for each side to present its argument before the judge (the instructor). Questions to be argued include the following.

- Was the food unfit to serve?
- Did the food cause the plaintiff harm?
- Did the restaurant or establishment violate the "warranty of sale"?
- Did the restaurant or establishment exhibit "reasonable care"?

After both sides have been presented, give the jury five minutes to deliberate and determine the fate of the restaurant or establishment. No mistrials or appeals!

2. Knowledge and Application of Food Safety

Objective: *After completing this activity, class participants should be able to identify information that every manager should know and apply at his or her establishment, as recommended by the FDA.*

Directions: The FDA recommends that local and state regulatory agencies hold the person in charge of a restaurant or establishment responsible for knowing and applying the information listed below. Assign teams of two or three people. Depending upon the number of teams, assign each team one or more topics from the list. Ask the teams to prepare a discussion of current food-safety knowledge related to their assigned topics. After the teams have had adequate time to prepare, ask them to present their information to the class. After the presentation, solicit group discussion. You might ask class members how well the management teams in their own establishments follow the FDA recommendations.

The FDA recommends that the person in charge of a restaurant or establishment know and apply the following information.

○ The diseases that are carried or transmitted by food and the symptoms of these diseases

○ Points in the flow of food where hazards can be prevented, eliminated, or reduced, and how procedures meet the requirements of the local code

○ The relationship between personal hygiene and the spread of disease, especially concerning cross-contamination, hand contact with ready-to-eat foods, and handwashing

○ How to keep injured or ill employees from contaminating food or food-contact surfaces

○ The need to control the length of time that potentially hazardous foods remain at temperatures where disease-causing microorganisms can grow

○ The hazards involved in the consumption of raw or undercooked meat, poultry, eggs, and fish

○ Safe cooking temperatures and times for potentially hazardous foods, such as meat, poultry, eggs, and fish

○ Safe temperatures and times for the safe refrigerated storage, hot holding, cooling, and reheating of potentially hazardous foods

○ Correct procedures for cleaning and sanitizing utensils and food-contact surfaces of equipment

○ The types of poisonous and toxic materials used in the operation, and how to safely store, dispense, use, and dispose of them

○ The need for equipment that is sufficient in number and capacity, and is properly designed, constructed, located, installed, operated, maintained, and cleaned

○ The sources of the operation's water supply and the importance of keeping it clean and safe

○ The rights, responsibilities, and authorities the local code assigns to employees, managers, and the local health department

Training Tips on the Job

1. Developing a Food-Safety Complaint Form for Your Establishment

Purpose: *To involve your management team in the development of a standard complaint questionnaire form for your establishment. Note: involving the management team will help clarify your policies regarding how to handle a possible complaint of foodborne illness.*

Directions: Every establishment should have a standard complaint questionnaire form on hand to record customer reports of possible foodborne illness. These forms are essential in order to ensure that complaints are handled in a professional manner, information is fully documented, and that all appropriate questions are included.

Ask members of the management team to help write and review questions for the form. It is a good idea to look at similar forms from other establishments. Ask your local regulatory agency or state restaurant association for input. If possible, you may want to have a lawyer review the form. When the content of the form is approved, print up copies and make them accessible to the management team.

Inform your team that it is essential to use these forms whenever a customer has a complaint about foodborne illness. Let the team know how you want a complaint to be handled. They need to know what they should say to the customer, and what steps to take once the form has been filled out. It may be a good idea to role-play scenarios regarding a customer complaint. This will allow managers to practice, and can expose any weaknesses that may exist in your program.

2. Marketing Your Safe Food

Purpose: *To involve your management team in a brainstorming session to determine different ways to market the food safety of your establishment to both employees and customers.*

Directions: Conduct a brainstorming session with your management team to determine different ways by which you can market your food-safety systems and procedures to employees and customers. If you already market your systems and procedures, use the brainstorming session to improve your current marketing practices.

Bring up the following questions in the session:

Employees

○ How do we demonstrate to our employees that we are committed to food safety?

○ What food-safety systems are currently in place? Do foodhandlers understand these systems?

○ What training commitment have we made regarding food safety? How is this training made available to employees?

○ What documentation do we keep related to food safety? Do we monitor Critical Control Points? If so, do employees understand how to accurately monitor and document Critical Control Points?

Customers

○ How is our commitment to food safety visible to our customers?

○ What food-safety principles do our customers see being put into practice by our staff?

○ What written materials (signs, plaques, posters, decals, statements on packaging, etc.) are in place to inform our customers of our commitment to food safety?

Based on the input received from your management team, come up with a plan to market your food-safety systems and procedures to employees and customers, or to improve your current marketing practices. Implement the plan and evaluate it at management meetings. If necessary, modify your plan based on the evaluation.

DISCUSSION QUESTIONS

1. What is the difference between a foodborne illness and an outbreak of foodborne illness?

2. What are the potential costs and liabilities associated with foodborne-illness outbreaks?

3. Give two reasons why the elderly are at higher risk for contracting foodborne illnesses.

4. List the three major types of hazards to food safety.

5. A chef cuts up a salmon on a cutting board, then thoroughly rinses the cutting board and knife in warm water. She then uses the same cutting board and knife to slice washed, fresh parsley. Is this an acceptable foodhandling practice? Why or why not? On what key food-safety practice does this example focus?

MULTIPLE-CHOICE STUDY QUESTIONS

1. Why do people who live with chronic illnesses have a higher risk of contracting a foodborne illness?

   A. They are likely to eat food prepared in large quantities while they are in a hospital.

   B. Their internal disease-fighting systems are likely to be weaker than normal.

   C. Their allergic reactions to chemicals used in food production might be greater than normal.

   D. They are likely to have diminished appetites and do not want to cook for themselves.

2. Which type of food would be the most likely to cause a foodborne illness?

   A. Tomato juice
   B. Baked potatoes
   C. Stored whole wheat flour
   D. Dry powdered milk

3. Your restaurant is closed Sunday and Monday. Tuesday morning you open the restaurant and notice that the refrigerator is not running. When you check the internal thermometer, it reads 50°F (10°C). What should you do with the fresh ground beef in the refrigerator?

   A. Cook and serve it within two hours.
   B. Freeze it right away.
   C. Discard it.
   D. Re-chill it immediately to below 41°F (5°C).

4. Which of the following is not a common characteristic of potentially hazardous foods?

   A. They are moist.
   B. They are dry.
   C. They are neutral or slightly acidic.
   D. They are high in protein.

5. When a foodborne illness occurs, it is usually caused by a combination of three factors. These factors are

   A. time-temperature abuse, unacceptable personal hygiene, and physical hazards.

   B. time-temperature abuse, chemical hazards, and physical hazards.

   C. time-temperature abuse, cross-contamination, and chemical hazards.

   D. time-temperature abuse, cross-contamination, and unacceptable personal hygiene.

6. In order for a foodborne illness to be considered an "outbreak," how many people must experience the illness after eating the same food?

   A. 1   B. 2   C. 10   D. 20

7. An effective food-safety system, such as HACCP, will

   A. react to food-safety hazards quickly after they occur.

   B. prevent food-safety hazards before they occur.

   C. be based on standards set up by establishments like yours.

   D. be based on standards mandated by the government.

8. The largest threat to food safety comes from which of the following contaminants?

   A. Pesticides   C. Microorganisms

   B. Hair   D. Food additives

9. To prevent cross-contamination, foodhandlers should not

   A. touch raw meat and then touch cooked or ready-to-eat food.

   B. allow food to remain at temperatures above 41°F (5°C).

   C. take food temperatures when receiving food.

   D. hold food at temperatures below 140°F (60°C).

10. Foodhandlers must practice all of the following hygienic practices except

   A. proper handwashing.

   B. daily bathing.

   C. wearing clean clothing to work.

   D. getting periodic AIDS tests.

ADDITIONAL RESOURCES

Books and Periodicals

Jay, J. M. (1996). *Modern food microbiology.* New York: Chapman & Hall.

McCoy, J. J. (1990). *How safe is our food supply?* New York: F. Watts.

National Research Council. (1998). *Ensuring safe food: From production to consumption.* Washington DC: National Academy of Sciences/Author.

Olson, D. G. (1998). Irradiation of food. *Food Technology, 52*(1), 56-65.

Web Sites

1999 FDA Model Food Code

http://vm.cfsan.fda.gov/~dms/fc99-toc.html

Complete outline of the FDA's latest code for regulating operations that provide food directly to consumers. Also includes a quick synopsis of changes from the 1997 Food Code.

American Public Health Association (APHA)

http://www.apha.org

Information on disease prevention and health promotion as well as other resources for public health professionals.

Centers for Disease Control and Prevention (CDC)

http://www.cdc.gov

Data and statistics, publications, and other health information from the United States Centers for Disease Control.

FightBac!

http://www.fightbac.org

The Web site for the Partnership for Food Safety Education, a coalition of government, industry, and consumer groups.

Food Marketing Institute

http://www.fmi.org

This nonprofit association conducts programs in research, education, industry relations, and public affairs for its membership of food retailers and wholesalers. Their Web site offers a wealth of food-safety information.

Institute of Food Technologists (IFT)

http://www.ift.org

IFT is a nonprofit scientific society with 28,000 members working in food science, food technology, and related professions in industry, academia, and government. This site includes information on its policies and publications, education and industry news, and meeting and convention locations.

International Association of Milk, Food, and Environmental Sanitation (IAMFES)

http://www.iamfes.org

IAMFES keeps members informed of the latest scientific, technical, and practical developments in food safety and sanitation. Their Web site includes booklets and links to other food-safety sites.

International Commission on Microbiological Specifications for Food (ICMSF)

http://www.dfst.csiro.au/icmsf.htm

ICMSF provides timely, science-based guidance to government and industry on appraising and controlling the microbiological safety of foods.

Morbidity and Mortality Weekly Report (MMWR)

http://www.cdc.gov/epo/mmwr/mmwr_ss.html

The MMWR page on the Centers for Disease Control's Web site allows you to search the CDC's publications for information on foodborne illness.

National Center for Infectious Diseases

http://www.cdc.gov/ncidod

The CDC's National Center for Infectious Diseases offers information on the efforts to prevent and control diseases caused by bacteria and fungi, as well as fact sheets and articles on foodborne and diarrheal diseases.

National Environmental Health Association (NEHA)

http://www.neha.org

The Web site for the National Environmental Health Association features information, resources, and publications; also offers information on credentialing for sanitarians.

National Food Processors Association (NFPA)

http://www.nfpa-food.org

NFPA is a scientific and technical trade association for the food industry. Their site provides industry news and educational materials on issues of food safety, research, and food science.

National Food Safety Database

http://www.foodsafety.org

A compilation of food-safety database information from government, consumer, and public health organizations, this is a one-stop Web site for food-safety information on the Internet.

National Restaurant Association

http://www.restaurant.org

The National Restaurant Association Web site provides information on government agencies affecting the restaurant industry, the latest training and certification updates, and links to state restaurant associations and hospitality schools and universities.

United States Department of Agriculture Food Safety and Inspection Service

http://www.fsis.usda.gov

The USDA's Food Safety and Inspection Web site offers the latest food-safety news, educational materials, and HACCP-implementation materials.

World Atom

http://www.iaea.or.at/worldatom

Web site for the International Atomic Energy Agency provides information on efforts by the International Consultative Group on Food Irradiation to advise and evaluate international activities of food irradiation.

Chapter 2
The Microworld

TEST YOUR FOOD-SAFETY KNOWLEDGE

1. **True or False:** Proper cooking will destroy all pathogens in food. *(See Multiple Barriers for Controlling the Growth of Microorganisms, page 2-8.)*

2. **True or False:** A foodborne intoxication is caused by eating food that contains a pathogen. *(See Foodborne Infection vs. Foodborne Intoxication, page 2-20.)*

3. **True or False:** Most foodborne illnesses are caused by viruses. *(See Bacteria, page 2-2.)*

4. **True or False:** Bacteria will not grow at refrigeration temperatures. *(See Temperature, page 2-6.)*

5. **True or False:** Highly acidic foods such as tomato sauce provide a perfect environment for bacteria to grow. *(See Acidity, page 2-6.)*

Learning Objectives

After completing this chapter, you should be able to:

○ Identify the four basic types of microbial contaminants; give examples, and describe preventive actions for each.

○ Differentiate between foodborne infection and foodborne intoxication and identify the major causes of each.

○ Identify the microbial risks associated with various types of food.

○ Explain the conditions conducive to bacterial growth.

Key Terms

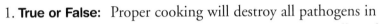

Microorganisms
Pathogens
Toxins
Bacteria
Virus
Parasite
Fungi
Spoilage microorganism
Spores
pH (Acidity)
Lag phase

Vegetative microorganism
Log phase
Stationary phase
Death phase
FAT-TOM
Temperature danger zone
Aerobic
Anaerobic
Facultative

Water activity (a_w)
Host
Mold
Yeast
Infection
Intoxication
Toxin-mediated infection
Vehicle
Food irradiation

In the previous chapter, you learned that microorganisms pose the greatest threat to food safety, and that disease-causing microorganisms are responsible for the majority of foodborne illness outbreaks. In this chapter, you will learn about the microorganisms that cause foodborne illness, as well as the conditions they require in order to grow. When you understand these conditions, you will begin to see how the growth of microorganisms can be controlled, a topic that will be covered in greater detail in later chapters.

Microorganisms are small, living beings that can be seen only with a microscope. While not all microorganisms cause disease, some do. These are called **pathogens.** Eating food contaminated with pathogens or their **toxins** (poisons) is the leading cause of foodborne illness.

MICROBIAL CONTAMINANTS

There are four types of microorganisms that can contaminate food and cause foodborne illness: **bacteria, viruses, parasites,** and **fungi.**

These microorganisms can be arranged into two groups: **spoilage microorganisms** and pathogens. Mold is an example of spoilage microorganism. While moldy food has an unpleasant appearance, smell, and taste, it seldom causes illness. However, pathogens such as *Salmonella, E. coli* O157:H7, and the virus that causes Hepatitis A cannot be seen, smelled, or tasted, but food contaminated by these pathogens often causes some form of illness when ingested.

BACTERIA

Of all microorganisms, bacteria are of greatest concern to the manager. They are more commonly involved in foodborne illness than any of the other foodborne microorganisms or contaminants. Knowing what bacteria are, and understanding the environment in which they grow, is the first step in controlling them.

Basic Characteristics of Bacteria that Cause Foodborne Illness

Bacteria that cause foodborne illness have some basic characteristics.

○ They are living, single-celled organisms.

○ They may be carried by a variety of means: food, water, humans, and insects.

○ Under favorable conditions, they can reproduce very rapidly.

○ Some can survive freezing.

○ Some form into **spores,** a change that protects the bacteria from unfavorable conditions.

○ Some can cause food spoilage; others can cause disease.

○ Some cause illness by producing toxins as they multiply, die, and break down.

Bacterial Growth

To grow and reproduce, bacteria need the following:

○ Adequate time

○ Proper temperature

○ Ample moisture

○ Food

○ Appropriate **pH (acidity)**

○ The necessary level of oxygen

Their growth can be broken down into four progressive stages (phases): lag, log, stationary, and death. *(See Exhibit 2a.)*

When bacteria are first introduced to a food, they go through an adjustment period, called the **lag phase.** In this phase, their numbers are stable and they are preparing for growth. To control their number, it is important to prolong the lag phase as long as possible. You can accomplish this by controlling the bacteria's requirements for growth, such as time, temperature, moisture, and pH. If these conditions are not controlled, bacteria can enter the next phase, the log phase, where they will grow remarkably fast.

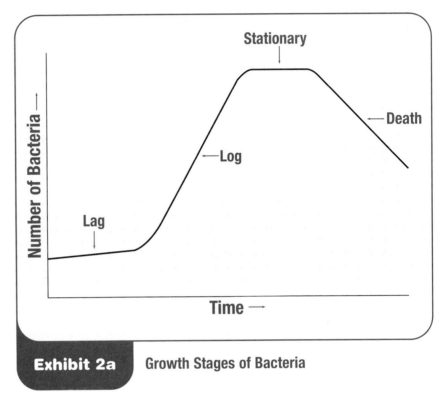

Exhibit 2a **Growth Stages of Bacteria**

Bacteria reproduce by splitting in two. Those that are in the process of reproduction are called **vegetative microorganisms.** As long as conditions are favorable, bacteria can grow and multiply very rapidly, doubling their number as often as every twenty minutes *(see Exhibit 2b).* This is called exponential growth, and it occurs in the **log phase.** Foods will become unsafe rapidly during the log phase.

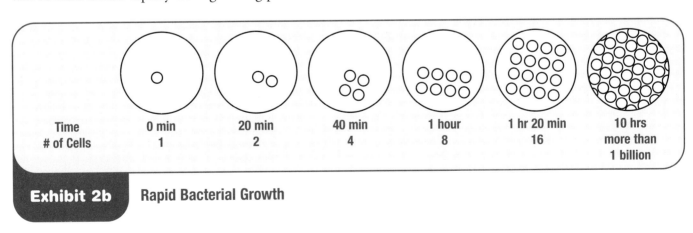

Time	0 min	20 min	40 min	1 hour	1 hr 20 min	10 hrs
# of Cells	1	2	4	8	16	more than 1 billion

Exhibit 2b **Rapid Bacterial Growth**

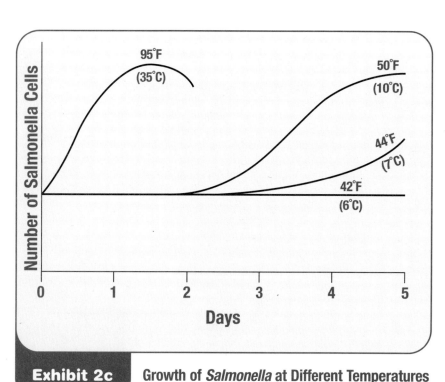

Exhibit 2c   Growth of *Salmonella* at Different Temperatures

Bacteria can continue to grow until nutrients and moisture become scarce, or conditions become unfavorable. Eventually, the population reaches a **stationary phase,** where just as many bacteria are growing as are dying. When the number of bacteria dying exceeds the number that are growing, the population declines. This is called the **death phase.**

The time required for bacteria to adapt to a new environment (lag phase), and to begin a rapid rate of growth (log phase), depends on several factors. *Exhibit 2c* shows how different temperatures affect the growth rate of *Salmonella* bacteria. As the graph shows, at warmer temperatures (95°F [35°C]), *Salmonella* grows more quickly than at colder temperatures (44°F and 50°F [7°C and 10°C]). At even colder temperatures (42°F [6°C]), *Salmonella* doesn't grow at all, but notice that it doesn't die either (prolonging the lag phase).

Vegetative Stages and Spore Formation

Although vegetative bacteria may be resistant to low—even freezing temperatures—they can be killed by high temperatures. For example, pathogenic bacteria can be killed during proper cooking. *(This will be discussed in Chapter 7.)* Some types of bacteria, however, have the ability to change into a different form, called a spore. The spore's thick wall protects the bacteria against unfavorable conditions, such as high or low temperature, low moisture, and high acidity.

While a spore cannot reproduce, it is capable of turning back into a vegetative organism when conditions become favorable again. For example, bacteria in food may form a spore when exposed to freezer temperatures, allowing the bacteria to survive. As the food thaws and conditions change, the spore can turn back into a vegetative cell and begin to grow in the food.

Since spores are so difficult to destroy, it is important to cook, cool, and reheat foods properly in order to keep bacteria that may be in food from growing to harmful levels.

Conditions that Support the Growth of Microorganisms

All microorganisms, except viruses, are able to grow in foods if the conditions are right. To grow, microorganisms need food, water, appropriate temperatures, oxygen, and proper pH. Therefore, you must control these necessities in order to control the growth of microorganisms.

FAT-TOM: What Microorganisms Need to Grow

The conditions that favor the growth of most foodborne microorganisms (except viruses) can be remembered by the acronym **FAT-TOM.** *(See Exhibit 2d.)* Each of these conditions for growth will be explained in more detail in the next several paragraphs.

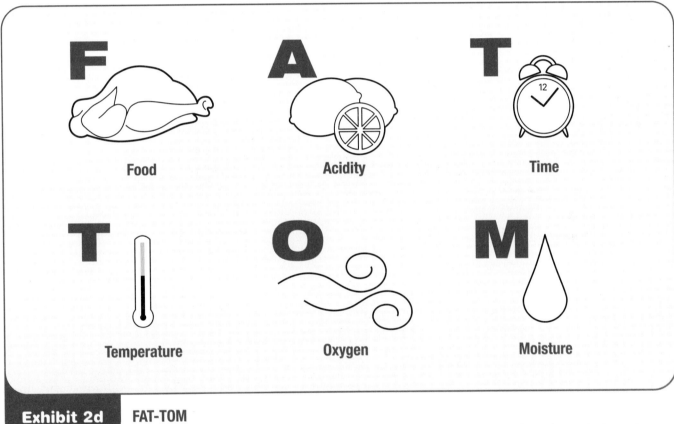

F — Food	A — Acidity	T — Time
T — Temperature	O — Oxygen	M — Moisture

Exhibit 2d **FAT-TOM**
The conditions that favor the growth of most foodborne microorganisms (except viruses) can be remembered by the acronym FAT-TOM.

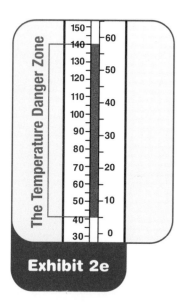

Exhibit 2e

**Temperature and
Bacterial Growth**
Most foodborne micro-
organisms grow well at
temperatures between 41°F
and 140°F (5°C and 60°C).

Key Point

Pathogenic
bacteria
grow well in foods with a
pH between 4.6 and 7.5.

Key Point

*Listeria
monocyto-
genes* and *Yersinia
enterocolitica* are able to
grow at refrigeration
temperatures.

Food

To grow, microorganisms need nutrients, specifically proteins and carbohydrates. These substances are commonly found in foods such as meat, poultry, dairy products, and eggs.

Acidity

Microorganisms typically do not grow in foods that are highly acidic or highly alkaline. They grow best in foods that have a neutral pH. The ph of a substance tells how acidic or alkaline it is. The pH scale ranges from 0 to 14.0. A food with a pH between 0 and 7.0 is acidic while a food with a pH between 7.0 and 14.0 is alkaline. A pH of 7.0 is neutral. Some microorganisms grow better than others in foods with a specific pH. *(See Exhibit 2f on the next page.)*

Yeasts and mold tend to grow in acidic foods (with a pH below 4.6) such as fruits, fruit juices, and jams. Pathogenic bacteria grow well in foods with a pH between 4.6 and 7.5. Foods with a pH higher than 7.0 typically do not support the growth of foodborne microorganisms. However, there are exceptions. A few bacteria have been known to grow in foods with a pH below 4.6 and above 7.5. For example, *E. coli* O157:H7 has been found to reproduce in unpasteurized apple juice, which has a pH below 4.0.

Temperature

Most foodborne microorganisms grow well between the temperatures of 41°F and 140°F (5°C to 60°C). *(See Exhibit 2e.)*

This range is known as the **temperature danger zone.** However, exposing microorganisms to temperatures outside the danger zone does not necessarily kill them. Refrigeration temperatures, for example, may only slow their growth. Some bacteria such as *Listeria monocytogenes* and *Yersinia enterocolitica* are able to grow well at refrigeration temperatures. Bacterial spores can often survive extreme heat and cold.

Time

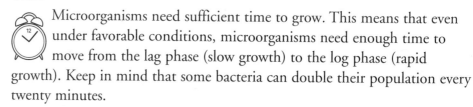

Microorganisms need sufficient time to grow. This means that even under favorable conditions, microorganisms need enough time to move from the lag phase (slow growth) to the log phase (rapid growth). Keep in mind that some bacteria can double their population every twenty minutes.

If contaminated food remains in the temperature danger zone for four hours or more, pathogenic microorganisms can grow to levels high enough to make someone ill. Therefore, it is important to control the amount of time potentially hazardous foods remain in this temperature danger zone.

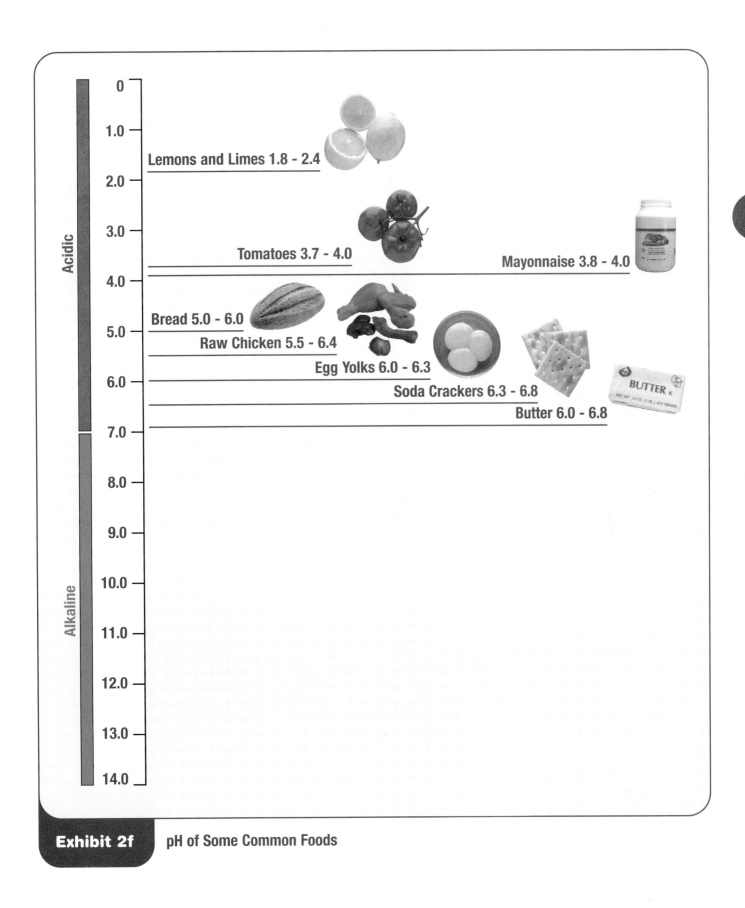

0

1.0

Lemons and Limes 1.8 - 2.4

2.0

3.0

Tomatoes 3.7 - 4.0   Mayonnaise 3.8 - 4.0

4.0

Bread 5.0 - 6.0

5.0
Raw Chicken 5.5 - 6.4

Egg Yolks 6.0 - 6.3

6.0
Soda Crackers 6.3 - 6.8

Butter 6.0 - 6.8

7.0

8.0

9.0

10.0

11.0

12.0

13.0

14.0

Acidic

Alkaline

Exhibit 2f   pH of Some Common Foods

Oxygen

Different microorganisms have different oxygen requirements for growth. They can be categorized as aerobic, anaerobic, or facultative. **Aerobic** microorganisms require oxygen to grow. **Anaerobic** microorganisms can grow only when oxygen is absent. **Facultative** microorganisms can grow either with or without the presence of oxygen. Most microorganisms that cause foodborne illness are facultative.

Moisture

Most foodborne microorganisms grow well in moist foods. The amount of moisture in a food is called its **water activity (a_w).** Water activity (a_w) is measured on a scale from 0 through 1.0, with distilled water having a water activity of 1.0. Most microorganisms that cause foodborne illness grow best in foods with water activities between 0.85 and 0.97, although some can grow in foods with lower water activity levels. Potentially hazardous foods typically have a water activity of 0.85 or above. *Exhibit 2g* presents the water activity levels of various foods.

Multiple Barriers for Controlling the Growth of Microorganisms

FAT-TOM is the key to controlling the growth of microorganisms. Multiple barriers need to be put in place to deny the microorganism as many as possible of the conditions that support growth. The list below provides some simple barriers that can be combined to create multiple barriers to *control* the growth of pathogens. To kill or reduce the level of pathogens, foods must be cooked to their recommended minimum internal temperatures.

○ **Make the food more acidic.** Add vinegar, lemon juice, lactic acid, or citric acid.

○ **Raise or lower the temperature of the food.** Move food out of the temperature danger zone by cooking it to the proper temperature, or refrigerating it to 41°F (5°C) or below, or by freezing it.

○ **Lower the water activity (a_w) of the food.** Dry food by adding sugar, salt, alcohol, or acid. Food can also be air-dried or freeze-dried to remove water.

○ **Limit the amount of time the food is in the temperature danger zone.** Prepare food as close to service as possible.

The major foodborne illnesses caused by bacteria are listed in *Exhibit 2h.* *(Biological toxins are discussed in Chapter 3.)*

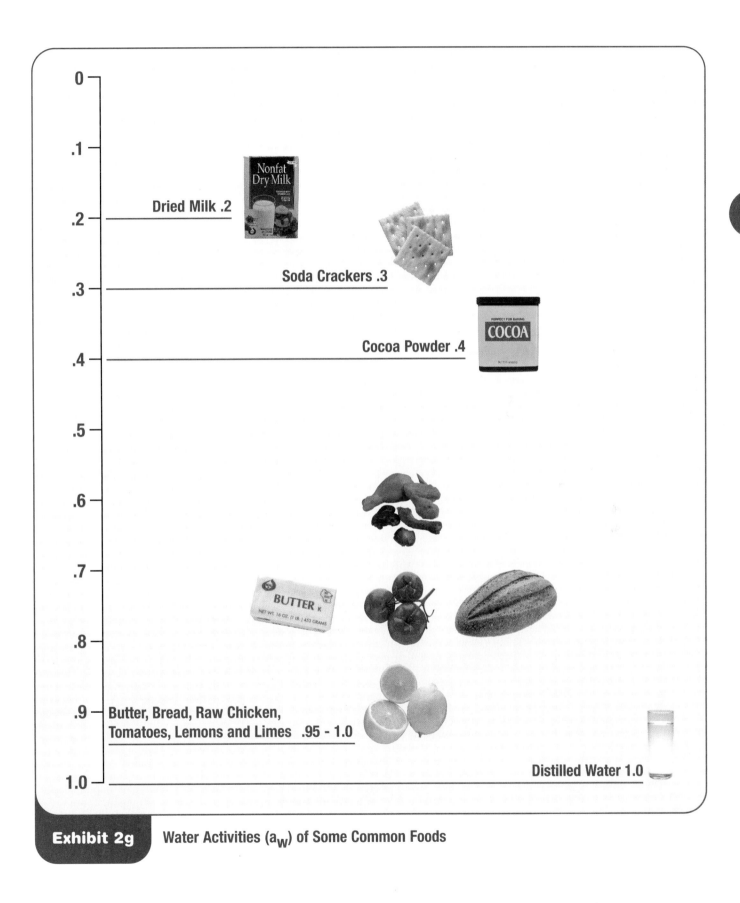

0

.1

Dried Milk .2

.2

Soda Crackers .3

.3

Cocoa Powder .4

.4

.5

.6

.7

BUTTER
NET WT. 16 OZ. (1 LB.) 453 GRAMS

.8

.9

Butter, Bread, Raw Chicken,
Tomatoes, Lemons and Limes   .95 - 1.0

Distilled Water 1.0

1.0

Exhibit 2g   Water Activities (a$_w$) of Some Common Foods

Exhibit 2h   Major Foodborne Illnesses Caused by Bacteria

Foodborne Illness	Salmonellosis (nontyphoid)	Shigellosis	Listeriosis
Bacteria	Salmonella spp.	Shigella spp.	Listeria monocytogenes
Characteristics of Bacteria	Does not form spores; facultative; some can survive pHs below 4.5	Does not form spores; facultative; some produce shiga toxin	Does not form spores; facultative; resists freezing, drying, and heat; can grow at refrigeration temperatures
Type of Illness	Infection (possibly toxin-mediated)	Toxin-mediated infection	Infection
Symptoms	Abdominal cramps, headache, nausea, fever, diarrhea, and sometimes vomiting; may cause severe dehydration in infants and elderly; may cause arthritic symptoms 3 to 4 weeks later	Diarrhea (may be bloody), abdominal pain, fever, nausea, cramps, vomiting, chills, fatigue, dehydration	Nausea, vomiting, diarrhea, headache, persistent fever, chills, backache, meningitis, encephalitis, septicemia, and intrauterine or cervical infections in pregnant women, which may result in spontaneous abortion or stillbirth; most often affects fetuses, infants, and pregnant women
Incubation Period	6 to 48 hours; usually 12 to 36 hours	12 to 50 hours; usually 1 to 3 days	3 to 70 days; usually about 3 weeks
Duration of Illness	1 to 2 days (sometimes longer)	Usually 4 to 7 days; can be indefinite (depends on treatment)	Indefinite (depends on treatment); high fatality rates in immunocompromised people
Source	Water, soil, insects, domestic and wild animals, and the human intestinal tract, especially as carriers	Human intestinal tract; flies; frequently found in water polluted by feces	Soil, water, and damp environments; humans, domestic and wild animals, especially fowl (intestinal tracts)
Foods Involved in Outbreaks	Poultry and poultry salads; meat and meat products; fish; shrimp; milk; shell eggs and egg products, such as custards, sauces, and pastry creams; tofu and other protein foods; sliced melons, sliced tomatoes, raw sprouts, and other fresh produce	Salads (potato, tuna, shrimp, chicken, and macaroni); lettuce; raw vegetables; milk and dairy products; poultry; moist and mixed foods	Unpasteurized milk and cheese; ice cream; frozen yogurt; raw vegetables; poultry and meats; seafood; prepared and chilled ready-to-eat foods (e.g., soft cheese, deli, foods, pâté)
Preventive Measures	Avoid cross-contamination; refrigerate food; thoroughly cook poultry to at least 165F (74C) for at least 15 seconds and cook other foods to minimum internal temperatures; properly cool cooked meats and meat products within 6 hours; avoid pooling eggs; ensure that employees avoid contaminating food and food-contact surfaces by practicing good personal hygiene	Avoid cross-contamination; ensure that foodservice employees practice good personal hygiene; use sanitary food and water sources; control flies; rapidly cool foods	Use only pasteurized milk and dairy products; cook foods to proper internal temperatures; avoid cross-contamination; clean and sanitize surfaces, keep facilities clean and dry

Exhibit 2h—Major Foodborne Illnesses Caused by Bacteria (Continued)

Foodborne Illness	Staphylococcal Food Poisoning (Enterotoxicosis)	Clostridium perfringens Enteritis	Bacillus cereus Gastroenteritis
Bacteria	Staphylococcus aureus	Clostridium perfringens	Bacillus cereus
Characteristics of Bacteria	Does not form spores; facultative; can survive high acidity (to pH 2.6); resists drying and freezing	Forms spores; anaerobic	Forms spores; facultative
Type of Illness	Intoxication	Toxin-mediated infection	Intoxication (emetic); toxin-mediated infection (diarrheal)
Symptoms	Nausea, retching, abdominal cramps, diarrhea; in severe cases—headache, muscle cramping, changes in blood pressure and pulse rate	Abdominal pain, diarrhea, nausea, dehydration (fever, headache, and vomiting usually absent)	Nausea and vomiting, sometimes abdominal cramps or diarrhea (emetic); watery diarrhea, abdominal cramps, pain, nausea (diarrheal)
Incubation Period	1 to 7 hours; usually 2 to 4 hours	8 to 22 hours; usually 10 to 12 hours	30 minutes to 6 hours (emetic); 6 to 15 hours (diarrheal)
Duration of Illness	1 to 2 days	Usually 24 hours; may last 1 to 2 weeks	Less than 24 hours (emetic); 24 hours (diarrheal)
Source	Skin, hair, nose; throat, infected sores; animals	Humans and animals (intestinal tracts), soil; soil contaminated with feces	Soil and dust; cereal crops
Foods Involved in Outbreaks	Reheated foods; ham and other meats; poultry; egg products and other protein foods; sandwiches; milk and dairy products; potato salads; custards; cream-filled pastries; salad dressings	Cooked meat; meat products; poultry; stew; gravy; beans that have been improperly cooled	Rice products; starchy foods (potato, pasta and cheese products); sauces; puddings; soups; casseroles; pastries; salads (emetic); meats; milk; vegetables; and fish (diarrheal)
Preventive Measures	Avoid contamination from unwashed bare hands; practice good personal hygiene; exclude foodservice employees with skin infections from foodhandling and preparation tasks; properly refrigerate food; rapidly cool prepared foods	Use careful time and temperature control in cooling and reheating cooked foods	Careful time and temperature control and quick-chilling methods to cool foods; adequate cooking of food

Exhibit 2h—Major Foodborne Illnesses Caused by Bacteria (Continued)

Foodborne Illness	Botulism	Campylobacteriosis	E. coli O157:H7 Enterohemorrhagic (EHEC)
Bacteria	Clostridium botulinum	Campylobacter jejuni	Escherichia coli
Characteristics of Bacteria	Forms spores; anaerobic	Does not form spores	Does not form spores; facultative; has survived freezing and high acidity (pHs below 4.0); can grow at refrigerator temperatures
Type of Illness	Intoxication (botulism)	Infection	Toxin-mediated infection
Symptoms	Vomiting and constipation or diarrhea may be present initially; progresses to fatigue, weakness, vertigo, blurred or double vision, difficulty speaking and swallowing, dry mouth; eventually leading to paralysis and death	Diarrhea (watery or bloody); fever, nausea, and vomiting; abdominal pain, headache, and muscle pain	Diarrhea (watery, may become bloody); severe abdominal cramps and pain; vomiting, and mild or no fever; may cause kidney failure in some people; symptoms more severe in the very young
Incubation Period	4 hours to 8 days; usually 12 to 36 hours	Usually 2 to 5 days, with a range of 1 to 10 days	3 to 8 days; usually 3 to 4 days
Duration of Illness	Several days to 1 year	2 to 5 days, usually no more than 10 days (relapses common); self-limiting	2 to 9 days
Source	Present on almost all foods of either animal or vegetable origin; soil; water	Domestic animals (sheep, pigs, cattle, poultry, and pets); the intestinal tracts of wild animals	Animals; particularly found in the intestinal tracts of cattle and humans
Foods Involved in Outbreaks	Foods that were underprocessed or temperature abused in storage; canned low-acid foods, untreated garlic-and-oil products, sautéed onions in butter sauce; leftover baked potatoes; stews; meat/poultry loaves; potential risks for MAP (modified-atmosphere packaging) and sous vide products	Unpasteurized milk and dairy products; raw poultry; nonchlorinated or fecal-contaminated water	Raw and undercooked ground beef; imported cheeses; unpasteurized milk and apple cider/juice; roast beef; dry salami; commercial mayonnaise; lettuce; nonchlorinated water
Preventive Measures	Do not use home-canned products; use careful time and temperature control for sous vide items and all large, bulky foods; purchase only acidified garlic-and-oil mixtures and keep refrigerated; sauté onions to order; rapidly cool leftovers	Thoroughly cook food to minimum safe internal temperatures; pasteurize milk; use treated water; avoid cross-contamination	Thoroughly cook ground beef to at least 155°F (68°C) for 15 seconds; avoid cross-contamination; practice good personal hygiene

Exhibit 2h—Major Foodborne Illnesses Caused by Bacteria (Continued)

Foodborne Illness	*Vibrio* spp. (noncholerae) Gastroenteritis/Septicemia	Yersiniosis
Bacteria	*Vibrio parahaemolyticus* and *Vibrio vulnificus*	*Yersinia enterocolitica*
Characteristics of Bacteria	Does not form spores; more common in warmer months	Does not form spores; facultative; can survive at pH below 4.5; can grow at low temperatures
Type of Illness	Infection	Infection
Symptoms	Diarrhea, abdominal cramps, nausea, vomiting, headache; severe cases of *V. vulnificus* include sudden chills, fever, blistering skin lesions, decreased blood pressure, and septicemia; fatality from *V. vulnificus* is high in immuno-compromised people	Fever and severe abdominal pain (mimics appendicitis); possible diarrhea, headache, sore throat, or vomiting
Incubation Period	*V. parahaemolyticus:* 4 to 96 hours, usually 12 to 24 hours; *V. vulnificus:* 12 hours to several days, usually 38 hours	1 to 11 days; usually 24 to 48 hours
Duration of Illness	*V. parahaemolyticus:* 1 to 8 days; *V. vulnificus:* days to weeks; death within a few days for immunocompromised people	Days to weeks; chronic in some individuals
Source	Oysters and other shellfish, (shrimp, lobsters, clams, mussels, and scallops), especially from the Gulf of Mexico	Domestic pigs primary reservoir; soil; water; wild animals; rodents
Foods Involved in Outbreaks	Raw or partially cooked oysters; raw or under-cooked shellfish (clams, mussels, and scallops)	Meats (pork, beef, lamb); oysters; fish; raw milk; contaminated pasteurized milk; tofu; nonchlorinated water
Preventive Measures	Avoid eating raw or undercooked seafood (particularly oysters); avoid cross-contamination; freezing does not completely destroy these bacteria; purchase from approved sources	Minimize cross-contamination from pork; thoroughly cook foods to minimum safe internal temperatures; ensure that facilities and equipment are properly sanitized; follow proper storage procedures; use only sanitary, chlorinated water supplies

Key Point

Viruses cannot reproduce outside a living cell.

Key Point

Viruses do not require a potentially hazardous food as a medium for transmission.

Health Alert

Practicing good personal hygiene and minimizing hand contact with ready-to-eat foods is an important defense against foodborne viruses.

VIRUSES

Viruses are the smallest of the microbial contaminants. They consist of genetic material wrapped with an outer layer of protein. While a virus cannot reproduce outside a living cell, once inside, it will reproduce more viruses. Viruses are responsible for several foodborne illnesses such as Hepatitis A, and infections caused by Norwalk virus and rotavirus. Rotavirus infections are the leading cause of severe gastroenteritis among infants and young children worldwide.

Basic Characteristics of Viruses

Viruses share some basic characteristics.

○ They rely on a living cell to reproduce.

○ They do not require a potentially hazardous food as a medium for transmission.

○ They are not complete cells.

○ They do not reproduce in food.

○ Some may survive freezing and cooking.

○ They can be transmitted from person to person, from people to food, and from people to food-contact surfaces.

○ They usually contaminate food through a foodhandler's improper personal hygiene.

○ They can contaminate both food and water supplies. An example is shellfish harvested from sewage-contaminated waters.

Practicing good personal hygiene is an important defense against foodborne viruses. It is especially important to minimize hand contact with ready-to-eat foods. Information about major foodborne illnesses caused by viruses, and methods of prevention is in *Exhibit 2i*.

Exhibit 2i   Major Foodborne Illnesses Causes by Viruses

Foodborne Illness	Hepatitis A	Norwalk Virus Gastroenteritis	Rotavirus Gastroenteritis
Virus	*Hepatovirus* or Hepatitis A virus	Norwalk and Norwalk-like viral agents	Rotavirus
Type of Illness	Infection	Infection	Infection
Symptoms	Mild or no illness, then sudden onset of fever, general discomfort, fatigue, headache, nausea, loss of appetite, vomiting, abdominal pain, and jaundice after several days	Nausea, vomiting, diarrhea, abdominal cramps, headache, and mild fever	Vomiting and diarrhea, abdominal pain, and mild fever (illness more common in children than adults)
Incubation Period	10 to 50 days	Usually 1 to 2 days; ranges from 10 to 50 hours	Usually 1 to 3 days
Duration of Illness	1 to 2 weeks; severe cases may last several months	Usually 1 to 3 days	Usually 4 to 8 days
Source	Human intestinal and urinary tracts; contaminated water	Human intestinal tract and contaminated water	Human intestinal tract; contaminated water
Foods Involved in Outbreaks	Water; ice; shellfish; salads; cold cuts and sandwiches; fruits and fruit juices; milk and milk products; vegetables; any food that will not receive a further heat treatment	Water; shellfish (especially raw or steamed); raw vegetables; fresh fruits and salads; contaminated water	Water and ice; raw and ready-to-eat foods (such as salads, fruits, and hors d'oeuvres); contaminated water
Preventive Measures	Obtain shellfish from approved sources; prevent cross-contamination from hands; ensure foodhandlers practice good personal hygiene; clean and sanitize food-contact surfaces; and use sanitary water sources	Obtain shellfish from approved sources; prevent cross-contamination from hands; ensure foodhandlers practice good personal hygiene; thoroughly cook foods to minimum safe internal temperatures; and use sanitary, chlorinated water	Ensure foodhandlers practice good personal hygiene; thoroughly cook foods to minimum safe internal temperatures; and use sanitary, chlorinated water

PARASITES

Parasites are organisms that need to live in or on a host organism in order to survive. A **host** is a person, animal, or plant on which another organism lives and takes nourishment. Parasites can live inside many animals that humans use for food, such as cattle, poultry, swine, and fish. Foodborne parasites include protozoa, roundworms and flatworms. Parasites require many different hosts to carry out their life cycles. However, they are typically passed to humans through an animal host. To prevent foodborne illness caused by parasites, use proper freezing and cooking techniques; avoid cross-contamination; use sanitary water supplies; and follow proper handwashing procedures, especially after using the restroom.

Basic Characteristics of Foodborne Parasites

Parasites share some basic characteristics, as follows.

○ They are living organisms that need a host to survive.

○ They grow naturally in many animals such as swine and can be transmitted to humans.

○ Most are very small, often microscopic, but larger than bacteria.

○ They may be killed by proper cooking or freezing.

○ They pose hazards to both food and water.

Major foodborne illnesses caused by parasites and methods of prevention are presented in *Exhibit 2j.*

FUNGI

Fungi range in size from microscopic, single-celled organisms to very large, multicellular organisms. They are found naturally in air, soil, plants, animals, water, and some foods. **Molds, yeasts,** and mushrooms are examples of fungi. The fungi of concern to establishments are molds and yeasts. *(Mushroom toxins are discussed in Chapter 3.)*

Molds

Individual mold cells can usually be seen only with a microscope. However, fuzzy or slimy mold colonies, consisting of a large number of cells, are often visible to the naked eye. Bread mold is an example. The spores produced by molds are not the same as the spores produced by bacteria. Molds use spores for reproduction.

Molds are responsible for the spoilage of many foods. They cause spoilage with discoloration, formation of odors, and off-flavors. Molds are able to grow on almost any food, at almost any storage temperature. They can also grow in

Key Point

Parasites are typically passed to humans through an animal host.

Exhibit 2j Major Foodborne Illnesses Caused by Parasites

Foodborne Illness	Trichinosis	Anisakiasis	Giardiasis
Parasite	*Trichinella spiralis*	*Anisakis simplex*	*Giardia duodenalis* (formerly *G. lamblia*)
Characteristic of Parasites	Roundworm	Roundworm	Protozoan
Type of Illness	Infection	Infection	Infection
Symptoms	Nausea, diarrhea, abdominal pain, occasionally vomiting, swelling around the eyes, fever; later, muscle soreness, thirst, extreme sweating, chills, hemorrhaging (bleeding), and fatigue	Tingling or tickling sensation in throat, vomiting or coughing up worms; in severe cases, severe abdominal pain, cramping, vomiting, and nausea	Fatigue, nausea, intestinal gas, weakness, weight loss, and abdominal cramps
Incubation Period	2 to 28 days; depends on number of larvae ingested	A few hours to 2 weeks	3 to 25 days; usually about 7 to 10 days
Duration of Illness	Several days to more than 30 days; depends on treatment and health status of individual	Up to 3 weeks	Usually 1 to 2 weeks; infectivity may last months
Source	Domestic pigs; wild game, such as bears and walruses	Marine fish, especially bottom-feeders	Wild animals; especially beavers and bears; domestic animals (dogs and cats); intestinal tract of humans
Foods Involved in Outbreaks	Undercooked pork or wild game; pork and nonpork sausages (ground meats may be contaminated by meat grinders)	Raw, undercooked, or improperly frozen seafood, especially cod, haddock, fluke, Pacific salmon, herring, flounder, monkfish, and fish used in sushi and sashimi	Water; ice; salads; (possibly) other raw vegetables
Preventive Measures	Cook pork and other meats to minimum internal cooking temperatures; larvae may be killed by freezing pork products less than 6 inches thick at 5°F (-15°C) for 30 days or 30°F (-1°C) for 12 days; wash, rinse, and sanitize equipment such as sausage grinders and utensils used in the preparation of raw pork and other meats	Obtain seafood only from certified sources; properly freeze fish; avoid eating raw or partly cooked fish and shellfish; fish intended to be eaten raw should be frozen at -4°F (-20°C) or lower for 7 days in a freezer, or at 31°F (-1°C) or lower for 15 hours in a blast chiller; *Anisakis* parasites are not killed by marinades	Use sanitary, chlorinated water supplies; ensure that foodhandlers practice good personal hygiene; wash raw produce carefully.

Exhibit 2j—Major Foodborne Illnesses Caused by Parasites (Continued)

Foodborne Illness	Toxoplasmosis	Intestinal Cryptosporidiosis	Cyclosporiasis
Parasite	*Toxoplasma gondii*	*Cryptosporidium parvum*	*Cyclospora cayetanensis*
Characteristic of Parasites	Protozoan; not directly passed from person to person	Protozoan; sporocysts are resistant to most chemical disinfectants	Protozoan
Type of Illness	Infection	Infection	Infection
Symptoms	Often, there are no symptoms. When symptoms occur they include enlarged lymph nodes in head and neck; severe headaches, severe muscle pain; rash; individuals with compromised immune systems, such as HIV-infected people, and pregnant women and their fetuses are most at risk.	Severe, watery diarrhea; may have no symptoms	Watery diarrhea, loss of appetite, weight loss, bloating, gas, abdominal cramps, nausea, vomiting; muscle aches, mild or no fever; fatigue
Incubation Period	5 to 20 days; depends on immune system of individual	1 to 12 days; average is 7 days	Days to weeks; usually about 1 week
Duration of Illness	Depends on treatment (relapses common)	4 days to 3 weeks; indefinite and life-threatening in immunocompromised people	A few days to a month, or longer; usually several weeks (relapses spanning 1 to 2 months are common)
Source	Animal feces (especially felines), mammals, birds	The intestinal tract of humans, cattle, and other domestic animals; drinking water contaminated with runoff from farms or slaughterhouses	The intestinal tract of humans; contaminated water supplies
Foods Involved in Outbreaks	Raw or undercooked meat contaminated with this parasite, especially pork, lamb, venison, and hamburger meat	Water; salads and raw vegetables; milk; raw foods, such as apple cider; ready-to-eat foods	Water; marine fish; raw milk; raw produce
Preventive Measures	Avoid raw or undercooked meats; thoroughly cook meats to the minimum internal temperature, so there is no pink inside; properly wash hands that come in contact with soil, raw meat, cat feces, or raw vegetables	Ensure that foodhandlers practice good personal hygiene; thoroughly wash produce; use sanitary water	Ensure that foodhandlers practice good personal hygiene; thoroughly wash produce; use sanitary water

environments that are moist or dry, have a high or low pH, and are salty or sweet. They typically prefer to grow in and on sweet, acidic foods that have a low water activity. Molds often spoil fruits, vegetables, meats, cheeses, and breads because of the water activity and pH of these types of foods. *(See Exhibit 2k.)*

Some molds produce toxins, which can cause allergic reactions, nervous system disorders, kidney damage, and liver damage. For example, aflatoxin, which is produced by the molds *Aspergillus flavus* and *Aspergillus parasticus,* can cause liver diseases.

Foods such as corn and corn products; peanuts and peanut products; cottonseed; milk; and tree nuts, such as Brazil nuts, pecans, pistachio nuts, and walnuts have been associated with aflatoxins.

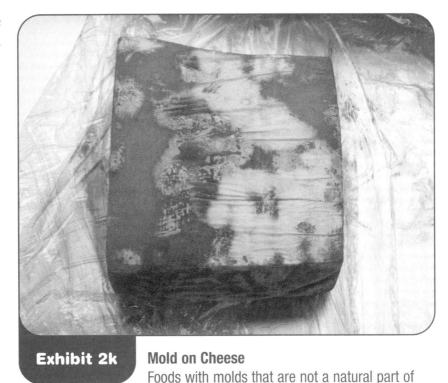

Exhibit 2k

Mold on Cheese
Foods with molds that are not a natural part of the product should always be discarded.

Basic Characteristics of Foodborne Molds

Molds share some basic characteristics.

○ They spoil food, and sometimes cause illness.

○ They grow under almost any condition, but grow well in sweet, acidic foods with low water activity.

○ Freezing temperatures prevent or reduce the growth of mold but do not destroy it.

○ Some molds produce toxins called aflatoxins.

○ Mold cells and spores may be killed by heating, but mold toxins may not be destroyed.

○ To avoid illnesses caused by mold toxins, throw out moldy food, unless the mold is a natural part of the food (e.g., cheeses such as Gorgonzola, bleu cheese, Brie, and Camembert. These are specific types of molds which are grown in very specific conditions.)

Although mold cells and spores can be killed by heating them, toxins that may be present are not destroyed by normal cooking methods. Foods with molds that are not a natural part of the product should always be discarded.

Exhibit 2l   **Yeast on Jelly**
Foods that have been spoiled by yeast should
be discarded.

Yeasts

Some yeasts are known for their ability to spoil food rapidly. Carbon dioxide and alcohol are produced as yeast slowly consumes food. Yeast spoilage may therefore produce a smell or taste of alcohol. Yeast may appear as a pink discoloration or slime and may bubble.

Yeasts are similar to molds in that they grow well in sweet, acidic foods with low water activity, such as jellies, jams, syrup, honey, and fruit juice. *(See Exhibit 2l.)* Foods that have been spoiled by yeast should be discarded.

FOODBORNE INFECTION VERSUS FOODBORNE INTOXICATION

Foodborne diseases are classified as infections, intoxications, or toxin-mediated infections.

○ Foodborne **infections** result when pathogens grow in the intestines of someone who has eaten food contaminated by those pathogens. Typically, the symptoms of a foodborne infection, such as fever, do not appear immediately.

○ Foodborne **intoxications** are caused by eating food containing poisonous toxins, such as enterotoxins (intestinal toxins that are produced by pathogens). A person does not need to eat live microorganisms to become ill, just the toxins produced by them. Typically, the symptoms of a foodborne intoxication appear quickly, within a few hours. Many toxins are not destroyed by cooking.

○ Foodborne **toxin-mediated infections** result from eating food that contains pathogens. These pathogens grow in the intestines and produce toxins that can cause illness.

RECENT TRENDS IN FOODS ASSOCIATED WITH FOODBORNE ILLNESS

In the past, the foods involved in outbreaks of foodborne illness were usually unpasteurized milk or undercooked meat, poultry, or seafood. Today, some foods previously considered safe have become the vehicles for foodborne pathogens.

For example, it was assumed for centuries that the internal contents of an egg were sterile and safe to eat raw. Now, however, there are clear indications that eggs can be contaminated internally with *Salmonella*. Many outbreaks of salmonellosis have been traced to undercooked, contaminated eggs included in traditional recipes, such as eggnog, Caesar salad, lightly cooked omelets, French toast, and even meringue.

Many other foods such as produce and highly acidic foods, that were once considered safe are now being seen as potential **vehicles** for foodborne pathogens. Therefore, it is important for the industry to understand that food safety is a growing science. The industry must be kept informed of the latest findings and recommendations for food safety.

HAZARDS ASSOCIATED WITH FRESHER AND HEALTHIER FOODS

Many changes are occurring that can greatly affect the safety of food. Natural or organic foods without additives or preservatives are becoming more available. Foods from all over the world are becoming more accessible. In response to the public's health consciousness, many products commonly seen on grocery store shelves are now being altered.

Altering the makeup of a food is not without risk to the safety of the food. If the stability of a product was traditionally due to the water activity or pH (i.e., the product is too dry or acidic to allow microorganisms to grow), changing the makeup of the food may eliminate that barrier, causing the food to become unsafe.

Organic products may also be at risk to food safety. Organic foods are grown and processed without the use of pesticides, additives, and preservatives. Lacking these traditional barriers to microbial growth, organic products may have reduced shelf lives, and may allow spoilage microorganisms to grow.

EMERGING PATHOGENS AND ISSUES

Progress has been made in preventing foodborne diseases. For example, typhoid fever (caused by *Salmonella typhii*) was very common earlier this century. Typhoid is now almost forgotten in the United States due to improved methods of treating drinking water, better sewage treatment, milk sanitation and pasteurization, and shellfish sanitation. Similarly, cholera and trichinosis, once very common, are now also rare in the United States.

While some pathogens have been controlled, new foodborne pathogens have emerged. One of the first were nontyphoid strains of *Salmonella*, which have been increasing since World War II. In fact, in the United States, from 1988 to 1992, non-typhoid *Salmonella* strains caused 69 percent of the bacterial outbreaks whose origins were identified.

In the past, the foods involved in outbreaks of foodborne illness were often raw or inadequately cooked meat, poultry, seafood, or unpasteurized milk. In recent years, however, new foods have been recognized as vehicles of foodborne pathogens.

Eggs, for example, have been consumed raw or lightly cooked for centuries. Recently, a bacteria has adapted to survive inside the reproductive system of chickens and can contaminate the interior of the eggs. Even though this is rare, eggs must now be cooked thoroughly or heat treated or an illness may result.

In other instances, foods not typically associated with foodborne illness have recently been recognized as vehicles of foodborne pathogens. Produce, for example, has recently been associated with foodborne illness outbreaks. As detection techniques and foodborne illness investigations improve, new pathogens that cause foodborne illness are being recognized.

These trends have come to pass for a variety of reasons. As the population continues to grow, more food is needed. Food is produced in greater quantities and at centralized sources. More food, including fresh food, is shipped from greater distances across state and international borders. It remains important for establishments to purchase food from reputable suppliers and to keep food safe by using proper storage, preparation, and serving practices.

TECHNOLOGICAL ADVANCEMENTS IN FOOD SAFETY

Several new technologies are emerging to reduce pathogenic and spoilage microorganisms in foods. These include food irradiation, electron pasteurization, high-pressure processing, bacteriocins, and various light treatments, such as laser light, ultraviolet light, and pulsed light.

Food irradiation, often called cold pasteurization, involves exposing food to an electron beam or gamma rays, similar to the way microwaves are used to heat food. *(See Exhibit 2m.)* The FDA has approved the use of irradiation on certain foods in the United States. To date, irradiation is allowed for pork (to control the parasite *Trichinella spiralis*), poultry (to control bacterial

Key Point

Foods not typically associated with foodborne illness have recently been recognized as vehicles of foodborne pathogens.

Exhibit 2m

Radura Symbol
This is the international symbol for irradiated foods.

contaminants); fruits and vegetables (to prevent premature maturation and to control insect pests); and on herbs, spices, teas, and other dried vegetable substances (to control various microbial contaminants).

The benefits of food irradiation include:

○ Reduction or elimination of pathogens and spoilage microorganisms

○ Replacement of chemical treatments of foods

○ Extended shelf life of foods

Irradiated foods are only recently becoming widely available in the United States, despite regulatory approval and repeated endorsements by professional health and nutrition organizations. There is no evidence that harmful effects result from the amount or types of radiation used, and nutritional value, appearance, and taste are virtually not affected. However, some consumers are still apprehensive about foods exposed to radiation, due to their limited knowledge of the irradiation process. It has been shown that, once the process and benefits are explained to them, consumers are more receptive to the idea of irradiation and would purchase irradiated products.

It is important to note that food irradiation does not replace proper food storage and handling practices, since the food could still become unsafe if cross-contaminated. Therefore, when handling irradiated foods, you should follow the same food-safety practices that you would follow with other foods.

SUMMARY

Microbial contaminants are responsible for the majority of foodborne illness. Because bacteria, viruses, parasites, and fungi can be introduced at any point in the flow of food, food must be monitored and controlled throughout. Understanding how these microorganisms grow, reproduce, contaminate food, and infect humans is critical to understanding how to prevent the foodborne illnesses they cause.

Of all foodborne microorganisms, bacteria are of greatest concern to the manager. They are more commonly involved in foodborne illness than all other foodborne microorganisms. Bacteria are living, single-celled organisms that can be carried by a variety of means. Some cause food spoilage; others cause disease. Other bacteria cause illness by producing toxins as they multiply, die and break down. Under favorable conditions, bacteria can reproduce very rapidly. Although vegetative bacteria may be resistant to low or freezing temperatures, they can be killed by high temperatures such as those reached during cooking. Some types of bacteria, however, have the ability to form spores, which protect the bacteria from unfavorable conditions. Since spores are so difficult to destroy it is important to properly cook, cool, and reheat foods in order to keep bacteria from growing to harmful levels.

The acronym FAT-TOM, which stands for Food, Acidity, Time, Temperature, Oxygen, and Moisture, may help you remember the conditions that promote the growth of microorganisms. Microorganisms need nutrients in order to grow, specifically proteins and carbohydrates. They grow best in foods that have a neutral pH. Most foodborne microorganisms grow well between the temperatures of 41°F and 140°F (5°C to 60°C). They also need sufficient time within these temperatures to grow. If contaminated food remains in the temperature danger zone for four hours or more, pathogenic microorganisms can grow to levels high enough to make someone ill. Most microorganisms that cause foodborne illness can grow either with or without the presence of oxygen, and grow well in moist foods. FAT-TOM is the key to controlling the growth of microorganisms. Multiple barriers need to be put in place, which will deny the microorganism as many of the conditions that support growth as possible. These barriers include making food more acidic, lowering its temperature, and lowering the water activity of the food.

Viruses are the smallest of the microbial contaminants. While a virus cannot reproduce in food, once ingested it will cause illness. Viruses can be transmitted from person to person, from people to food, and from people to food-contact surfaces. They can contaminate both food and water supplies. Some may survive freezing and cooking. Practicing good personal hygiene and minimizing hand contact with ready-to-eat foods is an important defense against foodborne illness from viruses.

Parasites are organisms that need to live in or on a host organism to survive. They can live inside many animals that humans eat, such as cattle, poultry, swine, and fish. They may be killed by proper cooking and freezing.

Molds and yeasts are examples of fungi, which are a concern to establishments. Molds are mostly responsible for the spoilage of many foods but can produce toxins, which can harm us. Molds are able to grow in a variety of environments, but they typically prefer to grow in and on sweet, acidic foods that have a low water activity. Yeasts are known for their ability to spoil food rapidly. They are similar to molds in that they grow well in sweet, acidic foods with low water activity.

Foodborne diseases are classified as infections, intoxications, or toxin-mediated infections. Foodborne infections result when pathogens grow in the intestines of people who have eaten food contaminated by those pathogens. Typically, the symptoms of a foodborne infection, such as fever, do not appear immediately. Foodborne intoxications are caused by eating food containing poisonous toxins. A person does not need to eat live microorganisms to become ill, just the toxins produced by them. Typically, the symptoms of a foodborne intoxication appear quickly, within a few hours. Foodborne toxin-mediated infections result from eating food that contains pathogens. These pathogens grow in the intestines and produce toxins that can cause illness.

Our ever-changing world is opening new doors for microorganisms. Many new pathogens are emerging. New formulations for products are creating niches for microorganisms that may not have been able to grow in the product previously. However, many new technologies are available and others are being developed to prevent the growth of pathogenic microorganisms.

A CASE IN POINT

Case Study

A daycare center is serving stir-fried rice for lunch. The rice was properly cooked at 1:00 p.m. the day before; that is, it was heated to the minimum required temperature for at least fifteen seconds. The covered rice was then placed on the countertop and allowed to cool to room temperature. At 6:00 p.m., the cook placed it in the refrigerator. At 9:00 a.m. the following day, the rice was combined with the other ingredients for stir-fried rice and cooked to 165°F (74°C) for at least fifteen seconds. The cook covered the stir-fried rice and left it on the range until she gently reheated it at noon. Within an hour of eating the stir-fried rice, however, several of the children began vomiting, and a few had diarrhea. Samples from some of the children revealed that the rice, not the other ingredients, was considered to be the cause of the outbreak.

Based on the information given, was the illness caused by a bacteria, virus, parasite, or fungi? What is the name of the microorganism most likely to have caused the outbreak? Is this illness an infection or intoxication?

TRAINING TIPS

Training Tips for the Classroom

1. Twenty Questions

Objective: *After completing this activity, class participants should be able to identify the major foodborne illnesses caused by bacteria, viruses, and parasites, and be able to identify characteristics, symptoms, sources, foods involved, and preventive measures for each of these pathogens.*

Directions: After discussing Chapter 2, break your class into several groups. Assign each group a pathogenic bacteria, virus, or parasite that was discussed in the chapter.

Have each group study their "bug" for ten minutes. Ask them to focus on the following:

○ characteristics

○ symptoms of illness caused by the bug

○ source

○ foods involved

○ preventive measures

Then, one at a time, ask each group to stand in front of the class and solicit up to twenty questions from the other groups. Each group in the audience asks one question and makes a guess as to what bug that group represents. When a group correctly identifies the bug, points are awarded based on the number of unasked questions remaining. For example, if four questions were asked before a correct guess was made, the group making the correct guess would be awarded sixteen points (20 - 4 = 16).

In the unlikely event that no team can figure out what bug a group represents, the group presenting receives twenty points. Play until every team has gone to the front. The team with the most points wins!

Note: Allow up to forty minutes for this game. If time is an issue, select the six or so most common bugs, and assign larger groups.

2. Bug Scramble

Objective: *After completing this activity, class participants should be able to identify the major foodborne illnesses caused by bacteria, viruses, and parasites.*

Directions: Create a word scramble game for your class using the different types of bacteria, viruses, parasites and fungi discussed in Chapter 2. Scramble the name of each bug and write it on a transparency (one bug per transparency).

Example: suerec sullicab *(Bacillus cereus)*

Then, break your class into four or five groups, and give each group a bell. Ask each group to ring the bell when they have unscrambled the bug projected on the screen. If they are correct, they get twenty points. (If you choose, you can assign a certain number of points for each bug, based on its perceived difficulty. Otherwise, to keep it simple, you can just give twenty points for each one.)

To enhance the learning process, you could require the group that unscrambles the word correctly to give two or three characteristics of the bug, and give bonus points for each correct answer.

As usual, the team with the most points wins!

DISCUSSION QUESTIONS

1. Why should potentially hazardous foods not be left in the temperature danger zone for more than four hours? Refer to some of the phases of bacterial growth in your explanation.

2. Jams and jellies have water activity levels of about 0.75. Based on this information alone, do they provide an environment that favors bacterial growth? Explain your thinking.

3. Compare and contrast viruses and bacteria.

4. Explain the differences between foodborne infection and foodborne intoxication.

5. Explain why it is important to use multiple barriers to prevent bacterial growth. Give examples in your explanation.

MULTIPLE-CHOICE STUDY QUESTIONS

1. To what does the acronym FAT-TOM refer?
   A. The major types of microorganisms that can cause foodborne illness
   B. The conditions that support the growth of microorganisms
   C. Human health hazards associated with foods that are high in fat and sodium
   D. The federal agency responsible for monitoring food safety

2. All of the following are conditions that support the growth of microorganisms except
   A. moisture.
   B. a protein or carbohydrate food source.
   C. high acidity.
   D. temperatures between 41°F and 140°F (5°C and 60°C).

3. A person who has a foodborne infection has most likely eaten a food containing .
   A. a ciguatoxin.   C. a plant toxin.
   B. histamine.   D. a live microorganism.

4. Which food is more likely to transmit parasites to humans?
   A. Improperly cooked eggs
   B. Improperly frozen sushi
   C. Improperly refrigerated milk
   D. Unpasteurized apple juice

5. Which of the following is not a basic characteristic of foodborne mold?
   A. It grows well in sweet acidic foods with low water activity.
   B. Freezing temperatures prevent or reduce its growth but do not destroy it.
   C. Its cells and spores may be killed by heating, but the toxins it produces may not be destroyed.
   D. It needs a host to survive.

6. Which of the following statements regarding foodborne intoxication is true?

   A. The symptoms of intoxication often appear days after exposure.
   B. The medical treatment for intoxication can be painful.
   C. Foodborne intoxication is more common than foodborne infection.
   D. The symptoms of intoxication appear quickly, within a few hours.

7. Which of the following microorganisms is most likely to cause a foodborne infection?

   A. *Listeria monocytogenes*
   B. *Bacillus cereus*
   C. *E. coli* O157: H7
   D. *Clostridium botulinum*

8. Which of the following microorganisms is most likely to cause a foodborne intoxication?

   A. *Staphylococcus aureus*
   B. *Shigella*
   C. *Campylobacter jejuni*
   D. *Salmonella*

9. Which of the following foods is commonly associated with an outbreak of *Bacillus cereus?*

   A. Cooked rice
   B. Blanched vegetables
   C. Raw chicken
   D. A hamburger cooked rare

10. Which of the following foods is commonly associated with an outbreak of *Vibrio?*

    A. Raw oysters
    B. Unpasteurized milk
    C. Apple cider
    D. Sliced melons

ADDITIONAL RESOURCES

Books and Periodicals

American Public Health Association. (1995). *Control of communicable diseases in man.* Washington DC: Author.

Buchanan, R. L. & Doyle, M. P. (1997). Foodborne disease significance of *Escherichia coli* O157:H7 and other enterohemorrhagic *E. coli. Food Technology, 51 (*10), 69-76.

Bush, R. K. (1995). Seafood allergy and allergens: A review. *Food Technology, 49*(10), 103-116.

Carrera, N. (1997). Avoiding cross-contamination. *Pizza Today, 16*(5), 84.

Cliver, D. O. (1997). Virus transmission via food. *Food Technology, 51 (*4), 71-78.

Cliver, D. O. (Ed.). (1990). *Foodborne diseases.* San Diego, CA: Academic Press.

Dulen, J. (1998). Under the microscope. *Restaurants & Institutions, 108*(21), 80.

Durocher, J. (1997). Germ warfare. *Restaurant Business, 96*(22), 123.

Farkas, D. (1996). Creating awareness. *Restaurant Hospitality, 80*(9), 112.

Featsent, A. W. (1998). Food fright! Consumers' perceptions of food safety versus reality. *Restaurants USA, 18*(6), 30.

Grossbauer, S. (1998). Food safety on the Web. *FoodService Director, 11*(2), 140.

Hernandez, J. (1997). Eliminating the *E. coli* threat. *Food Management, 32*(12), 67.

How to prevent foodborne illness. (1995). *FoodService Director, 8*(2), 150.

ICMSF (International Commission on Microbiological Specifications for Food). (1988). *Microorganisms in foods 1:*

Their significance and methods of enumeration (Rev. ed.). Toronto: University of Toronto Press.

ICMSF. (1997). *Microorganisms in foods 3: Vol. 1. Factors affecting life and death of microorganisms; Vol. 2. Food commodities; Vol. 3. Microbial ecology of foods.* New York: Academic Press.

ICMSF (1996). *Microorganisms in foods 5: Characteristics of microbial pathogens.* London: Blackie Academic & Professional.

Jay, J. M. (1996). *Modern food microbiology.* New York: Chapman & Hall.

Longree, K. & Armbruster, G. (1996). *Quantity food sanitation* (5th ed.). New York: Wiley.

Marriott, N. G. (1994). *Principles of food sanitation.* New York: Chapman & Hall.

The National Restaurant Association. (1996). *Sanitation survival kit.* Washington, DC: Author.

Neumann, R. (1998). The eight most frequent causes of foodborne illness. *Food Management, 33*(6), 28.

Puzo, D. P. (1997). USDA estimates financial costs of foodborne illness. *Restaurants & Institutions, 107*(16), 80.

Rehe, S. (1990). *Preventing foodborne illness: A guide to safe foodhandling.* Washington DC: United States Department of Agriculture Food Safety & Inspection Service.

Rubin, K.W. (1995). What to look for, how to prevent it: Food poisoning. *FoodService Director, 8*(11), 72.

Salmonella and food safety. (1995). *National Culinary Review, 19*(11), 26.

Spake, A. (1997). O is for outbreak. *United States News & World Report, 123*(20), 70.

Study considers seven foodborne illnesses. (1997). *FoodReview, 20*(3), 37.

Web Sites

1999 FDA Model Food Code

http://vm.cfsan.fda.gov/~dms/fc99-toc.html

Complete outline of the FDA's latest code for regulating operations that provide food directly to consumers. Also includes a quick synopsis of changes from the 1997 Food Code.

American Public Health Association (APHA)

http://www.apha.org

Information on disease prevention and health promotion as well as other resources for public health professionals.

American Society of Microbiology

http://www.asmusa.org

Comprehensive site includes publications, archives, and links to other valuable microbiology sites.

The Bad Bug Book

http://vm.cfsan.fda.gov/~mow/intro.html

This on-line handbook provides basic facts regarding foodborne pathogenic microorganisms and natural toxins. It brings together in one place information from the Food and Drug Administration, the Centers for Disease Control and Prevention, the USDA Food Safety Inspection Service, and the National Institutes of Health.

Centers for Disease Control and Prevention (CDC)

http://www.cdc.gov

Data and statistics, publications, and other health information from the United States Centers for Disease Control.

CDC Division of Bacterial and Mycotic Diseases

http://www.cdc.gov/ncidod/dbmd

This division of the CDC is dedicated to preventing and controlling the many emerging, re-emerging, drug-resistant, and other important bacterial and mycotic diseases in the United States and around the world. From this site you can access fact sheets on many foodborne and mycotic diseases.

FDA Center for Food Safety and Applied Nutrition (CFSAN)

http://vm.cfsan.fda.gov/list.html

Comprehensive site from CFSAN offers a wealth of food-safety information, from foodborne illness to food labeling. CFSAN strives to be a leader in food safety, and to protect consumers from economic fraud, promote sound nutrition, and encourage innovation.

FightBac!

http://www.fightbac.org

The Web site for the Partnership for Food Safety Education, a coalition of government, industry, and consumer groups.

Food and Drug Administration (FDA)

http://www.fda.gov

Web site for the Food and Drug Administration features link to the FDA's Center for Food Safety and Applied Nutrition.

Institute of Food Technologists (IFT)

http://www.ift.org

IFT is a nonprofit scientific society with 28,000 members working in food science, food technology, and related professions in industry, academia, and government. This site includes information on its policies and publications, education and industry news, and meeting and convention locations.

International Association of Milk, Food, and Environmental Sanitation (IAMFES)

http://www.iamfes.org

IAMFES keeps members informed of the latest scientific, technical, and practical developments in food safety and sanitation. Their Web site includes booklets and links to other food-safety sites.

International Commission on Microbiological Specifications for Food (ICMSF)

http://www.dfst.csiro.au/icmsf.htm

ICMSF provides timely, science-based guidance to government and industry on appraising and controlling the microbiological safety of foods.

Microbionet

http://www.sciencenet.com.au/

This site provides information on some of the thousands of forms of bacteria. The site profiles each organism's particular characteristics in short paragraphs.

Morbidity and Mortality Weekly Report (MMWR)

http://www.cdc.gov/epo/mmwr/mmwr_ss.html

The MMWR page on the Centers for Disease Control's Web site allows you to search the CDC's publications for information on foodborne illness.

National Center for Infectious Diseases

http://www.cdc.gov/ncidod

The CDC's National Center for Infectious Diseases offers information on the efforts to prevent and control diseases caused by bacteria and fungi, as well as fact sheets and articles on foodborne and diarrheal diseases.

National Environmental Health Association (NEHA)

http://www.neha.org

The Web site for the National Environmental Health Association features information, resources, and publications; also offers information on credentialing for sanitarians.

National Food Safety Database

http://www.foodsafety.org

A compilation of food-safety database information from government, consumer, and public health organizations, this is a one-stop Web site for food-safety information on the Internet.

National Restaurant Association

http://www.restaurant.org

The National Restaurant Association site provides information on government agencies affecting the restaurant industry, the latest training and certification updates, and links to state restaurant associations and hospitality schools and universities.

National Shellfish Sanitation Program Manual of Operation

http://vm.cfsan.fda.gov/~ear/nsspman.html

An outline of the FDA's Center for Food Safety and Applied Nutrition's National Shellfish Sanitation Program Manual of Operation, which ensures safe molluscan shellfish.

United States Department of Agriculture (USDA)

http://www.usda.gov

The United States Department of Agriculture's Web site features information, publications, and other educational materials about the nation's agriculture.

United States Department of Agriculture's Food Safety and Inspection Service

http://www.fsis.usda.gov

The USDA's Food Safety and Inspection Web site offers the latest food-safety news, educational materials, and HACCP implementation materials.

Chapter 3
Contamination, Food Allergies, and Foodborne Illness

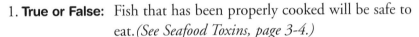

TEST YOUR FOOD-SAFETY KNOWLEDGE

1. **True or False:** Fish that has been properly cooked will be safe to eat.*(See Seafood Toxins, page 3-4.)*

2. **True or False:** Cooking can destroy fungal toxins in certain varieties of wild mushrooms. *(See Mushroom Toxins, page 3-5.)*

3. **True or False:** Food additives listed on the government's GRAS (Generally Regarded As Safe) list will not cause allergic reactions. *(See Food Allergies, page 3-9.)*

4. **True or False:** Unopened cleaning products may be stored with unopened cans and packages of food. *(See Chemicals and Pesticides, page 3-7.)*

5. **True or False:** Most biological toxins found in food occur naturally and are not caused by the presence of microorganisms. *(See Biological Contamination, page 3-2.)*

Table of Contents

Learning Objectives

After completing this chapter, you should be able to:

○ Identify the three types of contamination (biological, chemical, and physical), and give examples of each.

○ Identify ways in which food can become contaminated.

○ Identify methods to prevent biological, chemical, and physical contamination.

Key Terms

Biological contaminant
Biological toxins
Ciguatera
Scombroid poisoning
Histamine
Chemical contaminant
Physical contaminant
Food allergy

Food additive
Preservatives
Nitrites
Sulfites
Monosodium glutamate (MSG)
GRAS (Generally Regarded as Safe)

Food is considered contaminated when it contains hazardous substances. These substances may be biological, chemical, or physical. The most common food contaminants are the **biological contaminants** that belong to the microworld—bacteria, parasites, viruses, and fungi, which were discussed in Chapter 2. Most foodborne illnesses are the result of microbial contamination, but biological and chemical toxins are responsible for many foodborne illnesses as well. Each year in the United States, it is estimated that biological and chemical toxins are responsible for millions of foodborne illnesses. While biological and chemical contamination pose a significant threat to food, the danger from physical hazards should also be recognized.

TYPES OF FOODBORNE CONTAMINATION

Ensuring the safety of food is the most important job of the manager. A thorough understanding of the causes and prevention of the various types of contamination can help you keep food safe.

Biological Contamination

As we learned in Chapter 2, a foodborne intoxication occurs when a person eats food containing a **biological toxin,** which was produced by a plant or produced or ingested by an animal. Toxins in seafood, plants, and mushrooms are responsible for many cases of foodborne illness annually in the United States. Most of these biological toxins occur naturally and are not caused by the presence of microorganisms. Some occur in fish as a result of their diet.

Seafood Toxins

The **ciguatera** toxin occurs in certain predatory tropical reef fish such as amberjack, barracuda, grouper, and snapper *(see Exhibit 3a)*. Ciguatera accumulates in the tissue of these large, predatory fish after they eat smaller fish that have fed upon certain species of algae. When a person eats fish containing this toxin, an illness may result (that requires weeks or months for recovery). Symptoms of ciguatera intoxication include vomiting, severe itching, nausea, dizziness, hot and cold flashes,

Exhibit 3a   **Ciguatera Toxin**
Because cooking doesn't destroy the ciguatera toxin, predatory tropical reef fish such as red snapper should be purchased only from approved suppliers.

temporary blindness, and sometimes hallucinations. Cooking does not destroy the ciguatera toxin. Therefore, it is very important to purchase predatory tropical reef fish only from approved suppliers.

Shellfish may contain toxins that occur because of the algae upon which they feed. The illnesses caused by shellfish poisoning vary and are specific to the type of toxin consumed. Paralytic shellfish poisoning (PSP), the most serious type, may lead to respiratory failure and even death if respiratory support is not provided. PSP is generally associated with mussels, clams, cockles, and scallops. Since cooking may not destroy shellfish toxins, it is important to purchase shellfish from approved suppliers who can certify that they are harvested from safe waters *(see Exhibit 3b).*

Exhibit 3b

Shellfish Toxins
Purchase shellfish from approved suppliers who can certify that they are harvested from safe waters.

Scombroid poisoning is one of the more common forms of illness caused by fish toxin in the United States. It occurs when scombroid species of fish such as tuna, mackerel, bluefish, skipjack, swordfish, and bonito are time-temperature abused. Under these conditions, the bacteria associated with the fish produce the toxin **histamine.** This odorless, tasteless chemical causes scombroid intoxication when it is consumed. Symptoms of the illness include flushing and sweating, a burning peppery taste in the mouth, dizziness, nausea, and headache. Sometimes a facial rash, hives, edema, diarrhea, and abdominal cramps will follow. Scombroid poisoning, also known as histamine poisoning, has been associated with other fish species as well, such as mahi-mahi, marlin, and sardines. The histamine toxin is not destroyed by cooking or freezing. Since time-temperature abuse during the harvesting process may cause scombroid fish to become unsafe, it is important to purchase these fish from reputable suppliers who practice strict time-temperature controls.

Some fish toxins are systemic, that is, they occur as a natural part of the fish. An example of a potentially toxic fish is pufferfish. Pufferfish contains tetrodotoxin in its liver, skin, and other organs. Consuming the toxin in this fish can produce rapid and violent death. Few cases of this type of poisoning have been reported in the United States. Cooking may not destroy systemic fish toxins. Pufferfish should be eaten only if it is handled and prepared by a trained and licensed chef.

Key Point

Purchase scombroid fish such as tuna and swordfish from reputable suppliers who practice strict time-temperature controls.

The following general procedures should be practiced to guard against a seafood-specific foodborne illness.

○ Purchase fish from reputable suppliers who maintain strict time-temperature controls.

○ Refuse fish that have been thawed and refrozen.

○ Thaw frozen fish at refrigerator temperatures below 41°F (5°C).

○ Check temperature. The temperature of fresh fish must be 41°F (5°C) or lower upon arrival. However, this temperature standard will not prevent a scombroid intoxication if the fish has been temperature-abused during the harvesting process or prior to being received. This is why it is critical to purchase fish from reputable suppliers.

Remember that toxins are not living organisms, so cooking or freezing may not destroy them. Therefore, most of the preventive measures taken against biological toxins must be implemented when purchasing and receiving fish. Chapter 5 will discuss the receiving process in greater detail.

Plant Toxins

Toxins are a natural part of some plants. Most poisonings caused by plants result when toxic plants have been used in medicinal home remedies. Foodborne illnesses have occurred after people have consumed rhubarb leaves, jimsonweed, and the root of water hemlock. Some illnesses have occurred after animals have eaten toxic plants, and people have consumed the by-products of those animals. For example, people have become ill after they consumed milk from cows that had eaten snakeroot. People have become ill from consuming honey produced by bees that had gathered nectar from mountain laurel or rhododendrons.

Some plants may be toxic in their raw state, but safe when properly cooked. For example, fava beans and red kidney beans may be unsafe if eaten when raw or undercooked. In general, toxic plant species and products prepared with them should be avoided. Only commercially processed honey and properly cooked beans should be used.

Mushroom Toxins

Outbreaks of foodborne illness associated with mushrooms are almost always caused by the consumption of wild mushrooms that have been collected by amateur mushroom hunters. Most cases occur when toxic mushroom species are confused with edible species. The symptoms of intoxication vary depending upon the species consumed. Some mushroom toxins will destroy humans' internal organs. Others cause convulsions,

Key Point

Toxic plant species and products prepared with them should not be used in an establishment.

hallucinations, and coma. Still others produce nausea, vomiting, abdominal cramping, and diarrhea.

Cooking or freezing will not destroy the toxins produced by toxic mushrooms. Establishments should not use wild mushrooms or products made with them. All mushrooms should be purchased from approved suppliers *(see Exhibit 3c).*

Exhibit 3d on the next page summarizes information about common biological toxins, the sources of these toxins, the foods with which they are often associated, and preventive measures that can be taken to keep these toxins from causing foodborne illness.

Exhibit 3c **Mushroom Toxins**
Establishments should not use wild mushrooms or products made with them.

Chemical Contamination

Chemical contaminants are responsible for many cases of foodborne illness. Contamination can come from a variety of substances normally found in establishments. These include toxic metals, pesticides, and chemicals.

Toxic Metals

Utensils and equipment that contain toxic metals such as lead, copper, brass, zinc, antimony, and cadmium can cause a foodborne illness from chemical contamination. If acidic foods are stored in or prepared with this type of equipment, they can leach these metals from the item and contaminate the food. For example, storing tomato sauce in a copper pot or lemonade in a pewter pitcher or a galvanized (zinc-coated) tub could lead to a foodborne illness from chemical contamination. Only food-grade utensils and equipment should be used to prepare and store food.

Improperly installed carbonated beverage dispensers can also create a hazard. If carbonated water is allowed to flow back into the copper supply lines, it could leach copper from the line and contaminate the beverage. Beverage dispensing systems should be installed and maintained only by professionals.

Key Point

Only food-grade utensils and equipment should be used to prepare and store food.

Exhibit 3d	Biological Contamination (Biological Toxins)

Biological Toxin	Source Of Contamination	Associated Foods	Preventive Measures
Seafood Toxins			
Ciguatera Toxin	Fish that have eaten algae that contain the toxins	Predatory tropical reef fish such as amberjack, barracuda, grouper, and snapper	Cooking does not destroy these toxins; purchase tropical reef fish only from approved suppliers
Scombroid Toxin (Histamine)	Histamine produced by bacteria in some fish when they are time-temperature abused	Primarily occurs in tuna, bluefish, mackerel, skipjack, roundfish, and bonito; other fish, such as mahi-mahi, marlin, and sardines, have also been implicated in histamine poisoning	Cooking does not destroy histamine; because time-temperature abuse during the harvesting process may cause the fish to become unsafe, it is important to purchase from reputable suppliers
Shellfish Toxins	Shellfish that have eaten a type of algae that contains the toxin	Shellfish, especially mollusks, such as mussels, clams, cockles, and scallops	Cooking may not destroy this toxin; purchase these shellfish from approved suppliers who can certify that they are harvested from safe waters
Systemic Fish Toxins	Toxins that are a natural part of some fish	Pufferfish, moray eels, and freshwater minnows	Cooking may not destroy systemic fish toxins; pufferfish should be handled and prepared by properly trained chefs
Plant Toxins	Toxins that are a natural part of some plants	Fava beans, rhubarb leaves, jimsonweed, water hemlock, and apricot kernels; honey from bees that have gathered nectar from mountain laurel; milk from cows that have eaten snakeroot, jimsonweed, and other toxic plants	Cooking may not destroy these toxins; avoid these plant species and also products prepared with them
Fungal Toxins	Toxins that are a natural part of some varieties of fungi	Poisonous varieties of mushrooms and other fungi	Cooking does not destroy these toxins; do not use wild mushrooms; purchase mushrooms only from approved suppliers

Chemicals and Pesticides

Chemicals such as cleaning products, polishes, lubricants, and sanitizers can contaminate food if they are improperly used or stored. Follow the directions supplied by the manufacturer when using these chemicals. Use caution when using chemicals during operating hours, to prevent contamination of food and food-preparation areas. Store chemicals away from food packaging, utensils, and equipment used for food. Keep them in a locked storage area in their original container. If chemicals must be transferred to smaller containers or spray bottles, label each container appropriately.

Pesticides are often used in kitchens and in food-preparation and storage areas to control pests such as rodents and insects. If used, pesticides should be applied only by a trained Pest Control Operator (PCO). All food should be wrapped or stored prior to application of the pesticide. Pesticides should be stored with the same care as other chemicals used in the establishment.

Exhibit 3f on the next page summarizes information about some common chemical contaminants, their sources, the foods with which they are often associated, and preventive measures that can be taken to keep them from causing foodborne illness.

Physical Contamination

Physical contamination results from the accidental introduction of foreign objects into foods. Common **physical contaminants** may include items such as metal shavings from cans *(see Exhibit 3e)*, staples from cartons, glass from broken light bulbs, blades from plastic or rubber scrapers, fingernails, hair, bandages, and dirt. The contaminant may even be a natural part of the food itself, such as the bones in chicken or fish. Closely inspect the foods you receive and take steps to ensure that food will not be physically contaminated during the flow of food in your operation.

FOOD ALLERGIES

Many people have food allergies. A **food allergy** is the body's negative reaction to a particular food or foods. Depending on the person, allergic reactions may occur immediately after

Key Point

Chemicals should be stored away from food.

Key Point

Pesticides should be applied only by a trained Pest Control Operator (PCO).

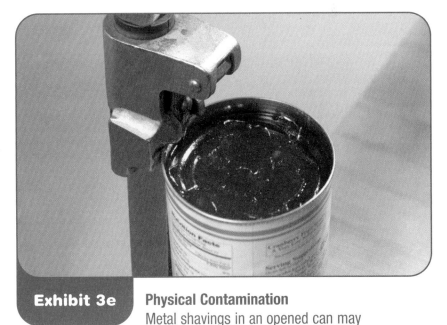

Exhibit 3e **Physical Contamination**
Metal shavings in an opened can may contaminate the food inside.

| | Exhibit 3f | Chemical Contamination | | |

Chemical Toxin	Source	Associated Foods	Preventive Measures
Toxic Metals	Utensils and equipment containing potentially toxic metals, such as lead, copper, brass, zinc, antimony, and cadmium	Any foods, but especially high-acid foods such as sauerkraut, tomatoes, and citrus products; the acidity of these foods can cause metal ions to leach into their liquids	Use only food-grade storage containers
			Use metal and plastic containers only for their intended use
			Use only foodservice brushes on food; do not use paintbrushes or wire brushes
		Carbonated beverages; the carbonated water used to make the beverage can leach copper ions from copper pipes into the water	Do not use enamelware, which may chip and expose the underlying metal
			Do not cook high-acid foods in metal utensils
			Do not use equipment or utensils made of materials that contain lead (such as pewter) for food preparation
			Do not use zinc-coated (galvanized) equipment or utensils in food preparation
			Use a backflow prevention device to prevent carbonated water in soft drink dispensing systems from flowing back into the water intake system
Chemicals	Cleaning products, polishes, lubricants, and sanitizers		Follow manufacturers' directions for storage and use; use only recommended amounts
			Store away from food packaging, utensils, and equipment used for food
			Store in a dry cabinet in original labeled containers, apart from other chemicals that may react with them
			Tools used for dispensing chemicals should never be used on foods
			If chemicals must be transferred to smaller containers or spray bottles, label each container appropriately
			Use only food-grade lubricants or oils on kitchen equipment or utensils
Pesticides	Used in kitchens and food-preparation and storage areas to control pests, such as rodents and insects		Wash fruits and vegetables before preparation or consumption
			Pesticides should be applied only by a trained professional; wrap or store all food before applying pesticides

the food is eaten or several hours later. The reaction could include some or all of the following symptoms.

○ Itching in and around the mouth, face, or scalp

○ Tightening in the throat

○ Wheezing or shortness of breath

○ Hives

○ Swelling of the face and eyes

○ Loss of consciousness

○ Death

Employees should be aware of the most common food allergens, which include milk and dairy products, egg and egg products, fish, shellfish, wheat, soy and soy products, peanuts, and other nuts.

Some people are also allergic to common **food additives** and **preservatives** such as sulfites, nitrites, and monosodium glutamate (MSG). **Nitrites** are preservatives used by the meat industry. Nitrites have been linked to cancer, especially when nitrite-treated meats are over-browned or burned, which produces cancer-causing substances.

Many products contain **sulfites,** which are used to preserve the freshness and/or color of certain foods such as dried and preserved fruits and vegetables. Sulfites can cause allergic reactions in some people, especially those with asthma. These reactions include nausea, diarrhea, asthma attacks, and in some cases, loss of consciousness.

Monosodium glutamate (MSG) is used to enhance flavor in many packaged foods and is included on the federal government's **GRAS (Generally Regarded as Safe)** list of chemicals that are safe to use in food. However, in some people, monosodium glutamate can cause flushing, dizziness, headache, a dry burning throat, and nausea. Use only the recommended amount of MSG in recipes.

Your employees should be able to inform customers of menu items that contain these potential allergens. Designate one person per shift to answer customers' questions regarding your menu items. Keep the following points in mind to help customers with allergies enjoy a safe meal at your establishment.

○ Be able to fully describe each of your menu items when asked. Tell customers how the item is prepared and identify any "secret" ingredients that are used.

○ If you don't know if an item is free of an allergen, tell the customer. Urge the customer to order something else.

○ When preparing food for a customer with allergies, ensure that the food makes no contact with the ingredient that the customer is allergic to. Make sure all cookware, utensils, and tableware are allergen free to prevent food contamination.

○ Serve menu items as simply as possible to customers with allergies. Sauces and garnishes are often the source of allergic reactions. Serve these items on the side.

SUMMARY

Biological and chemical toxins are responsible for many outbreaks of foodborne illness. Most occur naturally and are not caused by the presence of microorganisms. Some occur in the animal as a result of its diet. Since toxins are not living organisms, cooking or freezing typically will not destroy them. Most measures taken to prevent foodborne intoxications center on proper purchasing and receiving.

There are several things that a manager can do to help prevent seafood-specific toxins from causing foodborne illness. Purchase seafood from reputable suppliers who maintain strict time-temperature controls and can certify that the seafood has been harvested from safe waters. Make sure that fish is received at 41°F (5°C) or lower, and refuse fish that has been thawed and refrozen. Thaw fish at refrigeration temperatures.

Toxins are a natural part of some plants. Some plants may be toxic in their raw state, but safe when properly cooked. In general, toxic plant species and products prepared with them should be avoided. Foodborne-illness outbreaks associated with mushrooms are almost always caused by the consumption of wild mushrooms which have been misidentified. Establishments should not use wild mushrooms or products made with them. All mushrooms should be purchased from approved suppliers.

Chemical contaminants are responsible for many cases of foodborne illness. Contamination can come from a variety of substances normally found in the establishment, such as toxic metals, pesticides, and cleaning chemicals. Only food-grade utensils and equipment should be used to prepare and store food. Cleaning products, polishes, lubricants, and sanitizers should be used as directed. Use caution when using these chemicals during operating hours and store them properly. If used, pesticides should be applied only by a trained Pest Control Operator (PCO).

Physical contamination results from the accidental introduction of foreign objects into foods. Closely inspect the foods you receive, and take steps to ensure that food will not become physically contaminated during its flow through your operation.

Some people are allergic to common food additives and preservatives such as sulfites, nitrites, and monosodium glutamate (MSG). You should be able to inform customers of these and other potential food allergens that may be included in the food that is served at your establishment.

A CASE IN POINT I

Case Study

Roberto receives a shipment of frozen mahi-mahi steaks. The steaks are frozen solid at the time of delivery, and the packages are sealed and contain a lot of ice crystals. Roberto accepts the mahi-mahi steaks and thaws them at a refrigeration temperature of 38°F (3°C). The thawed fish steaks are then held at this temperature during the evening shift, and are cooked to order. The cooks follow all appropriate guidelines for preparing, cooking, holding, and serving the fish, monitoring time and temperature throughout the process. Unfortunately, these fish steaks are implicated in an outbreak of scombroid intoxication. Explain why this may have happened.

A CASE IN POINT II

Case Study

A busboy, Mark, has been asked to spray for ants in the dry storage room. He makes sure that all foods are covered, then sprays along the ants' trail. He then places the can of spray on the top shelf in the dry-storage room, along with the other pesticides and foodservice chemicals, and returns to busing tables. What has Mark done correctly? What unsafe practices are evident?

TRAINING TIPS

Training Tips for the Classroom

1. Pick Your Poison

Objective: *After completing this activity, class participants will be able to identify foods often associated with biological toxins, and determine preventive measures that can be taken to keep these toxins from causing foodborne illness.*

Directions: Using *Exhibit 3d* as a reference, create a restaurant menu that includes (but is not limited to) the list of foods often associated with biological toxins. Break the class into teams of two. Pass out a copy of the menu to each team and ask them to do the following:

1. Circle the menu items which are often associated with a biological toxin.
2. Determine preventive measures that can be taken to keep these toxins from causing a foodborne illness.

Ask the teams to present their ideas to the rest of the class. Discuss these ideas, and create a list of preventive measures for each menu item.

2. "You Found What In Your Soup?"

Objective: *After completing this activity, class participants will be able to identify common physical and chemical contaminants, and determine methods to prevent them from occurring.*

Directions: Have your students take out a blank sheet of paper. Ask them to list as many examples of physical and chemical contamination of food as they can think of in exactly two minutes.

When time has expired, randomly solicit examples of physical contamination. After each example is given, ask if anyone in the class has ever experienced this type of physical contamination before, as a foodhandler or customer. If so, ask the following questions:

○ How do you think it might have happened?
○ Could it have caused illness or injury?
○ If not, what other type of "damage" did it cause?
○ How would you have responded to such an incident?
○ How could it have been prevented from happening?

After the examples of physical contamination have been exhausted, repeat the process with examples of chemical contamination. Create a checklist of safe foodhandling practices and policies that will prevent both physical and chemical contamination from occurring in an establishment.

TRAINING TIPS

Training Tips on the Job

1. Developing a "Customer Complaint" Form

Purpose: *To document customer complaints regarding food items. Information on this form can then be used to identify corrective actions that may need to be made in the establishment's processes, procedures, or facility.*

Directions: Create a customer complaint form with the following headings:

Date   **Item**   **Reason Sent Back**   **Corrective Action**

Require employees to document incidents that resulted in the customer sending food back to the kitchen. Analyze customer complaints to determine if the complaint was the result of a hazard that requires a change in current processes or procedures.

Share customer complaints with employees, and get their input regarding corrective actions and preventive measures that should be taken. Explain to employees the importance of documenting customer complaints. Make sure that employees are not penalized for telling the truth, especially since the complaint may be the result of a mistake they made.

2. Performing a Physical or Chemical Hazard Inspection

Purpose: *To determine how food might become physically or chemically contaminated in your facility and to identify corrective actions that must be taken.*

Directions: Using the sources of contamination identified in *Exhibits 3e* and *3f,* conduct an inspection of your facility to identify any possible problems. Solicit help from your foodservice team to do this. Look at the layout of facility itself and the foodhandling procedures that are followed during receiving and storage, preparation and cooking, and holding and serving. Identify places where food can become physically or chemically contaminated. Look for possible "worst-case scenarios."

Next, identify preventive measures that can be taken. Let the preventive measures identified in *Exhibit 3f* serve as a guide. Some corrective actions can be made immediately (such as simple changes in procedures), while others may require more time (equipment changes).

Keep your foodservice team involved in this process, especially if changes will affect their work habits or current procedures. Everyone must be notified of any changes in processes or procedures, and the reasons for these changes must be clearly explained.

DISCUSSION QUESTIONS

1. How do the causes of scombroid intoxication differ from the other biological toxins listed in *Exhibit 3d*?

2. What preventive measures should be taken to guard against a seafood-specific foodborne illness?

3. A restaurant is serving broiled red snapper, along with side dishes of asparagus and rice with wild mushrooms. Which of these foods are at risk for containing biological toxins, and why?

4. Explain why serving orange juice from an enamelware pitcher is inadvisable.

5. Mike is serving drinks and cannot find the ice scoop. He uses a clean, sanitary glass tumbler as temporary ice scoop. Is this a safe practice?

MULTIPLE-CHOICE STUDY QUESTIONS

1. You have ordered some frozen tuna steaks for your restaurant. When the delivery arrives, you notice that there is excessive frost and ice in the package. You refuse the delivery. Why?

   A. You suspect that the steaks may have been contaminated with a cleaning compound.

   B. You suspect the fish may contain ciguatoxins.

   C. You suspect that the steaks have been temperature-abused and may cause scombroid intoxication if you serve them.

   D. You believe the supply company may have treated the steaks with an unauthorized preservative.

2. Which food ingredient could cause stew to become chemically contaminated?

   A. Canned tomatoes   C. Dry mustard
   B. Cubed beef   D. Wild mushrooms

3. Your pest-control company sprayed during the overnight closing hours. When you arrived the next morning, you noticed that some apples and grapes had been left out on the counter that night. What should you do?

   A. Discard the fruit.

   B. Wash the fruit under running water before serving it.

   C. Put the fruit in the refrigerator immediately.

   D. Call the pest control company to find out if they had sprayed near the fruit.

4. You scoop some ice from your restaurant's ice storage bin and notice a drinking glass inside. The glass is missing a large piece of the rim. You find the chunk at the bottom of the storage bin. What should you do?

   A. Inspect each cube carefully by hand before serving the ice in beverages.

   B. Discard all of the ice and clean out the bin before restarting the icemaker.

   C. Transfer the ice to a clean container and flush out the bin before putting the ice back.

   D. No action is needed since you found both pieces of the glass.

5. Which situation could result in the physical contamination of a food-preparation area?

   A. A foodservice worker forgets to wear a hat while working in the area.

   B. The exhaust hood above the area is made of stainless steel.

   C. MSG, which was used to prepare a previous item, was spilled in the area.

   D. The cutting board used in the area is made of wood.

6. Which of the following fish are not associated with ciguatera?

   A. Barracuda   C. Snapper

   B. Grouper   D. Puffer fish

7. Which action should an establishment take to prevent the contamination of food?

   A. Do not use fresh fruits and vegetables that have been chemically treated while growing.

   B. Purchase food products from approved suppliers only.

   C. Keep high-acid foods separate from other types of foods.

   D. Store foods only in clean glass containers.

8. Which of the following items is an example of a physical food contaminant?

   A. A wild mushroom
   B. Dirt on a head of lettuce
   C. A temperature-abused tuna steak
   D. Sulfites used to preserve a package of dried apricots

9. All of the following can lead to the contamination of food except

   A. cooking tomato sauce in a copper pot.
   B. storing orange juice in a pewter pitcher.
   C. using a backflow prevention device on a carbonated beverage dispenser.
   D. serving fruit punch in a galvanized tub.

10. Which of the following statements is true about fish containing ciguatera or scombroid (histamine) toxins?

    A. Freezing will destroy these toxins.
    B. Cooking will not destroy these toxins.
    C. Proper receiving will prevent these toxins from making a person sick.
    D. Cooking will destroy these toxins.

ADDITIONAL RESOURCES

Books and Periodicals

American Academy of Allergy, Asthma and Immunology & The Food Allergy Network. (n.d.). *What you need to know about food allergies.* [Brochure]. Washington DC: National Restaurant Association.

Benenson, A. (Ed.). (1995). *Control of communicable diseases in man* (16th ed.). Washington, DC: American Public Health Association.

Bush, R. K. (1995). Seafood allergy and allergens: A review. *Food Technology, 49*(10), 103-116.

Dulen, J. (1998). Under the microscope. *Restaurants & Institutions, 108*(21), 80.

Grossbauer, S. (1998). Food safety on the Web. *FoodService Director, 11*(2), 140.

How to prevent foodborne illness. (1995). *FoodService Director, 8*(2), 150.

ICMSF (1996). *Microorganisms in foods 5: Characteristics of microbial pathogens.* London: Blackie Academic & Professional.

The National Restaurant Association. (1996). *Sanitation survival kit.* Washington, DC: Author.

Neumann, R. (1998). The eight most frequent causes of foodborne illness. *Food Management, 33*(6), 28.

Puzo, D. P. (1997). USDA estimates financial costs of foodborne illness. *Restaurants & Institutions, 107*(16), 80.

Rehe, S. (1990). *Preventing foodborne illness: A guide to safe foodhandling.* Washington DC: United States Department of Agriculture Food Safety & Inspection Service.

Rubin, K. W. (1995). What to look for, how to prevent it: Food poisoning. *FoodService Director, 8*(11), 72.

Spake, A. (1997). O is for outbreak. *U.S. News & World Report, 123*(20), 70.

Web Sites

1999 FDA Food Model Code

http://vm.cfsan.fda.gov/~dms/fc99-toc.html

Complete outline of the FDA's latest code for regulating operations that provide food directly to consumers. Also includes a quick synopsis of changes from the 1997 Food Code.

American Public Health Association (APHA)

http://www.apha.org

Information on disease prevention and health promotion as well as other resources for public health professionals.

The Bad Bug Book

http://vm.cfsan.fda.gov/~mow/intro.html

This on-line handbook provides basic facts regarding foodborne pathogenic microorganisms and natural toxins. It brings together in one place information from the Food and Drug Administration, the Centers for Disease Control and Prevention, the USDA Food Safety Inspection Service, and the National Institutes of Health.

Centers for Disease Control and Prevention (CDC)

http://www.cdc.gov

Data and statistics, publications, and other health information from the United States Centers for Disease Control.

FDA Center for Food Safety and Applied Nutrition (FDA-CFSAN)

http://vm.cfsan.fda.gov/list.html

Comprehensive site from CFSAN offers a wealth of food-safety information, from foodborne illness to food labeling. CFSAN strives to be a leader in food safety, and to protect consumers from economic fraud, promote sound nutrition, and encourage innovation.

FDA-CFSAN: Pesticides, Metals, and Chemical Contaminants

http://vm.cfsan.fda.gov/~lrd/pestadd.html

Technical information on pesticides, toxic metals, and chemicals that can adversely affect food.

FDA Seafood and Information Resources

http://vm.cfsan.fda.gov/seafood1.html

Web site for the FDA's Center for Food Safety and Applied Nutrition includes seafood information and resources, including seafood pathogens and contaminants.

FightBac!

http://www.fightbac.org

The Web site for the Partnership for Food Safety Education, a coalition of government, industry, and consumer groups.

The Food Allergy Network

http://www.foodallergy.org

Information about food allergies.

Institute of Food Technologists (IFT)

http://www.ift.org

IFT is a nonprofit scientific society with 28,000 members working in food science, food technology, and related professions in industry, academia, and government. This site includes information on its policies and publications, education and industry news, and meeting and convention locations.

International Association of Milk, Food, and Environmental Sanitation (IAMFES)

http://www.iamfes.org

IAMFES keeps members informed of the latest scientific, technical, and practical developments in food safety and sanitation. Their Web site includes booklets and links to other food-safety sites.

International Commission on Microbiological Specifications for Food (ICMSF)

http://www.dfst.csiro.au/icmsf.htm

ICMSF provides timely, science-based guidance to government and industry on appraising and controlling the microbiological safety of foods.

Morbidity and Mortality Weekly Report (MMWR)

http://www.cdc.gov/epo/mmwr/mmwr_ss.html

The MMWR page on the Centers for Disease Control's Web site allows you to search the CDC's publications for information on foodborne illness.

National Center for Infectious Diseases (NCID)

http://www.cdc.gov/ncidod

The CDC's National Center for Infectious Diseases offers information on the efforts to prevent and control diseases caused by bacteria and fungi, as well as fact sheets and articles on foodborne and diarrheal diseases.

National Food Safety Database

http://www.foodsafety.org

A compilation of food-safety database information from government, consumer, and public health organizations, this is a one-stop Web site for food-safety information on the Internet.

National Restaurant Association

http://www.restaurant.org

The National Restaurant Association site provides information on government agencies affecting the restaurant industry, the latest training and certification updates, and links to state restaurant associations and hospitality schools and universities.

National Shellfish Sanitation Program Manual of Operation

http://vm.cfsan.fda.gov/~ear/nsspman.html

An outline of the FDA's Center for Food Safety and Applied Nutrition's National Shellfish Sanitation Program Manual of Operation, which ensures safe molluscan shellfish.

United States Department of Agriculture (USDA)

http://www.usda.gov

The United States Department of Agriculture's Web site features information, publications, and other educational materials about the nation's agriculture.

United States Department of Agriculture's Food Safety and Inspection Service

http://www.fsis.usda.gov

The USDA's Food Safety and Inspection Web site offers the latest food-safety news, educational materials, and HACCP-implementation materials.

Chapter 4
The Safe Foodhandler

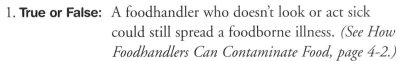

TEST YOUR FOOD-SAFETY KNOWLEDGE

1. **True or False:** A foodhandler who doesn't look or act sick could still spread a foodborne illness. *(See How Foodhandlers Can Contaminate Food, page 4-2.)*

2. **True or False:** Wearing gloves is an acceptable substitute for handwashing. *(See Use of Gloves, page 4-7.)*

3. **True or False:** Foodhandlers must always wash their hands after smoking. *(See Handwashing, page 4-6.)*

4. **True or False:** A foodhandler who has been diagnosed with Salmonellosis cannot continue to work at an establishment while he has the disease. *(See Policies for Reporting Illness and Injury, page 4-11.)*

5. **True or False:** If a foodhandler receives a minor cut on her hand, she should be sent home immediately. *(See Policies for Reporting Illness and Injury, page 4-11.)*

Learning Objectives

After completing this chapter, you should be able to:

○ Identify personal behaviors that can contaminate food.

○ Identify proper hand-washing procedures.

○ Respond properly to food-handler cuts, wounds, and sores to ensure food safety.

○ Identify procedures that foodhandlers must follow when using gloves.

○ Identify foodhandler health problems that are a possible threat to food safety and determine appropriate actions which should be taken.

○ Give examples of clothing, accessories, and other items worn by foodhandlers that pose a threat to food safety.

○ Identify policies that should be implemented at the establishment regarding eating, drinking, and smoking while working.

Key Terms

Gastrointestinal illness
Infected lesion
Hepatitis A
Carriers

Personal hygiene
Single-use paper towel
Hand sanitizers

Finger cot
Hair restraint
Jaundice

Customers frequently judge an establishment by observing the appearance and behavior of the foodhandler serving them. Good personal hygiene is a critical protective measure against foodborne illness, and customers expect it. By establishing a personal hygiene program that includes specific policies, and by training and enforcing those policies, you can minimize your risk of foodborne illness incidents and lost business.

It is ironic that *people* are the cause and the victims in foodborne illness incidents. At every step in the flow of food through the operation, from receiving through final service, foodhandlers can contaminate food and cause customers to become ill. The manager who wants to provide safe food must build a sanitary barrier between the product and the people who prepare, serve, and consume it. This requires trained foodhandlers who possess the knowledge, skills, and attitude necessary to operate a safe food system.

HOW FOODHANDLERS CAN CONTAMINATE FOOD

In previous chapters you learned that foodborne illness is caused by microorganisms, which can often be transmitted to food by foodhandlers. Foodhandlers can contaminate food when:

○ They have been diagnosed with a foodborne illness.

○ They show symptoms of **gastrointestinal illness** (an illness relating to the stomach or intestine).

○ They have **infected lesions** (infected wound or injury).

○ They live with or are exposed to a person who is ill.

○ They touch anything that may contaminate their hands.

Even an apparently healthy person may be hosting dangerous pathogens. With some diseases, such as Hepatitis A, an extremely hearty virus, an individual is at the most infectious stage of the disease for several weeks before symptoms appear. With others diseases, such as Salmonellosis, the *Salmonella* bacteria may remain in a person's system for months after all signs of infection have ceased. Some people are called **carriers** because they may carry pathogens, yet never become ill themselves.

The next three paragraphs provide examples which will help illustrate the routes by which food can become contaminated by employees.

A deli foodhandler who was diagnosed with Salmonellosis failed to inform his manager that he was ill for fear of losing wages. It was later determined that he was the cause of an outbreak that involved more than 200 customers through twelve different products.

A foodhandler suffering from diarrhea, a symptom of gastrointestinal illness, didn't wash his hands and made approximately five thousand people ill when he mixed a vat of buttercream frosting with his bare hands and arms. Typical symptoms of a gastrointestinal illness include diarrhea, fever, vomiting, jaundice, or sore throat with fever. Another large outbreak of foodborne illness was caused by a foodhandler who scratched an infected facial lesion and then handled a large amount of sliced pepperoni.

A foodborne illness outbreak was traced to a woman who prepared food for a dinner party. The investigation revealed that the woman was caring for her infant son, who had diarrhea. The woman could not recall washing her hands after changing the infant's diaper. As a result, twelve of her dinner guests became violently ill with symptoms that included diarrhea and vomiting.

Simple acts such as nose picking, rubbing an ear, scratching the scalp, touching a pimple or an open sore, or running fingers through the hair can contaminate food. Thirty to 50 percent of healthy adults carry staphylococci harmlessly in their noses, and about 20 to 35 percent carry them on their skin. If these microorganisms contaminate a foodhandler's hands which then touch food, the consequences can be severe. Because of these factors, foodhandlers must pay close attention to what they do with their hands and wash them often.

Key Point

Simple acts such as rubbing an ear, scratching the scalp, or touching a pimple or sore can contaminate food.

DISEASES NOT TRANSMITTED THROUGH FOOD

In recent years, the public has expressed growing concern over communicable diseases that are spread through intimate contact or by direct exchange of bodily fluids. Diseases such as AIDS (Acquired Immune Deficiency Syndrome) and Hepatitis B are not spread through food. These diseases are spread through blood transfusions, shared needles, and sexual contact, not through casual contact.

Although these diseases are not transmitted through food, as a manager you should be aware of the following laws concerning employees that are HIV- (Human Immunodeficiency Virus) positive or have Hepatitis B.

○ The Americans with Disabilities Act (ADA) provides civil rights protection to individuals who are HIV-positive or have Hepatitis B, and thus prohibits employers from firing people or transferring them out of foodhandling duties simply because they have these diseases.

○ Employers must maintain the confidentiality of employees who have any non-foodborne illness.

Key Point

Diseases such as AIDS or Hepatitis B are not spread through food.

Key Point

Managers must train their foodhandlers to wash their hands properly and when necessary.

COMPONENTS OF A GOOD PERSONAL HYGIENE PROGRAM

Good **personal hygiene** is key to the prevention of foodborne illness. Good personal hygiene includes:

○ Hygienic hand practices

○ Maintaining personal cleanliness

○ Wearing clean and appropriate uniforms and following dress codes

○ Avoiding unsanitary habits and actions

○ Maintaining good health

○ Reporting illnesses

Hygienic Hand Practices

Handwashing

While it may appear fundamental, many foodhandlers fail to wash their hands properly and as often as needed. As a manager, it is your responsibility to train your foodhandlers and then monitor them to make sure that they are washing their hands properly and when necessary. Never take this simple action for granted.

There are products on the market which help teach proper handwashing techniques. These products are applied to the hands, which are then washed normally. The employees' hands are then placed under illumination to reveal how well they were washed *(see Exhibit 4a)*.

To ensure proper handwashing in your establishment, train your foodhandlers to use the following steps *(See Exhibit 4b)*.

Step 1: Wet your hands with hot running water. The water should be as hot as the hands can comfortably stand, approximately 110°F (43°C).

Step 2: Apply soap. Apply enough soap to build up a good lather.

Step 3: Rub hands together for at least twenty seconds. Lather well beyond the wrists and the exposed portions of the arms.

Step 4: Clean under fingernails and between fingers. A nail brush is recommended.

Step 5: Rinse hands thoroughly under running water. Turn off the faucet using a sanitary **single-use paper towel** if available.

Step 6: Dry Hands. Use single-use paper towels or a warm-air hand dryer. Never use aprons or wiping cloths to dry hands after washing.

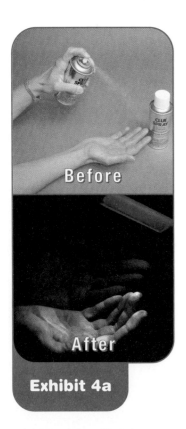

Before

After

Exhibit 4a

Handwashing Training Products

Pictures provided courtesy of Brevis Corporation, 3310 S. 2700 East, Salt Lake City, UT 84109, 800-383-3377.

1. Wet your hands with hot running water.

2. Apply soap.

3. Rub hands together for at least twenty seconds.

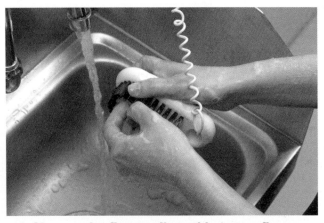

4. Clean under fingernails and between fingers.

5. Rinse hands thoroughly under running water.

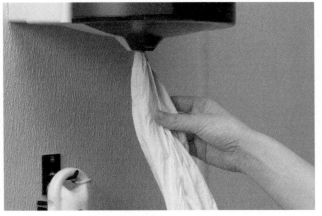

6. Dry hands.

Exhibit 4b   Proper Handwashing Procedure

Hand
sanitizers
should never be used
in place of proper
handwashing.

Food-
handlers
must keep fingernails
short and clean and
should not wear false
nails or nail polish.

Hand cuts
or sores
should be covered with
a clean bandage and a
glove or finger cot.

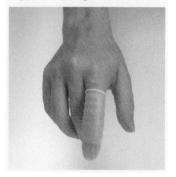

Approved **hand sanitizers** (a liquid used to lower the number of microorganisms on the surface of the skin) or hand dips may be used after washing, but should never be used in place of proper handwashing. If hand sanitizers are used, foodhandlers should never touch food or food-preparation equipment until the hand sanitizer has dried.

Foodhandlers must wash their hands after the following activities:

○ After using the restroom
○ Before and after handling raw foods
○ After touching the hair, face, or body
○ After sneezing, coughing, or using a handkerchief or tissue
○ After smoking, eating, drinking, or chewing gum or tobacco
○ After using any cleaning, polishing, or sanitizing chemical
○ After taking out garbage or trash
○ After clearing tables or busing dirty dishes
○ After touching clothing or aprons
○ After touching anything else that may contaminate hands, such as unsanitized equipment, work surfaces, or wash cloths

Hand Maintenance

In addition to proper washing, hands need other regular care to ensure that they will not transfer contaminants to food. Food, filth, and harmful substances can get caught under both long and short fingernails. Nail polish can disguise dirt under nails and may flake off into food. Long fingernails, false fingernails, and acrylic nails may be difficult to keep clean and can break off into food and therefore should not be worn while handling food. Fingernails should be kept short and clean. Nail polish should not be worn if the employee will be handling food.

Cuts and sores on hands, including hangnails, should be treated and kept covered with clean bandages. If hands are bandaged, clean gloves or **finger cots**, a protective covering, should be worn at all times to protect the bandage and to prevent it from falling off into food. You may need to move the foodhandler to another job, where he or she will not handle food or touch food-contact surfaces, until the injury heals. To ensure food safety, good hand maintenance must be learned and practiced without fail.

Use of Gloves

Gloves can protect hands from cuts and the effects of detergents and chemicals. They can also help ensure food safety by creating a barrier between hands and food (tongs or other utensils, or barriers such as deli tissue, can also be used to protect food). *See Exhibit 4c.* This is so important that some jurisdictions require the use of gloves for foodhandlers who work with ready-to-eat foods. However, other jurisdictions will allow an exception for the use of gloves if the establishment has a strong written policy about handwashing procedures that can be verified. Check with your local regulatory agency for requirements in your jurisdiction.

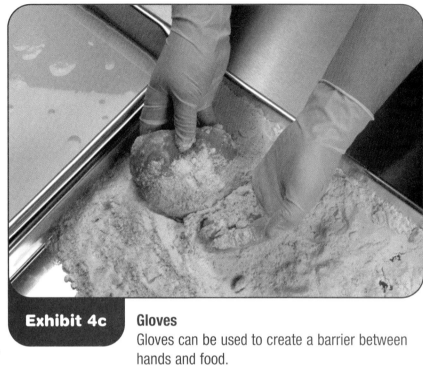

Exhibit 4c   **Gloves**
Gloves can be used to create a barrier between hands and food.

Gloves must never be used in place of handwashing. Foodhandlers must wash their hands before putting on gloves and when changing to a fresh pair. Gloves used to handle food are for single use only and should never be washed and re-used. Foodhandlers should change their gloves when necessary. Gloves should be changed:

○ As soon as they become soiled or torn

○ Before beginning a different task

○ At least every four hours during continual use, and more often when necessary

○ After handling raw meat and before handling cooked or ready-to-eat foods

Often, foodhandlers consider gloves more sanitary than bare hands. Because of this false sense of security, they might not change gloves as often as necessary. Training and supervision are the keys to avoiding this problem.

Managers must reinforce the habit of proper hand sanitation among foodhandlers. Proper hand sanitation may be hard to remember at first, but when practiced and reinforced, it can become a lifetime habit that will help prevent foodborne illness.

Key Point

Gloves must never be used in place of handwashing.

Key Point

Food-
handlers
must keep their hair
clean since oily hair
can harbor pathogenic
microorganisms.

Key Point

Food-
handlers
must always wear clean
clothes since dirty
clothes may harbor
disease-causing
microorganisms.

Key Point

Food-
handlers
must remove jewelry
prior to preparing or
serving food because
it can harbor
microorganisms.

Other Good Personal-Hygiene Practices

Personal hygiene can be a sensitive subject for some people, but because personal cleanliness and behaviors are vital to food safety, as a manager, you must address the subject with every foodhandler.

General Personal Cleanliness

In addition to following proper hand-hygiene practices, your foodhandlers must maintain personal cleanliness. Foodhandlers should bathe or shower before work. They must also keep their hair clean. Oily, dirty hair can harbor pathogenic microorganisms, and dandruff may fall into food or onto food-contact surfaces.

Proper Work Attire

A foodhandler's attire plays an important role in the prevention of foodborne illness, so it is essential that foodhandlers observe strict dress-code standards (see Exhibit 4d). Dirty clothes may harbor disease-causing microorganisms and give customers a bad impression of your establishment.

Managers should make sure that foodhandlers observe the following guidelines regarding their attire.

○ **Wear a clean hat or other hair restraint.** A **hair restraint** will keep hair away from food and keep the foodhandler from touching their hair. Foodhandlers with long beards should also wear beard restraints.

○ **Wear clean clothing daily.** Work clothes should be chosen to avoid contact with food or equipment, and to lessen the need for foodhandlers to touch their clothes to make adjustments. If possible, foodhandlers should put on work clothes at the establishment.

○ **Remove aprons when leaving food-prep areas.** For example, aprons should be removed prior to taking out garbage or using the restroom.

○ **Wear appropriate shoes.** Wear clean, closed-toed shoes with a sensible, non-slip sole.

○ **Remove jewelry prior to preparing or serving food or while around food-preparation areas.** Jewelry can harbor microorganisms, often tempts foodhandlers to touch it, and may pose a safety hazard around equipment. Remove rings (except for a plain wedding band), bracelets, watches, earrings, necklaces, and facial jewelry (such as nose rings, etc.).

Check with your local regulatory agency regarding requirements. These requirements should be written policies that are consistently monitored and enforced. All potential employees should be presented with these policies prior to employment.

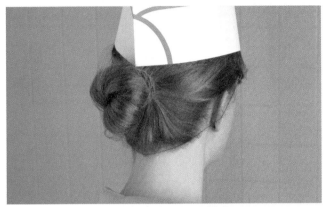

Hair properly restrained

Hair improperly restrained

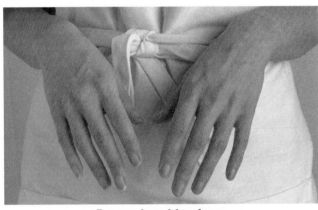

Proper hand hygiene:
clean, short fingernails; no jewelry or nail polish

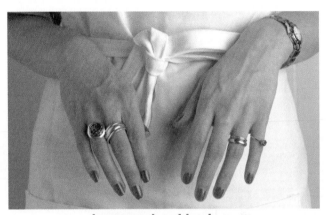

Improper hand hygiene:
long fingernails, jewelry, nail polish

Proper apron: clean

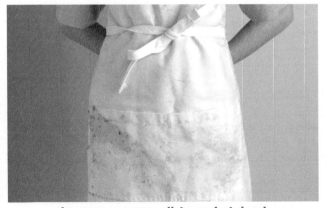

Improper apron: dirty and stained

Exhibit 4d   Proper and Improper Attire on the Job

Proper tasting process

Proper tasting process

Improper tasting process

Exhibit 4e   Safe and Unsafe Food Tasting Methods

Policies Regarding Eating, Drinking, Chewing Gum, and Tobacco

Small droplets of saliva can contain thousands of disease-causing microorganisms. In the process of eating, drinking, chewing gum and tobacco, and smoking, this saliva can be transferred to the foodhandler's hands or directly to the food that they are handling. For this reason, foodhandlers must follow strict policies regarding these activities.

Managers must implement the following policies in their establishments.

○ Foodhandlers must not smoke or chew gum or tobacco, while preparing or serving food, while in food-preparation areas, and while in areas used for equipment and utensil washing.

○ Foodhandlers must not eat or drink while in food-preparation areas or in areas used to clean utensils and equipment (with the exception of chefs properly tasting foods). Some jurisdictions allow employees to drink from a covered container with a straw. Check with your local regulatory agency.

Key Point

Food-handlers must not eat, drink, smoke, or chew gum or tobacco while preparing or serving food.

○ Foodhandlers should eat, drink, chew gum, or use tobacco products only in designated areas, such as an employee break room.

○ Foodhandlers should never spit in the establishment.

If chefs need to taste food during preparation, they must ladle a small amount into a separate dish and taste the food with a clean spoon. The dish and spoon should then be removed from the food-preparation area for cleaning and sanitizing *(see Exhibit 4e)*.

Policies for Reporting Illness and Injury

Foodhandlers must report health problems to the manager of the restaurant or establishment before working with food. If they become ill or are injured while working, they must report their condition immediately to their manager or supervisor. If the foodhandler's condition could possibly contaminate food or foodservice equipment, he or she must stop working and see a doctor. If the foodhandler must take medication while working, the medicine must be stored with his or her personal belongings away from areas where food is prepared, served, and stored.

According to the FDA Model Food Code, managers must exclude from the establishment foodhandlers who have been diagnosed with a foodborne illness, and notify the local regulatory agency. Managers must also exclude foodhandlers from working with or around food if they have the following symptoms:

○ Fever

○ Diarrhea

○ Vomiting

○ Sore throat

○ **Jaundice** (yellow skin and eyes)

Managers must work with local regulatory agencies to determine when foodhandlers can safely return to work.

Any cuts, burns, boils, sores, skin infections, or infected wounds should be covered with a bandage when the foodhandler is working with or around food or food-contact surfaces. Bandages should be clean, dry, and must prevent leakage from the wound. As mentioned previously, waterproof disposable gloves or finger cots should be worn over bandages on hands. Foodhandlers wearing bandages may need to be temporarily reassigned to duties which do not involve contact with food or food-contact surfaces.

Vaccination for Hepatitis A

Hepatitis A is a disease that causes inflammation of the liver and is transmitted to food by poor personal hygiene or contact with contaminated water. It infects many people each year, resulting in community-wide outbreaks. Of all foodborne illnesses facing the foodservice industry, Hepatitis A is the only one that can be prevented by vaccine. While effective handwashing is a critical practice to prevent contamination, vaccinating your foodhandlers for Hepatitis A can provide an additional barrier. This may be especially recommended in areas where Hepatitis A outbreaks are highly prevalent.

Health Alert

Managers must not allow foodhandlers to work at the establishment if they have been diagnosed with a foodborne illness.

Health Alert

Foodhandlers must not work with or around food if they have symptoms which include fever, diarrhea, vomiting, a sore throat, or jaundice

Health Alert

Vaccination and effective handwashing can help prevent an outbreak of Hepatitis A.

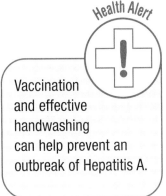

MANAGEMENT'S ROLE IN A PERSONAL HYGIENE PROGRAM

Management plays a critical role in the effectiveness of a personal hygiene program. Responsibilities include:

○ Modeling proper behavior for foodhandlers at all times

○ Establishing proper personal hygiene policies. Policies should be revised when laws and regulations change, and when changes are recognized regarding the science of food safety. Foodhandlers should be retrained as necessary

○ Training foodhandlers on personal hygiene policies

○ Continuously supervising sanitary practices, and retraining foodhandlers as necessary

Job Assignments

When job descriptions are developed and responsibilities are assigned, consider the risk of cross-contamination and plan tasks to prevent it. The risk may be higher if foodhandlers are required to perform several different duties than if specific foodhandlers are assigned to each duty. For example, an employee who is expected to prepare and wrap food, clear off tables, and then return to food preparation duties could more easily contaminate food than an employee who is assigned to just one of these tasks. By planning tasks to prevent cross-contamination, you will minimize the amount of time needed for supervision and enable employees to follow sanitation rules more easily.

SUMMARY

Foodhandlers can contaminate food at every step in its flow through the establishment. Good personal hygiene is a critical protective measure against contamination and foodborne illness. A successful personal hygiene program depends on trained foodhandlers who possess the knowledge, skills, and attitude necessary to maintain a safe food system.

Foodhandlers have the potential to contaminate food when they have been diagnosed with a foodborne illness, when they show symptoms of a gastrointestinal illness, when they have infected lesions, or when they touch anything that may contaminate their hands. Foodhandlers must pay close attention to what they do with their hands since simple acts such as nose picking or running fingers through the hair can contaminate food. Proper handwashing must also be practiced. This is especially important after using the restroom, before and after handling raw food, after sneezing and coughing, and after smoking, eating, or drinking. It is up to the manager to monitor handwashing to make sure it is thorough and frequent. In addition, hands need other care to ensure that they will not transfer contaminants to food. Fingernails should be kept short and clean. Cuts and sores should be covered with clean bandages, and covered with a glove or finger cot.

Gloves may be used to create a barrier between hands and food; however, they should never be used in place of handwashing. Foodhandlers must wash their hands before putting on gloves and when changing to a fresh pair. Gloves used to handle food are for single use and should never be washed or re-used. They must be changed when necessary.

Personal hygiene can be a sensitive subject for some people, but because it is vital to food safety, it must be addressed with every employee. All employees must maintain personal cleanliness. They should bathe or shower before work and keep their hair clean.

Prior to handling food, foodhandlers must put on a clean hair restraint, put on clean clothing, remove jewelry, and put on appropriate shoes. Aprons should always be removed when the employee leaves food-preparation areas.

Establishments should implement strict policies regarding eating, drinking, smoking, and chewing gum and tobacco. These activities should not be allowed when the foodhandler is preparing or serving food or working in food-preparation areas.

Foodhandlers must be encouraged to report health problems to management before working with food. If their condition could contaminate food or equipment, they must stop working and see a doctor. Managers must not allow foodhandlers who have been diagnosed with a foodborne illness to work, and must notify the local regulatory agency. Managers must also exclude foodhandlers from working with or around food if they have symptoms that include fever, diarrhea, vomiting, sore throat, and jaundice.

Management plays a critical role in the effectiveness of a personal hygiene program. By establishing a program that includes specific policies, and by training and enforcing those policies, managers can minimize the risk of causing a foodborne illness.

A CASE IN POINT I

Case Study

Chris works at a quick-service restaurant. She is suffering from seasonal allergies, so she carries a small pack of tissues with her. Her assigned responsibility is to make salads. She properly washes her hands and puts on single-use gloves before she starts her shift. When Chris needs to sneeze, she steps away from the food-preparation area, pulls a clean tissue out of her pocket, sneezes into it, then discards it. Because her medication gives her a dry mouth, Chris keeps a glass of water at her station. Her clearly labeled bottle of allergy medication has to be kept cold, so she stores it in the walk-in refrigerator.

Does this situation represent a threat to food safety? Explain why or why not, and tell what Chris is doing right and wrong in this scenario.

A CASE IN POINT II

Case Study

Marty works for a catering company. A few days ago, he was serving hot foods from chafing dishes at an outdoor music festival sponsored by the local community college. He did not wear gloves, because he used spoons and tongs to serve the food. His manager noticed that Marty made multiple trips to the bathroom during his four-hour shift. These trips did not interrupt service to customers because there were plenty of staff members and Marty hurried to and from the restroom.

The nearest restrooms had soap, separate hot and cold water faucets, and a working hot-air dryer, but no paper towels. Each time Marty used the restroom, he washed his hands quickly and then dried them on his apron. Throughout the following week, the manager of the catering company received several telephone calls from people who had attended the music festival and had eaten from the buffet. They each complained of diarrhea, fever, and chills. One call was from a mother of a young boy who was hospitalized for dehydration from the diarrhea. The doctor reported that the boy had shigellosis.

Explain how Marty might have caused an outbreak of shigellosis. What measures should have been taken to prevent such a foodborne illness outbreak?

TRAINING TIPS

Training Tips for the Classroom

1. Habits with Hands Role-Play

Objective: *After completing this activity, class participants should be able to determine how hands can cross-contaminate food, and identify the proper times to wash hands.*

Directions: Ask for three volunteers. Solicit the volunteers prior to teaching Chapter 4, and provide them with enough time to prepare for their role-play. Assign one volunteer to role-play a cook, another a server, and the third, a manager.

Volunteers are to design a three-minute role-play in which they portray their character performing various tasks. While doing so, they should also depict different ways to cross-contaminate food with their hands.

For example, the person playing the cook might handle raw poultry, and then prepare sandwiches without washing her hands, or the server may sneeze into his hands, wipe them on his apron, and then serve dinner rolls to a guest using his bare hands. The goal is to design scenes with as many violations as possible.

Prior to presenting Chapter 4 to the class, tell them that a role-play is going to be performed. Ask class participants to note every incident of cross-contamination that they see during the role-play. At the end of the role-play, discuss the incidents noted, and develop a list which identifies when hands should be washed. Then discuss the proper procedure for handwashing.

2. Yuck

Objective: *After completing this activity, class participants should recognize the importance of proper handwashing.*

Directions: Purchase some petri dishes and petri-agar mix from a science supply store. Follow the directions for mixing the agar for each dish. Issue one petri dish per class participant. Ask each participant to lightly press a finger or thumb into the agar. Cover the petri dishes and place them in a warm location. Allow the dishes to sit for at least forty-eight hours. As a variation of this activity, you might ask participants to wash their hands and make another impression in the agar. Separate these impressions from the first set.

After a couple of days, a growth should appear where the class participants left their finger or thumb impression in the agar. Compare the impressions left by the unwashed hands to the impression left by the washed hands.

This simple activity will show class participants that microorganisms, while invisible to the unaided eye, can be present on the body. It should also emphasize how important it is to properly wash hands.

3. Glo Germ™ Activity

Objective: *After completing this activity, class participants should recognize the importance of proper handwashing.*

Directions: Ask for a volunteer from the class to participate in an experiment. Apply some iridescent "germs" (harmless orange liquid) from a *Glo Germ* kit to the hands of the volunteer and ask him to rub it in. Have him wash his hands.

When the volunteer returns, turn off the lights, and hold a black light over his hands to see how effectively he washed them. Often, the hands will glow around the fingernails and around watches or jewelry. Walk the volunteer around the room and use the black light to show the rest of the class how many germs have survived the handwashing process.

Discuss the importance of proper handwashing with the class.

Training Tips on the Job

1. Gotcha

Purpose: *This activity may help make handwashing a habit at your establishment.*

Directions: Create a list of activities after which handwashing must be performed. Share this list with all of your foodhandlers, and designate a "gotcha" day. The "gotcha" day will be a handwashing monitoring day. Foodhandlers will monitor one another. Whenever a foodhandler is caught not washing his or her hands after one of the listed activities, fellow foodhandlers should shout "gotcha."

By the end of the day it will be evident to the staff that handwashing is easy to forget when they are busy. Handwashing must become a habit at your establishment. Ask the staff to continue the "gotcha" activity until handwashing becomes a habit. If necessary, offer incentives to keep the staff involved in this activity.

2. Creating the Safe Foodhandler

Purpose: *To make foodhandlers aware of personal hygiene standards necessary to keep food safe in your establishment.*

Directions: Develop a colorful wall poster with a drawing of "The Safe Foodhandler." Make the drawing life-size if possible. Noted on the drawing should be all of the aspects of good hygiene from head to toe, from clean and properly restrained hair to proper closed-toe shoes. Make a checklist of basic hygiene standards such as coming to work healthy, taking a shower or bath before work, etc. Attach the checklist to the poster.

Place your poster in a conspicuous place for all foodhandlers to see.

DISCUSSION QUESTIONS

1. Describe the proper procedure for washing hands.

2. Identify personal behaviors that can contaminate food.

3. How can sneezing and coughing transmit foodborne illness?

4. Identify procedures that foodhandlers must follow when using gloves.

5. Identify employee health problems that are possible threats to food safety and determine appropriate actions that should be taken.

MULTIPLE-CHOICE STUDY QUESTIONS

1. Which of the following personal behaviors can contaminate food?
   A. Touching a pimple   C. Nose picking
   B. Touching hair   D. All of the above

2. After you've washed your hands, which of the following items can you use to dry your hands?
   A. Your apron   C. A common cloth towel
   B. Single-use paper towels D. A wiping cloth

3. Which item of personal apparel would most likely cause food to become unsafe?
   A. Earrings   C. A baseball-type cap
   B. A dark-colored shirt   D. A pair of athletic shoes

4. Which of the following situations is most likely to protect the safety of food by minimizing human hand contact?
   A. A foodhandler uses metal tongs to place cooked beef onto a bun.
   B. A foodhandler sneezes into a tissue before handling food.
   C. A foodhandler uses hand lotion after washing her hands.
   D. A foodhandler drinks from a sanitary plastic cup while preparing food.

5. Which of the following is the proper procedure for washing your hands?
   A. Run hot water, moisten hands and apply soap, rub hands together, apply sanitizer, dry hands.
   B. Run hot water, moisten hands and apply soap, rub hands together, rinse hands, dry hands.
   C. Run cold water, moisten hands and apply soap, rub hands together, rinse hands, dry hands.
   D. Run cold water, moisten hands and apply soap, rub hands together, apply sanitizer, dry hands.

6. Becky has an unhealed sore on the back of one hand. Can Becky perform her regular foodhandling duties?
   A. Yes, if Becky's doctor provides a certificate that the sore is not contagious.
   B. Yes, if Becky agrees to use an antiseptic hand lotion between jobs.
   C. Yes, if the sore is bandaged and Becky wears a glove to protect the bandage.
   D. No, Becky should stay home from work until the sore has healed.

7. Kim wore disposable gloves while she formed raw ground beef into patties. When she was finished, she continued to wear the gloves while she sliced hamburger buns. What mistake did Kim make?
   A. She failed to change her gloves and wash her hands after handling raw meat and before handling a ready-to-eat food item.
   B. She failed to wash her hands before wearing the same gloves to slice the buns.
   C. She failed to wash and sanitize her gloves before handling the buns.
   D. She failed to wear reusable gloves.

8. Which human disease can be transmitted to others through food?
   A. Diabetes   C. AIDS
   B. Hepatitis A   D. Arthritis

9. A foodhandler who has been diagnosed with Shigellosis should be
   A. told to stay home.
   B. told to wear gloves while working with food.
   C. told to wash his hands every fifteen minutes.
   D. assigned to a non-foodhandling position until he is feeling better.

10. Which of the following diseases can be prevented by a vaccine?

    A. Salmonellosis   C. Hepatitis A

    B. Listeriosis   D. *E. coli* O157:H7

11. Management can play a key role in promoting proper personal hygiene by

    A. providing hand lotion in handwashing stations.

    B. providing reusable gloves for foodhandlers.

    C. modeling proper behavior at all times.

    D. permitting smoking in food-preparation areas.

12. Foodhandlers should be excluded from working with or around food if they are experiencing which of the following symptoms?

    A. Fever, itching, fatigue   C. Vomiting, diarrhea, itching

    B. Fever, vomiting, diarrhea D. Fatigue, vomiting, itching

13. Which of the following policies should be implemented at establishments?

    A. Employees must not smoke while preparing or serving food.

    B. Employees must not eat while in food-preparation areas.

    C. Employees must not chew gum or tobacco while preparing or serving food.

    D. All of the above

14. Stephanie has a small cut on her finger and is about to prepare chicken salad. How should Stephanie's manager respond to the situation?

    A. Send Stephanie home immediately.

    B. Cover the hand with a glove or finger cot.

    C. Cover the cut with a clean, dry bandage, and a glove or finger cot.

    D. Cover the cut with a clean bandage.

15. Hands should be washed after which of the following activities?

    A. Touching your hair   C. Sneezing

    B. Drinking   D. All of the above

ADDITIONAL RESOURCES

Books and Periodicals

Cliver, D. O. (1997). Virus transmission via food. *Food Technology, 51*(4), 71-78.

Frable, F., Jr. (1997). 10 common errors in kitchen planning and ways to avoid them. *Nation's Restaurant News, 31*(6), 22.

Hernandez, J. (1998). Food safety: Preparation and cooking. *Food Management, 33*(5), 90.

Longree, K. & Armbruster, G. (1996). *Quantity food sanitation* (5th ed.). New York: Wiley.

More than an ounce of prevention. (1997). *Best Practices, 1*(3), 6-9.

Web Sites

1999 FDA Model Food Code

http://vm.cfsan.fda.gov/~dms/fc99-toc.html

Complete outline of the FDA's latest code for regulating operations that provide food directly to consumers. Also includes a quick synopsis of changes from the 1997 Food Code.

American Public Health Association (APHA)

http://www.apha.org

Information on disease prevention and health promotion as well as other resources for public health professionals.

Centers for Disease Control and Prevention (CDC)

http://www.cdc.gov

Data and statistics, publications, and other health information from the United States Centers for Disease Control.

FDA Center for Food Safety and Applied Nutrition (CFSAN)

http://vm.cfsan.fda.gov/list.html

Comprehensive site from CFSAN offers a wealth of food-safety information, from foodborne illness to food labeling. CFSAN strives to be a leader in food safety, and to protect consumers from economic fraud, promote sound nutrition, and encourage innovation.

FightBac!

http://www.fightbac.org

The Web site for the Partnership for Food Safety Education, a coalition of government, industry, and consumer groups.

Gojo

http://www.gojo.com

The Web site of this manufacturer of hand sanitizers offers product and ordering information.

Handgards, Inc.

http://www.handgards.com

The Web site of this supplier of disposable gloves, aprons, and dispensers offers product and ordering information.

National Center for Infectious Diseases

http://www.cdc.gov/ncidod

The CDC's National Center for Infectious Diseases offers information on the efforts to prevent and control diseases caused by bacteria and fungi, as well as fact sheets and articles on foodborne and diarrheal diseases.

National Restaurant Association

http://www.restaurant.org

The National Restaurant Association site provides information on government agencies affecting the restaurant industry, the latest training and certification updates, and links to state restaurant associations and hospitality schools and universities.

SmithKline Beecham

http://www.sb.com

The Web site for this pharmaceutical company offers product information, health trends, and news.

KEEP IN MIND...

The safety of the food you serve at your establishment will depend largely on your understanding of food-safety concepts throughout the flow of food, as outlined in Chapters 5 through 8. Food safety also depends on your ability to develop a system that prioritizes, monitors, and verifies the most important food-safety practices. This system will be discussed in Chapter 9: Principles of HACCP.

As you read through this unit, it is important to keep in mind that the food-safety concepts discussed tell you **what** to do to keep food safe, and HACCP tells you **how** to consistently keep food safe. If you know what to do, but do not know how to do it, you will not have an effective and complete food-safety system.

Hazard Analysis Critical Control Point (HACCP) is a dynamic system that uses a combination of proper foodhandling procedures, monitoring techniques, and record keeping. The HACCP system enables you to consistently serve safe food by identifying and controlling possible hazards throughout the flow of food. Since HACCP is dynamic, it allows you to continuously improve your food-safety system. Because this is a preventive rather than reactive system, the National Restaurant Association and the Food and Drug Administration recommend that restaurants and other establishments develop and use a HACCP-based food-safety system. In fact, recognizing that the complete flow of food begins on the farm and continues through service, other government agencies and foodservice industries have accepted HACCP as the best food-safety system available. Industry segments such as canning plants, and meat, poultry, and seafood producers and processors are currently required by law to use HACCP.

Foodservice operators or managers who choose to develop a HACCP plan have found additional benefits, which can include improved quality, minimized waste, better product consistency, increased customer satisfaction, improved inspection scores, decreased customer complaints, and improved relations with the regulatory community.

Since your application of HACCP begins even before food arrives at your establishment, our presentation of the flow of food begins with purchasing and receiving (Chapter 5) and continues through storage (Chapter 6), preparation (Chapter 7), and service (Chapter 8). In Chapter 9, we pull the concepts from these chapters together to explain how the seven principles of HACCP are applied.

UNIT 2

THE FLOW OF FOOD THROUGH THE OPERATION

At Taco Bell, we are extremely dedicated to food safety. Serving 50 million customers weekly imposes an enormous responsibility to keep them safe. Ninety-percent of our managers are ServSafe certified. Food safety is one of those things you've got to get right the first time. You really only have one shot.

Gary DuBois
Director of Quality Assurance
And Food Safety
Taco Bell

Chapter 5
Purchasing and Receiving Safe Food

Knowledge

TEST YOUR FOOD-SAFETY KNOWLEDGE

1. **True or False:** Upon arrival, a delivery of fresh fish should be received at 41°F (5°C) or lower. *(See Fish, page 5-7.)*

2. **True or False:** *Sous vide* (vacuum-packed) products should be received at 41°F (5°C) or lower unless specified by manufacturer's directions. *(See MAP, Vacuum-Packed, and Sous Vide Foods, page 5-16.)*

3. **True or False:** A frozen food that is delivered in a frozen state is safe to use. *(See Refrigerated and Frozen Processed Foods, page 5-15.)*

4. **True or False:** If a sack of flour is dry upon delivery, the contents may still be contaminated. *(See Dry and Canned Products, page 5-16.)*

5. **True or False:** A delivery of hot roast beef and baked potatoes should be delivered at 140°F (60°C) or higher. *(See Hot Foods, page 5-18.)*

Key Terms

Temperature abuse
Shellstock identification tags
Modified atmosphere packaging (MAP)
Temperature danger zone
Bi-metallic stemmed thermometer
Time-temperature indicator (TTI)

Table of Contents

Learning Objectives

After completing this chapter, you should be able to:

○ Choose a safety-conscious food supplier.

○ Receive deliveries properly.

○ Determine when to accept or reject different types of foods during receiving.

○ Identify some of the microbial risks for different foods that can be minimized through proper receiving.

○ Calibrate thermometers and use them properly.

○ Check the temperature of different types of food during receiving.

Food-safety practices start long before you prepare or serve meals in your operation. To be sure the food you serve is safe, you must first control the quality and safety of food that comes in your back door.

Many things can happen to a product at different points within the flow of food. A frozen product that leaves the processor's plant in good condition, for example, may thaw en route to the distributor's warehouse, which will affect both product quality and safety.

The final responsibility for the safety of the food that enters your establishment rests with you. You can avoid many potential food-safety hazards by making sure products are received properly. Using approved suppliers and inspecting products when they are delivered are the first steps in the process.

CHOOSING A SUPPLIER

A number of factors go into selecting the right suppliers for a foodservice operation. While choosing a supplier who can deliver safe food is the ultimate goal, the level of service also needs to be considered. Before you accept any deliveries from a supplier, it is your responsibility to be sure the food you are purchasing comes from an approved source. Also check your suppliers to see if they meet or exceed the same food-safety standards you employ in your establishment.

Quality Standards

○ **Make sure your suppliers are getting their products from licensed, reputable purveyors and manufacturers.** Check with your regulatory agency to find out if your suppliers have had any food-safety problems or health code violations. Ask other operators what their experiences with a supplier have been.

○ **Inspect your supplier's warehouse or plant from time to time, if possible.** See if it is clean and well run.

○ **Ask your suppliers if they have a HACCP program in place.** (If they supply fresh produce, ask whether they have a Good Agricultural Practices Plan.) If not, ask what precautions or procedures they take to ensure the safety of the product.

○ **Find out if your supplier's employees are trained in food safety.**

○ **Check the condition of the supplier's delivery trucks.** Are they clean and well maintained? Do they hold refrigerated or frozen products at the proper temperatures? Are raw products separated from processed foods and fresh produce?

○ **Check your supplier's shipments for consistent product quality.** Inspect deliveries for unsafe packaging. Broken boxes, leaky packages, or dented cans are signs of careless handling.

○ **Request suppliers to deliver products when your staff has time to receive them properly.**

Rejecting Shipments

Remember, you have the right to refuse to accept any delivery.

You should have a company policy about returns, and your suppliers should be aware of it and agree to it. When foods don't meet acceptable standards, your employees should know what to do. If it is necessary to reject a product or shipment:

○ **Set the rejected product aside.** Keep it separate from other foods and supplies.

○ **Tell the delivery person exactly what is wrong with the product.** Use your purchase agreement and company standards to back up your decision to reject the product.

○ **Get a signed adjustment or credit slip from the delivery person before throwing the product away or letting the delivery person remove it.**

○ **Log the incident on the invoice or receiving document.** Note the food involved, including lot number and expiration date if appropriate, the standard that was not met, and the corrective action you took.

Once you have established a relationship with a supplier, continue to be a smart customer. Always inspect deliveries. Randomly check weights, break down cases, and take counts. Don't take anything for granted. You are buying not only products, but also service and food safety. Be sure you get what you pay for.

ACCEPTING DELIVERIES: QUALITY STANDARDS

When products are delivered to the back door of your establishment, you control the quality and safety of the foods you accept.

You can help prevent foodborne illnesses by refusing to accept any product that does not meet your standards. To control the receiving process, however, you must train staff members to closely follow a clearly established set of standards. If you set quality standards that products must meet before you accept deliveries, and establish procedures for inspecting products, you can reduce hazards before they enter your establishment.

Here are some general guidelines that can help you improve the way you receive deliveries.

HACCP Principle

For certain foods, such as shellfish, that will be eaten raw, receiving may be a Critical Control Point in a HACCP program.

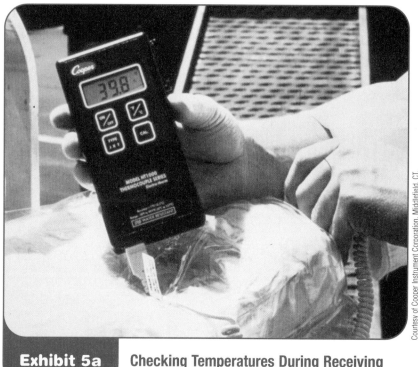

Courtesy of Cooper Instrument Corporation, Middlefield, CT.

Exhibit 5a

Checking Temperatures During Receiving
Take sample temperatures of refrigerated and frozen foods.

○ **Train employees to inspect deliveries properly.** Ideally, you should designate a dedicated receiving staff or trained employees whose sole responsibility is to inspect and receive deliveries. If this is impractical, employees who receive deliveries should be trained to judge product quality, check product for proper temperatures, know code dates, identify foods that have been thawed and refrozen, spot damage or insect infestation, and so on. They also should be authorized to accept, reject, and sign for deliveries.

○ **Inspect deliveries immediately.** Do a thorough visual inspection to count quantities, check for damaged product, and look for items that might have been repacked or mishandled. Spot-check weights and take sample temperatures of all refrigerated and frozen foods *(see Exhibit 5a)*. There is always a possibility that any food (even government-inspected products) may have been mishandled during shipment.

○ **If possible, receive only one delivery at a time.** Inspect and store each delivery before accepting another one, to avoid potential confusion and product abuse on the receiving dock. This helps ensure that employees are available to check the shipment properly and put it away quickly. The idea is to prevent refrigerated and frozen foods from sitting out any longer than necessary.

○ **Plan ahead for shipments.** Have clean hand trucks, carts, dollies, and containers available in the receiving area. Make sure that enough space is available in walk-ins and storerooms prior to receiving a shipment. Some operations use a refrigerator and freezer in the receiving area for temporary storage. If products need to be washed or broken down and rewrapped, make work space available as close to the receiving area as possible, to prevent dirt and pests from being brought into storage areas or the kitchen.

○ **Have the right information available.** Receivers should have a purchase order or order sheet ready to check against a supplier's invoice *(see Exhibit 5b* on the next page). The sheet should list quantities, quality specifications, and agreed-upon price. The sheet should also have room to record the date and time of delivery, product temperatures, and other notes.

○ **Correct any mistakes right away.** If any products are damaged, not at the correct temperature, or not delivered to specifications, do not accept them. Establish a return and credit policy with suppliers before placing your first order with them so there is no question about what to do when food does not meet your standards.

○ **Label all items for storage with the delivery date or the use-by date to ensure proper stock rotation.** Put products away as quickly as possible, especially products requiring refrigeration.

○ **Schedule deliveries during off-peak hours.** Arrange for deliveries when employees aren't busy so they have adequate time to inspect shipments properly.

○ **Keep the receiving area clean and well lighted to discourage pests.**

○ **Have a backup menu plan in case you have to return some food items.** If foods are not safe or not up to your standards, you may have to take the item off the menu, substitute another menu item, or try to arrange delivery from another supplier.

Exhibit 5b Check all orders that are received against the supplier's invoice.

Reprinted with permission from John DeWaele.

RECEIVING AND INSPECTING FOODS

Every food product delivered to your establishment should be inspected carefully for damage or potential contamination. Temperatures of products should be checked and recorded. While receiving temperatures for fresh foods are specific to the product, frozen foods should always be received frozen. In addition, note each food's appearance, as well as texture, smell, and taste in some cases. *(See Exhibit 5c on the next page.)*

Seafood

Fish and shellfish are very sensitive to rough treatment and **temperature abuse** and are easily damaged if mishandled. Both fresh and frozen seafood deteriorate quickly if handled improperly. Temperature abuse can cause the rapid growth of microorganisms and cause foodborne illness.

Exhibit 5c	**Acceptable and Unacceptable Conditions for Receiving Meat, Poultry, Seafood, and Eggs**
	Use this chart to determine whether to accept or reject deliveries.

Food	Accept	Reject
FRESH MEAT (such as beef, lamb, pork) Receive at 41°F (5°C) or lower	BEEF COLOR: bright cherry red LAMB COLOR: light red PORK COLOR: pink lean meat, white fat TEXTURE: firm and springs back when touched	COLOR: brown or greenish; brown, green, or purple blotches; black, white, or green spots TEXTURE: slimy, sticky, or dry PACKAGING: broken cartons, dirty wrappers, or torn packaging ODOR: sour odor
FRESH POULTRY Receive at 41°F (5°C) or lower	COLOR: no discoloration TEXTURE: firm and springs back when touched PACKAGING: should be surrounded by crushed, self-draining ice	COLOR: purple or green discoloration around the neck; dark wing tips (red wing tips are acceptable) TEXTURE: stickiness under the wings or around joints ODOR: abnormal, unpleasant odor
FRESH FISH Receive at 41°F (5°C) or lower	COLOR: bright red gills; bright shiny skin ODOR: mild ocean or seaweed smell EYES: bright, clear, and full TEXTURE: firm flesh that springs back when touched	COLOR: dull gray gills; dull dry skin ODOR: strong fishy or ammonia smell EYES: cloudy, red-rimmed, sunken TEXTURE: soft, leaves an imprint when pressed
FRESH SHELLFISH (such as clams, mussels, and oysters) Receive at 45°F (7°C) or lower for live shellfish	ODOR: mild ocean or seaweed smell SHELLS: closed and unbroken CONDITION: shipped alive; identified by shellstock identification tag. Retain tags for 90 days after product is used.	ODOR: strong fishy smell SHELLS: open shells that do not close when tapped (which means the shellfish are dead); broken shells CONDITION: dead on arrival TEXTURE: slimy, sticky, or dry
FRESH CRUSTACEAN (such as lobster, shrimp, and crabs) Receive at 45°F (7°C) or lower for live lobsters and crabs	ODOR: mild ocean or seaweed smell SHELL: hard and heavy for lobsters and crabs CONDITION: shipped alive; packed with seaweed and kept moist	ODOR: strong fishy smell SHELL: soft CONDITION: dead on arrival, tail fails to curl when lobster is picked up
FRESH EGGS Receive at an air temperature of 45°F (7°C) or lower	ODOR: none SHELLS: clean and unbroken CONDITION: firm, high yolks that are not easy to break and whites that cling to the yolk	ODOR: abnormal smell SHELLS: dirty or cracked

Fish

Fresh fish should be packed in self-draining crushed or flaked ice. Upon delivery, fresh fish is most acceptable at a temperature of 41°F (5°C) or lower. Fresh fish that is in good condition should meet the following standards.

○ Clear eyes

○ Bright red and moist gills

○ Firm flesh

○ Bright skin

○ Pleasant, mild scent of ocean or seaweed

Carelessly handled fish is not appetizing, let alone safe to eat. The following conditions are grounds for rejecting a shipment of fish.

○ Strong fishy smell or odor of ammonia

○ Cloudy, red-rimmed, sunken eyes

○ Dark, dull-red gills

○ Dry skin

○ Soft skin that leaves an imprint when pressure is applied with a finger

○ Tumors, abscesses, and cysts on the skin

See Exhibit 5d for examples of acceptable and unacceptable fish.

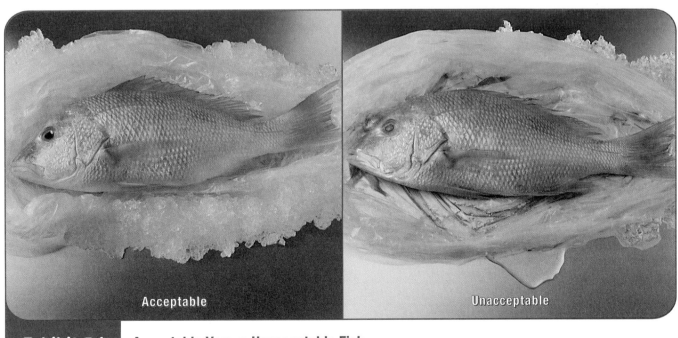

Acceptable   Unacceptable

Exhibit 5d   **Acceptable Versus Unacceptable Fish**

Frozen fish should be received frozen. If there is any indication that it has been allowed to thaw, do not accept it. Fish that has been thawed, then refrozen before getting to your establishment may have a sour odor and be off-color. Fillets often turn brown at the edges when they have been refrozen. A lot of ice or liquid in the bottom of the shipping box and moist, discolored, or slimy wrapping paper also are signs of refreezing.

Shellfish

Shellfish fall into two categories: crustacea, such as shrimp, crab, lobster; and molluscan bivalves (mollusks), such as clams, oysters, and mussels. Shellfish can be shipped live, fresh, frozen, in the shell, or shucked.

Interstate shipping of shellfish is monitored by the FDA. Shellfish must be bought only from suppliers on Public Health Service FDA lists of certified shellfish shippers or lists of state approved suppliers. Shucked shellfish must be packaged in nonreturnable containers clearly labeled with the name, address, and certification number of the packer. Packages containing less than one-half gallon must have a sell-by date. Packages containing more than one-half gallon should list the date the shellfish was shucked.

Both live and shucked molluscan shellfish may be received at temperatures of 45°F (7°C) or lower. When shipped live, shellfish must be delivered alive in nonreturnable containers. The FDA requires that live molluscan shellfish carry **shellstock identification tags** (see Exhibit 5e). Foodservice operators must write the date of delivery on the tags. The tags should remain attached to the containers the shellfish came in until the container is empty. Operators then must keep the tags on file for ninety days after the last shellfish has been used. Never mix shellfish from one shipment with another.

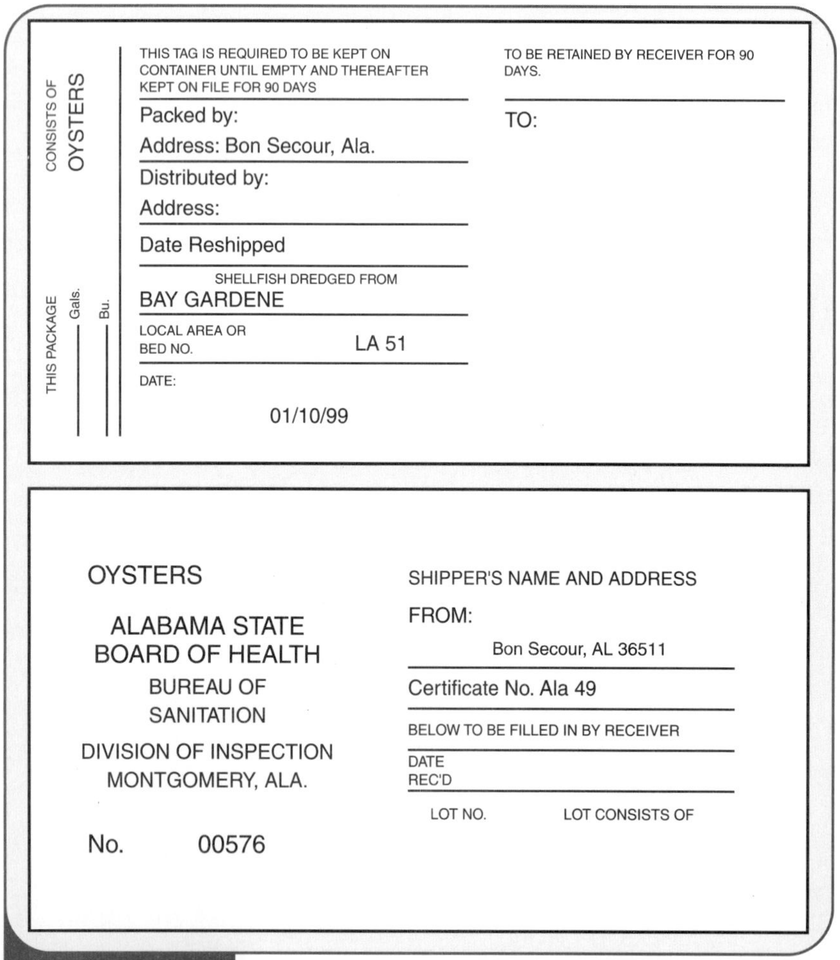

CONSISTS OF OYSTERS

THIS PACKAGE   Gals.   Bu.

THIS TAG IS REQUIRED TO BE KEPT ON CONTAINER UNTIL EMPTY AND THEREAFTER KEPT ON FILE FOR 90 DAYS

Packed by:

Address: Bon Secour, Ala.

Distributed by:

Address:

Date Reshipped

SHELLFISH DREDGED FROM
BAY GARDENE

LOCAL AREA OR BED NO.   LA 51

DATE:
01/10/99

TO BE RETAINED BY RECEIVER FOR 90 DAYS.

TO:

OYSTERS

ALABAMA STATE
BOARD OF HEALTH

BUREAU OF
SANITATION

DIVISION OF INSPECTION
MONTGOMERY, ALA.

No.   00576

SHIPPER'S NAME AND ADDRESS
FROM:
Bon Secour, AL 36511

Certificate No. Ala 49

BELOW TO BE FILLED IN BY RECEIVER

DATE
REC'D

LOT NO.   LOT CONSISTS OF

Exhibit 5e   **Shellstock Identification Tags**
Note that the tags must be dated when you receive the product, then kept on file for 90 days after the last shellfish has been used.

Shells of clams, mussels, and oysters will be closed if alive. Partly open shells may mean the clams, mussels, or oysters are dead. To find out, tap on the shells. If they close, the mollusks are still alive. If the shells do not close, or are badly cracked or broken, they should be discarded.

Fresh lobsters or crabs that are in good condition are easy to tell from those that have been poorly handled. A fresh lobster or crab will meet the following standards:

○ Show signs of movement

○ Have a hard and heavy shell

○ React when its eyes are pinched

○ Curl its tail under when turned on its back (lobsters)

Lobsters or crabs that show weak signs of life should be cooked right away. Dead ones must be discarded or returned to the vendor for credit.

Fresh Meat and Poultry

We often hear news stories of foodborne illnesses caused by *Salmonella* and *Campylobacter* from poultry or *E. coli* O157:H7 from beef or other meat. While fresh meat and poultry are carefully inspected by government agencies for wholesomeness, it is impossible to eliminate all microorganisms that are present during processing. Almost all cases of foodborne illness can be prevented, however, if food products are properly handled and prepared.

Fresh meat and poultry should be delivered at 41°F (5°C) or lower. When it arrives, inspect it closely, checking temperature, color, odor, texture, and packaging. Meat and poultry also must be purchased from plants inspected by the USDA or state department of agriculture. Meat products that have been inspected will be stamped with abbreviations for "inspected and passed" by the inspecting agency, along with the number identifying the processing plant *(see Exhibit 5f)*. Stamps will not appear on every cut of meat, but one should be present on every inspected carcass and on packaging. Suppliers must maintain written proof of government-inspected meats.

Meat and poultry inspection is mandatory. During the inspection process, products are checked for wholesomeness. USDA inspectors examine the carcass and viscera of each animal for possible signs of illness and check processing plants for sanitary conditions. The USDA inspection stamp means that both product and processing plant have met certain standards. It does not mean that the product is free of microorganisms that can cause foodborne illness. Operators still are responsible for handling and preparing these foods properly to be sure they are safe for consumers to eat.

Exhibit 5f

USDA Inspection Stamps for Meat, Poultry, and Eggs
Inspection is a mandatory process.

Exhibit 5g

USDA Grading Stamps for Meat, Poultry, and Eggs
Grading is a voluntary service that processors and packers pay for.

Most meats and poultry also carry a stamp that indicates their "grade" or level of quality. Grading is a voluntary service offered by the USDA and is paid for by processors and packers. Grades refer to the palatability and relative quality of the meat. USDA grades are printed inside a shield-shaped stamp *(see Exhibit 5g)*.

When receiving meat and poultry, consider the following factors:

Meats

○ **Beef should be a bright, cherry red.** Aged beef may be darker in color, and the type of packaging used sometimes affects color. Any beef that is turning brown or green should be rejected. Beef usually spoils on or near the surface first. An off color, slimy texture, or sour odor are signs that the meat has begun to deteriorate. Inspect ground beef very carefully. Ground beef spoils more easily than solid muscle cuts. Check packaging for broken cartons, dirty meat wrappers, and torn or leaking bags. Do not accept any product that arrives in this condition. *See Exhibit 5h* for examples of acceptable and unacceptable beef.

○ **Lamb is light red when it is fresh and properly exposed to air.** Do not accept any fresh lamb that is brown or has a whitish surface covering the lean meat.

○ **Fresh pork is light pink in color, with firm, white fat portions.** An excessively dark color, soft or rancid fat, and a sour odor all indicate the meat is spoiled and should be rejected.

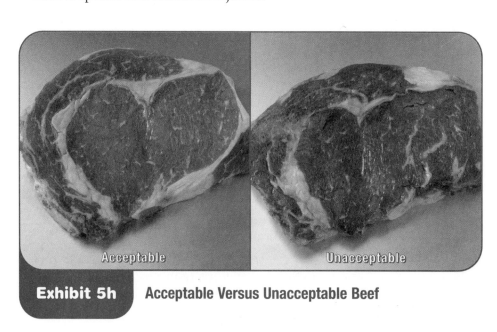

Acceptable   Unacceptable

Exhibit 5h   **Acceptable Versus Unacceptable Beef**

Poultry

Fresh poultry should be shipped in self-draining crushed ice and delivered at a temperature below 41°F (5°C) or chill packed. Poultry shipped and stored at temperatures of 28°F (-2°C) will likely have a significantly longer shelf life.

Mishandled poultry is easy to spot by its appearance *(see Exhibit 5i)*. Product that has started to spoil may have the following characteristics:

○ Purplish or greenish color

○ Abnormal odor

○ Stickiness under the wings and around joints

○ Dark wing tips (red tips are acceptable)

Poultry is inspected by federal or state agriculture agencies much the same way meat is. As with meat, the grading system is voluntary and paid for by processors. Processed poultry is inspected only at the processor's request.

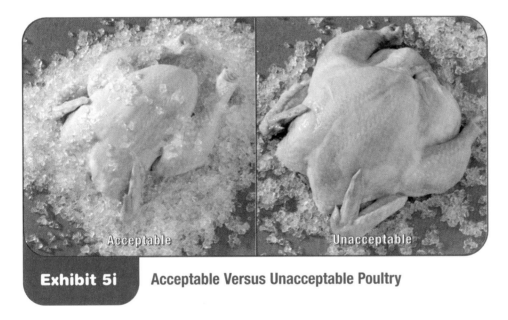

Exhibit 5i **Acceptable Versus Unacceptable Poultry**

Eggs

Purchase fresh eggs from approved, government-inspected suppliers. The USDA inspection shield on egg cartons indicates that federal regulations are enforced to maintain quality and reduce contamination. As with meat and poultry, grading is voluntary and is provided by the USDA. The official grade shield certifies that the eggs have been graded for quality under federal and/or state supervision.

Courtesy of American Egg Board, Park Ridge, IL.

Exhibit 5j

Acceptable Eggs
Eggs should be clean, dry, and free of cracks.

Choose suppliers who can deliver eggs within a few days of the packing date. Eggs must be delivered in refrigerated trucks. These trucks should be capable of documenting air temperature during transportation. When the eggs arrive, the air temperature of the truck should be 45°F (7°C) or lower, and the eggs must be stored immediately in refrigeration units that will hold them at 41°F (5°C). Check with your regulatory agency for temperature requirements in your jurisdiction.

Shells should be clean, dry, and free of cracks (see Exhibit 5j). To test freshness, break an egg into a flat dish. Acceptable eggs have firm, high yolks that are not easy to break, and the whites should cling to the yolks. In very high-quality eggs, the white will stand up on its own. Fresh eggs should have no odor. Immediately after inspection, refrigerate fresh eggs in their original containers. Eggs should be stored with small ends down.

Liquid, frozen, and dehydrated eggs must be pasteurized as required by law and bear the USDA inspection mark. They should be refrigerated or frozen at the proper temperature when delivered. Check packages for damage or indication of refreezing, and note use-by dates to be sure the product is in good condition and still usable.

Key Point

Purchase pasteurized dairy products only.

Dairy Products

Purchase pasteurized dairy products only. Unpasteurized milk and other dairy products are potential sources of microorganisms such as *Salmonella, Campylobacter,* and *Listeria,* all of which can cause serious illness. All milk and milk products should be labeled "Grade A." This means they meet standards for quality and sanitary processing methods set by the FDA and US Public Health Service. Dairy products with the Grade A label, such as cream, dried milk, cottage cheese, cream cheese, butter, cheese, and frozen products like ice cream, are made with pasteurized milk.

Like other refrigerated products, milk and dairy products should be received at 41°F (5°C) or lower, unless the law governing their distribution specifies a different temperature. *(See Chapter 6 for proper storage temperatures.)* Check with the proper regulatory agency for temperature requirements.

Fresh milk has a sweetish taste. Any milk that tastes sour, bitter, or moldy should be rejected. Milk has a sell-by date stamped on the container. Milk delivered after that date, as well as milk with any off odors, should be rejected.

Butter should have a sweet flavor, uniform color, and firm texture. Check for signs of mold, specks, or other foreign matter. Make sure that containers are clean and not damaged. Don't accept any butter that is rancid or has absorbed odors.

In the United States, all cheeses must meet certain standards of identity. In order for the product to be called "cheddar" or "mozzarella," for example, the government specifies ingredients that must be used, maximum moisture content, minimum fat content, and general characteristics. When cheeses are delivered, check to see that each type has its typical flavor, texture, and color. If the cheese has a rind, it should be clean and unbroken. Cheese should have no signs of mold or off odors. As always, check for proper temperature and clean, undamaged packaging.

Fresh Produce

Various fresh fruits and vegetables have different temperature requirements for transportation and storage, so they may be held at different temperatures. No specific temperature is mandated by regulation for the transportation and storage of fresh produce, with a few exceptions. Cut melons, a potentially hazardous food, must be held at 41°F (5°C) or below. Fresh-cut produce is best held at 33°F to 41°F (1°C to 5°C).

Fresh fruits and vegetables are highly perishable. They should be put into storage quickly. Most produce should not sit out at room temperature, let alone on a warm dock. Have a system in place to get fresh-cut and highly perishable items into a cooler.

In general, produce should not be washed before it is stored. While washing would not hurt leafy green items, many other products are likely to decay faster if washed before storage. This is especially true for mushrooms and berries. Wash produce just before preparation and serving.

All fruits and vegetables should be handled with care. If they are pinched, squeezed, or roughly handled, they will bruise and spoil more quickly.

Check products being delivered for signs of mishandling and insect infestation, including insect eggs and egg cases. Visually inspect produce for quality (see Exhibit 5k). Spoilage will show up in a variety of ways, including mold, blemishes, cuts, mushiness, discoloration, wilting, or dull appearance. What applies to one fruit or vegetable may not apply to another. For example, peaches with cuts in them could be considered poor quality, but

Key Point

In general, produce should not be washed before it is stored.

Exhibit 5k

Inspecting the Quality of Fruit
Spoilage will show up in a variety of ways including mold, blemishes, mushiness, discoloration, wilting, or dull appearance.

potatoes and carrots with cuts would be considered acceptable. Discoloration in produce may vary as well. For example, oranges may actually revert to a green color without affecting the quality of the orange or the juice.

Use smell and taste to help determine product quality. Unpleasant odors will tell you when a product is not acceptable. With fruits, sometimes taste is the best test. Outer peels or skins can be blemished without affecting flavor or quality. Be sure to wash or peel fruits and vegetables carefully before tasting them.

Since there are so many ways produce can show signs of spoilage, employees need to learn how to identify not only produce that is obviously unacceptable, but also produce that will spoil quickly in storage. *Appendixes A and B* list recommended storage temperatures for fresh fruits and vegetables.

Refrigerated and Frozen Processed Foods

More and more establishments are using prepared foods that are either refrigerated or frozen. These include precut meats, Individually Quick Frozen (IQF) poultry, frozen or refrigerated entrées that only require heating, and fresh-cut fruits and vegetables (including salads). Processed foods can save time and money, but only if they are treated with the same care given to other food products. Mishandled products that end up causing a foodborne illness can put an establishment out of business.

The temperature of refrigerated processed foods should be 41°F (5°C) when they are delivered. *(See Chapter 6 for proper storage temperatures.)* While these products are usually fully cooked or ready to eat, they still require careful handling. Inspect packaging for tears or holes and check use-by dates.

All frozen foods should be delivered frozen, with the exception of ice cream. Ice cream may be delivered and stored at temperatures of 6°F to 10°F (-14°C to -12°C) without affecting product safety or quality.

Key Point

All frozen foods, including meat, should be delivered frozen, with the exception of ice cream, which may be delivered and stored at temperatures of 6°F to 10°F (-14°C to -12°C).

Frozen foods should be checked for signs of thawing and refreezing. Simply because a product is frozen upon receipt does not mean it has not thawed during prior handling. Obvious signs are blocks of ice or liquid at the bottom of the case or large ice crystals on the product itself. Other signs are discoloration or dryness of the product and stains on the outer packaging. Foods should be wrapped in airtight packaging, and boxes and outer cartons should be clean and undamaged.

MAP, Vacuum-Packed, and Sous Vide *Foods*

MAP Foods

MAP stands for **modified atmosphere packaging**. By this method, air is removed from a food package and replaced with gases such as carbon dioxide and nitrogen. These gases help to extend the shelf life of the product. Many fresh-cut produce items are packaged using MAP methods.

Vacuum-Packed Foods

Vacuum packaging is the process of removing air from around a food product sealed in a package. Bacon is one example.

Sous Vide *Foods*

Sous vide (soo veed) is a French term meaning "under vacuum." Food processed by this method is vacuum-packed in individual pouches, partially or fully cooked, and then chilled. These foods are then heated for service in the establishment.

Sous vide products may be received either refrigerated or frozen, depending on the manufacturer. Some frozen, reduced-calorie meals are packaged using this method.

Receiving MAP, Sous Vide, *and* Vacuum-Packed Foods

Removing oxygen from packaged food can reduce or prevent the growth of some microorganisms that need oxygen to grow. However, the same conditions can promote the growth of anaerobic microorganisms, such as those that produce the botulism toxin.

Key Point

Check frozen food for signs of thawing and refreezing.

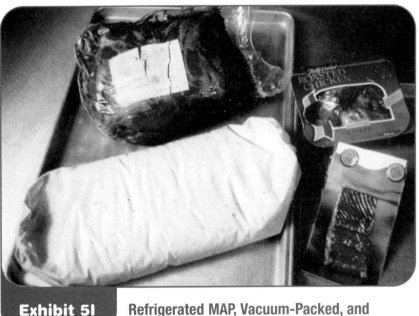

Exhibit 5I   **Refrigerated MAP, Vacuum-Packed, and** *Sous Vide* **Products**
Refrigerated MAP, vacuum-packed, and *sous vide* products should be received at 41°F (5°C) or below unless different temperatures are specified by the manufacturer.

Key Point

Reject dry foods if there are water stains on the packaging.

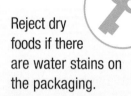

Key Point

Reject packages with holes, tears, or punctures.

For this reason, the FDA doesn't allow establishments to package foods on site using a MAP method except under certain conditions. To prepare *sous vide* or MAP foods, operators must have a HACCP plan in place and limit the foods being packaged to those that can't support the growth of *Clostridium botulinum.*

When you receive MAP, *sous vide,* and vacuum-packed foods, use the following guidelines.

○ Make sure the supplier has a HACCP plan in place.

○ Reject packages with leaks.

○ Reject product that appears slimy or has bubbles.

○ Reject product that is an unacceptable color.

○ Reject packages that contain product with an expired code date.

○ Refrigerated product must be delivered and stored at 41°F (5°C) or below unless the manufacturer provides different instructions.

○ Frozen products should be frozen when they arrive.

Dry and Canned Products

Dry and canned products seem to pose little threat to consumers. Most have a fairly long shelf life and are usually used long before they have a chance to spoil. However, canned products provide an ideal breeding ground for the microorganisms that cause deadly botulism. Dry foods can be contaminated by a variety of sources that can cause foodborne illnesses.

Dry foods must be kept dry. Most microorganisms need moisture to grow and multiply, which is why dry foods have a much longer shelf life than fresh foods. Check both outer cases and inner packaging for dampness or moisture. Reject the shipment if it is damp or shows signs of previous wetness (water stains).

Dry foods often attract pests. Because they can be stored at room temperature, dry foods may not be sealed and stored as securely as fresh, refrigerated, or frozen foods. Insects and rodents have an easier time getting into dry-food packages. Always inspect packages carefully for holes, tears, or punctures. Check products themselves for signs of infestation. You can spot insects or insect eggs in cereal or flour by sprinkling some of the product on brown paper. Rodents leave signs such as chewed packages and droppings in and around cartons.

Use sight and smell to inspect dry foods, too. Off colors and odors, spots of mold, or a slimy appearance is a sign of spoilage.

Canned foods must also be carefully checked for damage. Check can exteriors first, looking for the following signs of contamination *(see Exhibit 5m).*

○ **Swollen ends.** One or both ends of a can may bulge from gas produced by chemical or bacterial action inside. If one end bulges out when the other is pressed, discard it, because the can has not gone through the proper heat-treating process.

○ **Leaks and flawed seals.** If there are any leaks, flaws, or irregularities along the top or side seals, reject the can.

○ **Rust.** If a can is rusted, the contents may be too old, or the rust may have eaten holes in the can, which can lead to contamination.

○ **Dents.** Don't accept cans that have dents along side or top seams or dents large enough to make it impossible to open the can with a can opener. The seams may be broken. (Check with your regulatory agency regarding dented cans; some do not allow any dents.)

Any cans received without labels should be rejected. Once the exteriors of the received cans have been checked, spot-check the contents. Any foods that do not have a normal color, texture, or odor; that are foamy; or that have a milky-colored

Health Alert

Never taste canned foods that you are unsure of.

Exhibit 5m **Conditions for Rejecting Damaged Cans**
Bulging or swollen ends may be caused by gas from spoiled contents. Rust can perforate the can and may indicate the contents are old. Dents may cause seams to leak. Flawed top or side seals can let contaminants in.

liquid should be thrown out immediately. *Never* taste canned foods that you are unsure of. Botulism, a foodborne illness associated with canned foods, is so dangerous that people have died from just tasting and spitting out contaminated food.

Aseptically Packaged and Ultra High Temperature (UHT) Pasteurized Foods

Aseptically Packaged and Ultra High Temperature (UHT) pasteurized foods such as milk, juice, and puddings also can be received and stored at room temperatures. These products are pasteurized (heat-treated at very high temperatures for a short time) to kill microorganisms that can cause illness. Then they are packaged under sterile conditions to keep them from being contaminated. Once opened, however, the products should be refrigerated at 41°F (5°C).

Hot Foods

Occasionally, operators may receive a shipment of hot foods. Hot foods must be cooked properly as required by local or federal codes *(see Chapter 7 for guidelines)*. If you purchase hot foods, make sure the supplier has a HACCP plan or other means of documenting proper cooking methods and temperatures. Hot foods must be delivered at a temperature of 140°F (60°C) or above.

MONITORING TIME AND TEMPERATURE

Time and temperature play the most critical roles in the process of maintaining a safe food supply. They affect food quality and safety from the moment food arrives at the back door to the time a prepared meal is presented to the customer.

If food is kept in the **temperature danger zone** (41°F to 140°F or 5°C to 60°C) for longer than four hours, it must be discarded. This four-hour time period begins when the food is taken off the delivery truck, and continues through product storage, preparation, and cooking. The four-hour time period begins again after cooking. The more time food sits on a receiving dock before being refrigerated or put in the freezer, the less time that is left for preparation and cooking.

To manage both time and temperature, you need to monitor and control them. The thermometer may be the single most important tool you have to protect your food.

Choosing the Right Thermometer

There are many types of thermometers used in an establishment. Each is designed for a specific purpose. Some are used to measure the temperature of refrigerated or frozen storage areas. Others measure the temperature of equipment, such as ovens, hot holding cabinets, and warewashing machines.

Key Point

If food is kept in the temperature danger zone (41°F to 140°F or 5°C to 60°C) for longer than four hours, it must be discarded.

Key Point

The thermometer may be the single most important tool you have to protect your food.

Perhaps the most important types are the thermometers that measure the temperature of foods. The two most common types used in establishments are the bi-metallic stemmed thermometer and the digital thermometer. *(See Exhibit 5n.)*

Bi-Metallic Stemmed Thermometers

The most common and versatile type of thermometer in the foodservice industry is the **bi-metallic stemmed thermometer** *(see Exhibit 5o).* This type of thermometer measures temperature through a metal probe with a sensor in the end. Bi-metallic stemmed thermometers often have scales that measure temperatures from 0°F to 220°F (-18°C to 104°C). This makes them useful for taking temperatures of everything from incoming shipments of frozen foods to hot-holding cabinets and sanitizing solution in warewashing equipment. When you select this type of thermometer, it should have:

○ an adjustable calibration nut to keep it accurate;

○ easy-to-read numbered temperature markings;

○ a dimple to mark the end of the sensing area (which begins at the tip); and

○ accuracy to within ±2°F or ± 1°C.

Digital Thermometers

Digital thermometers measure temperatures through a metal probe or sensing area and display the results on a digital readout. These devices may use either a thermocouple or thermistor sensor to sense temperatures. They come in a wide variety of styles and sizes, from small pocket models to

Courtesy of Cooper Instrument Corporation, Middlefield, CT.

Thermometer Manufacturing, Atkins Technical, Gainesville, FL.

Exhibit 5n   **Types of Foodservice Thermometers**
Shown here from left to right are a bi-metallic stemmed thermometer and a digital thermometer.

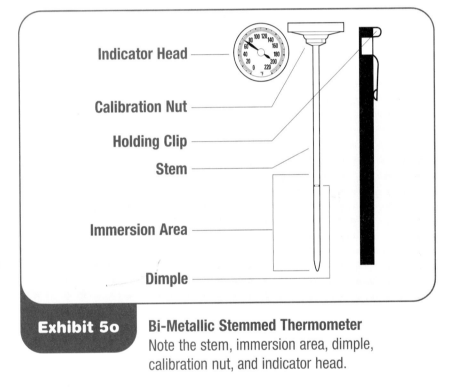

Indicator Head

Calibration Nut

Holding Clip

Stem

Immersion Area

Dimple

Exhibit 5o   **Bi-Metallic Stemmed Thermometer**
Note the stem, immersion area, dimple, calibration nut, and indicator head.

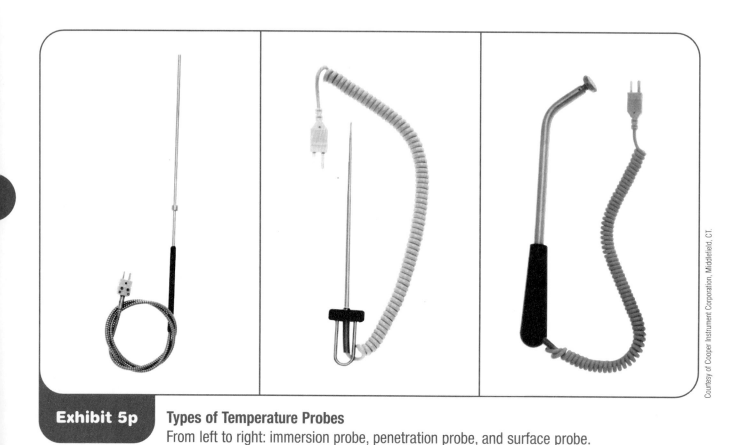

Courtesy of Cooper Instrument Corporation, Middlefield, CT.

Exhibit 5p   **Types of Temperature Probes**
From left to right: immersion probe, penetration probe, and surface probe.

panel-mounted displays. Many come with interchangeable temperature probes designed to measure the temperature of equipment and food.

Basic types of probes include surface, immersion, penetration, and air probes *(see Exhibit 5p)*. Immersion probes are designed to measure temperatures of liquids such as soups, sauces, or frying oil. Surface probes measure temperatures of flat cooking equipment like griddles. Penetration probes are used to measure the internal temperatures of foods. Air probes measure temperatures inside refrigerators or ovens.

Time-Temperature Indicators (TTI)

Some instruments are designed to monitor both time and product temperature. The **Time-Temperature Indicator** is one example. Some suppliers attach these self-adhesive tags or sticks to a food shipment to determine if the temperature of the product has exceeded safe limits during shipment or later storage. If the product temperature has exceeded these limits, the TTI provides an irreversible record of the incident. A change in color inside the indicators or windows of the TTI notifies the receiver that the product has experienced time and temperature abuse *(see Exhibit 5q)*.

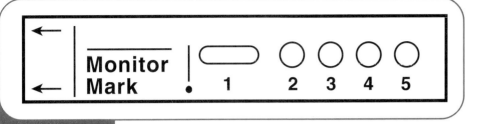

Exhibit 5q	**Time-Temperature Indicator**

A change in color in the windows of this TTI alerts the receiver that time and temperature abuse has occurred.

Rules for Using Thermometers

Employees should know what different thermometers are used for and how to care for them. They should follow a few simple rules with all thermometers.

○ **Keep thermometers and their cases clean.** Have an adequate supply of clean and sanitized thermometers on hand. Thermometers should be washed, rinsed, sanitized, and air dried before and after each use to prevent cross-contamination. Use an approved food-contact surface sanitizing solution to sanitize them.

○ **Measure internal temperatures of foods by inserting the end of the probe into the center of the product, usually the thickest part** (see *Exhibit 5r*). It is a good practice to take at least two readings, which should be in different locations because product temperatures may vary across the food portion. After inserting the sensing area of the thermometer into the food, wait at least fifteen seconds for the pointer or the digital readout to stop moving before reading the temperature. (Procedures for how to measure temperatures of packaged foods that are delivered by suppliers will be discussed later in this chapter.)

○ **Calibrate the thermometer regularly to make sure it is accurate.** This should be done before each shift or before each day's deliveries. Thermometers should also be recalibrated any time they have suffered a severe shock (for example, after being dropped or after an extreme change in temperature).

Cross-Contamination

Wash, rinse, sanitize, and air dry thermometers before and after each use to prevent cross-contamination. Use an approved food-contact surface sanitizing solution to sanitize all thermometers.

Exhibit 5r	**Measuring the Internal Temperature of Foods**

Place the thermometer probe in the geometric center when checking the temperature of food.

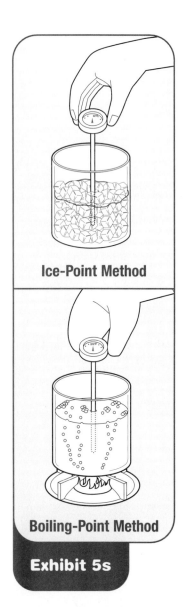

Ice-Point Method

Boiling-Point Method

Exhibit 5s

Using the Ice-Point and Boiling-Point Method

Thermometers that hang or sit in refrigerators, freezers, or ovens can easily be damaged by accident. To make sure that a thermometer is accurate, place a calibrated stemmed thermometer next to it and check the temperature. Hanging thermometers usually cannot be recalibrated and must be replaced if they are not accurate.

○ *Never* **use glass thermometers filled with mercury or spirits to monitor the temperature of food.** They can break and pose a serious danger to employees and customers.

How to Calibrate Thermometers

Most digital and bi-metallic stemmed probe thermometers can be calibrated easily. Two accepted methods of calibrating thermometers are the ice-point method and the boiling-point method. To properly calibrate your thermometers, follow these steps.

Ice-Point Method *(see Exhibit 5s)*

1. Fill a large glass with crushed ice. Add clean tap water until the glass is full and stir well.

2. Put the thermometer stem into the ice water so that the sensing area (from tip to about half an inch above the dimple) is completely submerged. Do not let the stem touch the sides or bottom of the glass. Wait at least thirty seconds until the indicator stops moving.

3. With the thermometer stem still in the ice water, hold the adjusting nut under the head of the thermometer securely with a wrench or other tool. Turn the thermometer head so that the pointer reads 32°F (0°C). Some digital thermometers have a reset button. Push it while the probe is in the ice water to automatically adjust the readout.

Boiling-Point Method *(see Exhibit 5s)*

For thermometers with scales that start reading above 32°F (0°C).

1. Bring clean water to a boil in a deep pan.

2. Put the thermometer stem into the boiling water so the sensing area (from the tip to about a half inch above the dimple) is completely submerged. Do not let the stem touch the sides or the bottom of the pan. Wait at least thirty seconds until the indicator stops moving.

3. With the thermometer stem still in the boiling water, hold the adjusting nut under the head of the thermometer securely with a wrench or other tool. Turn the thermometer head so that the pointer reads 212°F (100°C). Some digital thermometers have a reset button. Push it while the probe is in the boiling water to automatically adjust the readout.

Note that the boiling point of water changes based on atmospheric pressure and altitude above sea level. You might have to make adjustments depending on your location. The boiling-point is about 1°F (approximately 0.5°C) lower for each 550 feet above sea level. An establishment located 5,500 feet above sea level, for example, would have to adjust the pointer to 202°F (94°C) using this method.

How to Check Temperatures of Deliveries

Checking the temperature of foods that are delivered to your back door is one of the most important steps you can take to make sure the food you serve is safe. Temperature abuse is the most dangerous form of mishandling because it can affect both product quality and safety.

You can't control how food is handled before it reaches your facility, but you can decide if it meets your establishment's standards and determine whether or not to accept it. Through careful inspection, you can find out if a food product has been mishandled. You can also buy only from suppliers who use time-temperature indicators and temperature recording monitors in their delivery trucks.

More suppliers are using recording devices that continuously monitor temperatures at all times in their refrigerated delivery trucks. When delivered products appear to have suffered temperature abuse, the recording device can be checked to see if the temperature in the delivery truck changed at any time during transit.

To properly record the temperature of products arriving on your receiving dock, do the following.

○ **Have an adequate supply of clean, sanitized, and properly calibrated stemmed thermometers on hand.**

○ **Record the temperatures of all foods on a receiving log or copy of an invoice.**

○ **Remember that the temperature of all refrigerated foods should be 41°F (5°C) or below unless specified.** Various produce, though refrigerated, may be received and stored at higher temperatures for quality purposes.

○ **All frozen foods should be delivered frozen.** If you open a case to record a temperature, reseal the case, write the date on it, and initial it when you are finished. That will let other employees know it has been checked.

Key Point

Temperature abuse is the most dangerous form of mishandling because it can affect both product quality and safety.

Exhibit 5t   **Checking the Temperature of Packaged Foods**
Insert the thermometer between two packages to check their temperature.

○ **To check the temperature of packaged refrigerated or frozen foods, insert the thermometer probe between two packages in the center of a case** *(see Exhibit 5t).* Be careful not to puncture any wrapping on the product. Wait at least fifteen seconds and record the temperature.

○ **Check the temperature of milk by opening a carton and inserting the stem of the thermometer into the carton until at least two inches of the tip are submersed.** Do not let the tip touch the sides of the carton. Wait until the indicator stops moving and record the temperature. For milk in bulk packaging, fold the bag or pouch around the stem of the thermometer *(see Exhibit 5u).*

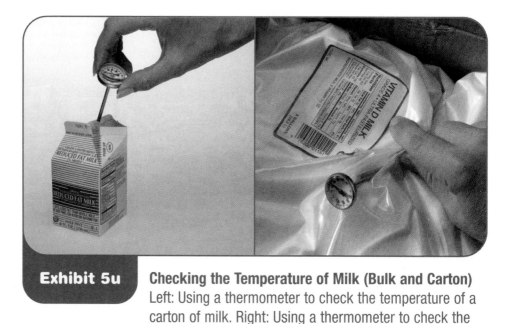

| Exhibit 5u | Checking the Temperature of Milk (Bulk and Carton) Left: Using a thermometer to check the temperature of a carton of milk. Right: Using a thermometer to check the temperature of bulk milk. |

○ **Temperature of products packed in ice, such as poultry or fresh fish, should be checked by inserting the probe into the product itself.** Use a very thin needle probe to prevent unsightly puncture marks, and insert it into the center of the product. Wait and record the temperature.

○ **To check the temperature of live molluscan shellfish, insert the probe into the middle of the carton or case, between the shellfish, for an ambient reading.** Check the temperature of shucked shellfish by inserting the probe into the container until the probe's sensing area is immersed.

○ **When receiving eggs, check the ambient air temperature of the delivery truck, and check the truck's temperature chart recorder for extreme**

temperature fluctuations during transport. Temperature fluctuations, high humidity, and warm (over 60°F or 15°C) temperatures may result in the growth of harmful microorganisms.

○ **Always be sure to use a clean, sanitized thermometer each time you take a temperature.** If you do not have extra thermometers at the back door, clean and sanitize your thermometer after each use. Keep a bucket of sanitizing solution on the dock, or use approved sanitizing wipes.

SUMMARY

The receiving area of your establishment is a control point in the flow of safe foods and is a Critical Control Point for certain foods. Even though federal and state government agencies regulate and monitor the production and transportation of foods such as meats, poultry, seafood, shellfish, eggs, dairy products, and canned goods, it is your responsibility to check the quality and safety of the foods that come into your establishment.

Receiving safe foods starts with the careful selection of approved and reputable suppliers. By working with suppliers, operators can take steps to ensure that the food they purchase is safe. They also can buy from suppliers who have HACCP plans and temperature recording devices that monitor how product is handled through processing, warehousing, and shipping.

Operators must plan delivery schedules so that products can be handled promptly and correctly. When deliveries arrive, shipments must be inspected carefully for the following.

○ **Completeness:** Invoices for deliveries should match purchase orders or order sheets. Deliveries should be checked for proper quantities, weights, and product specifications.

○ **Quality:** All products should meet agreed-upon standards. Packaging should be clean and undamaged. Code dates should be current. Foods should show no signs of mishandling.

○ **Safety:** Products must be delivered at the proper temperatures. All products, especially meat, poultry, and seafood, should be checked for color, texture, and odor. Live molluscan shellfish and crustacea must be delivered alive, or properly packed fresh or frozen. Eggs should be inspected for freshness and for dirty and cracked shells. Dairy products must be checked for freshness. Produce should be fresh and wholesome. Frozen foods should be inspected for signs of thawing and refreezing. Canned foods must be looked at carefully for signs of damage. Dry foods should be inspected for pest infestation and moisture. MAP, aseptic, and ultrapasteurized products should be checked to make sure their packaging is intact, recommended temperatures are correct, and code dates have not expired.

Employees assigned to receive deliveries should be trained to inspect foods properly. They should be able to identify damaged, contaminated, and spoiled products. They should be authorized to reject products that don't meet company standards and to sign for products that do.

Thermometers are the most important tool operators have to make sure that food products are not in the temperature danger zone any longer than necessary. Every facility should have an adequate supply of thermometers on hand. Thermometers must be cleaned and sanitized before and after each use. Managers should make sure that employees know what different thermometers are used for and can calibrate and use them properly.

A CASE IN POINT I

Case Study

On Monday, a number of foods were delivered to the Sunnydale nursing home during the busy lunch hour. All the food products were different: cases of frozen ground-beef patties, canned vegetables, frozen shrimp, fresh tomatoes, a case of potatoes, and fresh chicken.

Betty, the new assistant manager, thought the best thing to do was to put everything away and check it later, since she was very busy. She told Ed, who was in charge of receiving, to sign for the delivery and put the foods into storage. Ed asked her if it would be better to ask the delivery driver to come back later. Since she needed the chicken for that night's dinner, Betty asked Ed to accept the delivery anyway and went back to the front of the house.

Ed put the frozen shrimp and ground-beef patties in the freezer and the fresh chicken in the refrigerator. Then he put the fresh tomatoes, potatoes, and canned vegetables in dry storage. When he was finished, he went back to work in the kitchen.

What was done incorrectly? What could be the possible result?

A CASE IN POINT II

Case Study

ABC Seafood makes its usual Thursday afternoon delivery to The Fish House. John, a prep cook, is the only person in the kitchen when the driver rings the bell at the back door. The kitchen manager, who is in charge of receiving, is in a managers' meeting. The chef is out doing an errand, and the rest of the kitchen staff is on break, though some are still in the restaurant.

John follows the driver onto the dock where the driver unloads two crates of ice-packed fresh fish, a crate of live mussels, two crates of live oysters, a case of shucked oysters in plastic containers, and a case each of frozen shrimp and frozen lobster tails.

John goes back into the kitchen to get a stemmed pocket thermometer and remembers to look in the chef's office for a copy of the order form. He takes both out to the dock and begins to inspect the shipment.

John checks the product against both the order sheet and the invoice, then begins to check product temperatures. First he puts the thermometer probe between the packages inside the cases of frozen shrimp and lobster. Then he checks the temperature of the shucked oysters by taking the cover off one of the containers and inserting the probe into the container. The thermometer reads 45°F (7°C), which worries him. He remembers being taught that refrigerated products should be 41°F (5°C) or below. He decides not to say anything.

After wiping the thermometer probe off on his apron, John checks the internal temperature of one of the whole fish on top of one of the ice-pack crates. Finally, he reaches into one of the crates of live oysters with his hand to see if it feels cold. He notices a few mussels and oysters with open and broken shells. He removes those with broken shells, knowing the chef won't use them.

John records all his findings on the order sheet he took from the chef's office, signs for the delivery, and starts putting the products away. He puts away the live shellfish first, dumping the few that are still left in the refrigerator into the new crates so there is room. Then he puts the shucked oysters and fresh fish in the refrigerator and the shrimp and lobster into the freezer.

What was done wrong? What could be the result?

TRAINING TIPS

Training Tips for the Classroom

1.   Thermometer Calibration Demonstration

Objective: *After completing this activity, class participants should be able to identify the two acceptable methods for calibrating a bi-metallic stemmed thermometer and demonstrate how to calibrate one using each of these methods.*

Directions: Calibrate a bi-metallic stemmed thermometer for the class. The ice-point method is most practical, but if possible, verify the calibration using the boiling-point method after the ice-point method. Allow time for one or two members of the class to calibrate as well.

2.   Temperature Instrument Group Discussion

Objective: *After completing this activity, class participants should be able to explain how each temperature instrument functions and determine which ones would be best suited to their own establishments.*

Directions: Have your class discuss the primary use, as well as the advantages and disadvantages, of the following instruments for measuring temperature:

○ bi-metallic stemmed thermometer

○ thermistor

○ thermocouples

○ various hanging thermometers

3.   Receiving Game

Objective: *After completing this activity, class participants should be able to list signs of time or temperature abuse for all major food groups, and when given a specific picture or verbal description of a food product, determine if it should be accepted or rejected.*

Directions: Break the class into groups, so that you have four or five teams in the class, and give each group a bell.

Explain to the teams that you will show them a picture (or alternatively, give them a verbal description) of a product with certain characteristics (for example, a fish with sunken eyes, a dented can, discolored poultry, a properly packaged dry or canned product).

As soon as a team decides if they can accept or reject the product, they should ring their bell. They then state if they should accept or reject, giving their reasons. If correct, they receive a point. They double their points if they can state the "opposite" conditions which would make the product acceptable or unacceptable. (Give bonus points to any team that can identify a product where receiving is a Critical Control Point.)

4. Receiving Role-Play

Objective: *After completing this activity, class participants should be able to determine when to accept and reject a food product and be able to properly reject a food product from a delivery person.*

Directions: Ask two volunteers from the class to role-play a scene of a driver delivering several different types of food products to a manager in an establishment. Give the volunteers a list of specific food products that are to be delivered, and allow them five or ten minutes to prepare a role-play of how to refuse delivery of these products.

While the volunteers are preparing their scene, ask the rest of the class to each take out a blank piece of paper. When the role-play begins, members of the class are to write down as many "wrongs" in the receiving process as they see.

At the end of the role-play (after a round of applause!) ask the class how many different "wrongs" they have listed. The student with the most "wrongs" is to read his or her list. With the entire class participating, ask for the proper action that a foodservice manager or trained receiving employee should take when food products are delivered that are unacceptable.

Ask the group if anyone had additional "wrongs" that weren't discussed. Award prizes to both the role-play volunteers and the top "wrong-catchers" in class.

Training Tips on the Job

1. Field Trip to a Manufacturer or Supplier

Purpose: *After completing this activity, managers and supervisors will be able to identify the food-safety practices used by their manufacturers or suppliers to keep products safe. Note: Managers should become more knowledgeable about the products they handle and serve after this trip. It is always a good practice to go to the source and learn from the experts.*

Directions: Take your managers, supervisors, and any other interested employees on a field trip to a food manufacturer or distributor that you do business with, and get a tour of their facility.

Consider places such as a:

○ dairy plant   ○ seafood supplier

○ poultry plant   ○ bakery

○ meat-packing house   ○ food distribution warehouse

○ cannery

However, you should plan ahead. Group sizes may be limited, and access to different parts of these facilities is often restricted. (With the cooperation of a plant manager, these tours can be remarkably informative.)

2. Guest Lecturer: Thermometer Manufacturer Representative

Purpose: *After completing this activity, all staff members will be able to calibrate, properly use, and sanitize the thermometers utilized in their operation. Note: A manufacturer's rep will add credibility to the presentations, and because a new person is doing the presenting, this training may help hold the participants' attention.*

Directions: Bring in a knowledgeable and interesting manufacturer's rep from the company that manufactures the thermometers you use, and have him or her make a presentation on thermometers to your kitchen staff.

The rep should discuss the thermometers that are being used in your establishment, explaining how they are made, how they are calibrated, and how they are to be properly used and sanitized.

Make sure you are bringing someone in who is enthusiastic, is knowledgeable, and can present well. Provide information on who is in the audience, how long the presentation should be, what audiovisual equipment is available, how big and where the space is for the presentation, and so forth. Make it clear this is not to be a sales pitch.

Applications on the Job

3. Receiving Checklist

Purpose: *Having your receiving personnel work directly with suppliers to create a receiving checklist will create buy-in for both parties. Also, each group will have a better understanding of the other's needs.*

Directions: Working with both your receiving personnel and your suppliers, develop a checklist of acceptable and unacceptable conditions for the food items you normally receive.

Break down food items by category, then by specific item if necessary. Certain types of products may be listed as a general group, such as canned or dry foods.

For example:

Whole fresh chicken

Acceptable:
- shipped in crushed ice
- temperature of 41°F (5°C)
- free of discoloration
- free of off-odors
- free of stickiness
- USDA or state agriculture stamp

Unacceptable:
- shipped without ice
- temperature above 41°F (5°C)
- green, purple, or gray discoloration
- off-odors
- slime or stickiness on skin

4. Developing a Business Agreement with Your Suppliers

Purpose: *Developing a business agreement with your supplier, which identifies clearly defined standards, will help ensure that you receive safe products from them.*

Directions: Working with your receiving agents and your suppliers, develop a mutually acceptable agreement between your establishment and your food suppliers that clearly states your interest in receiving the safest food products possible.

Consider the following:

○ Work only with HACCP-based suppliers.

○ Receive deliveries at specified times and days.

○ Receive deliveries from clean, temperature-controlled trucks.

○ Receive foods at proper temperatures.

○ Receive only products that meet your checklist standards.

○ Receive foods in properly labeled, sealed packages or containers.

○ Receive foods free of pest infestation.

○ Specify that foods are to be inspected and signed for by authorized personnel only.

Having such an agreement with your suppliers helps establish a professional standard with the companies you work with, in order to help ensure a safe food product to your customers.

DISCUSSION QUESTIONS

1. List general guidelines for receiving foods safely.

2. Describe how to calibrate a thermometer using the ice-point method.

3. Describe the proper methods for measuring the receiving temperature of frozen foods, poultry delivered on ice, and bulk milk. What should the temperature be for each?

4. List three signs that fresh poultry has spoiled.

5. Name four types of external damage to cans that are cause for rejection.

MULTIPLE-CHOICE STUDY QUESTIONS

1. Which is most important in choosing a food supplier?

   A. It meets your food-safety standards.
   B. Its prices are the lowest.
   C. Its warehouse is close to your establishment.
   D. It offers a convenient delivery schedule.

2. When you are receiving a delivery of food for your establishment, it is important that you

   A. refrigerate it before checking it.
   B. check it before accepting it.
   C. put it with other recent deliveries.
   D. take it out of its original packaging before storing it.

3. How might you tell by looking at it that a food delivery should be rejected?

   A. You find a cracked egg in a box of eggs.
   B. You see frost on the outside of a box of frozen corn.
   C. The ice that was packed around fresh fish has melted, and the product is above 41°F (5°C).
   D. A box of lettuce contains some field soil.

4. How should fresh salmon be packaged for delivery and storage until cooking?

    A. Layered with salt

    B. Vacuum sealed

    C. Wrapped in dry, clean cloth

    D. Covered with crushed self-draining ice

5. What is the most important observation to make about the condition of a shipment of live oysters?

    A. Their shells are closed.

    B. They smell like fresh sea water.

    C. Their shells are open.

    D. They are a uniform color and size.

6. A box of sirloin steaks carries a USDA inspection stamp and also a USDA Choice grade stamp. What do these stamps tell you?

    A. The farm that supplied the beef uses only USDA-certified animal feed.

    B. The meat processor that packaged the steaks meets USDA sanitary standards, and the meat quality is acceptable.

    C. The meat wholesaler meets USDA quality grading standards.

    D. The steaks are free of disease-causing microorganisms.

7. How could you tell if a whole fresh chicken has been exposed to temperature abuse?

    A. It came from an unlicensed supplier.

    B. The wing tips are brown.

    C. It has been in your refrigerator for more than one day.

    D. The skin is dry.

8. To check the freshness of a delivery of shell eggs, you crack one onto a plate. What about the egg would tell you that you can accept the delivery?

    A. It smells like a newly lighted matchstick.

    B. The shell cracks in half perfectly.

    C. It comes out of its shell completely.

    D. The yolk is high and firm.

9. Statements from a dairy supplier's sales brochure are listed below. Which statement should tell you not to hire this supplier?

   A. We make our cheese with only the freshest unpasteurized milk.
   B. From farm to you, our milk is kept at temperatures below 41°F (5°C).
   C. Money-back guarantee—our prices are the lowest in the area.
   D. We deliver according to your schedule and needs.

10. Which of the following do not have to be received at 41°F (5°C) or below?

    A. Beef, poultry, shellfish, and lamb
    B. Eggs, dairy products, fruits, and vegetables
    C. Ham, luncheon meats, stuffing, and duck
    D. Broth, goose, giblets, and bacon

11. Upon delivery, what is one way to tell if a frozen food product has not been properly handled?

    A. The box is water stained.
    B. The contents of the box are frozen solid.
    C. The outside of the box is warmer than 0°F (-18°C).
    D. The box is delivered in a plastic bag.

12. When should you reject the delivery of a dry or canned food product?

    A. The label on a can of peaches is tearing.
    B. The top of a can of tomatoes is bulging out.
    C. A bag of oatmeal is delivered at a temperature higher than 41°F (5°C).
    D. A box of rice does not show a USDA inspection stamp.

13. Which is not a proper step in the ice-point method of calibrating a thermometer?

    A. First, insert the probe of the thermometer into a glass of ice water.
    B. Second, wait until the temperature indicator stops moving.
    C. Third, remove the probe from the ice water.
    D. Finally, adjust the pointer on the temperature indicator to read 32°F (0°C).

14. You are the manager of a restaurant with a soup and salad bar. How should you measure the temperature of the soup to ensure that it is safe to eat?

A. Insert an immersion probe into the center of each soup pot.

B. Attach a surface probe to the surface of each soup pot.

C. Ladle some soup from each pot into bowls and lay bi-metallic probes on the surface.

D. Observe each soup pot to see if steam is rising from the surface.

15. Your manager has asked you to purchase a new thermometer for the restaurant. Which would not be a proper choice?

A. A thermometer accurate to ±2°F (±1°C)

B. A digital thermometer with a thermistor sensor

C. A digital thermometer with a thermocouple sensor

D. A mercury-filled glass thermometer

ADDITIONAL RESOURCES

Books and Periodicals

Cichy, R. F. (1993). *Sanitation management.* East Lansing, MI: Educational Institute of the American Hotel & Motel Association.

Cross out cross-contamination. (1998). *Best Practices, 2*(1), 6-9.

Equipped for food safety. (1998). *Best Practices, 2*(4), 6-10.

Kotschevar, L. H. (1994). *Quantity food purchasing* (4th ed.). New York: Macmillan.

Loken, J. K. (1995). *The HACCP food-safety manual.* New York: Wiley.

Longree, K. (1996). *Quantity food sanitation* (5th ed.). New York: Wiley.

National Restaurant Association Educational Foundation. (1998). *A Practical Approach to HACCP: Coursebook.* Chicago: Author.

The right way to…accept deliveries. (1998). *Best Practices, 2*(1), 12-13.

The right way to…calibrate equipment. (1997). *Best Practices, 1*(3), 12.

The right way to…prepare beef. (1998). *Best Practices, 2*(3), 12-13.

The right way to…prepare chicken. (1998). *Best Practices, 2*(1), 10.

The right way to…prepare eggs. (1998). *Best Practices, 2*(2), 10-11.

The right way to…prepare pork. (1997). *Best Practices, 1*(3), 10-11.

The right way to…preparing fruits and vegetables. (1998). *Best Practices, 2*(4), 12-13.

The right way to…seafood safety. (1999). *Best Practices, 3*(1), 16.

Secrets of self-inspection, (1998). *Best Practices, 2*(2), 6-9.

United States Dept. of Health and Human Services, Public Health Service, & Food and Drug Administration. (1997). *Food Code: Recommendations of the United Public Health Service of the Food and Drug Administration.* Springfield, VA: United States Dept. of Commerce. (NTIS No. PB97-133656)

Warfel, M. C. (1996). *Purchasing for food service managers* (3rd ed.). Berkeley, CA: McCutchan.

Web Sites

1999 FDA Model Food Code

http://vm.cfsan.fda.gov/~dms/fc99-toc.html

Complete outline of the FDA's latest code for regulating operations that provide food directly to consumers. Also includes a quick synopsis of changes from the 1997 Food Code.

Alaska Seafood Marketing Institute

http://www.alaskaseafood.org

The Alaska Seafood Marketing Institute works to improve seafood quality by teaching fishermen, processors, retailers, and restaurateurs about proper handling of Alaska seafood products. Their Web site also includes industry news, suppliers, and recipes.

American Egg Board

http://www.aeb.org

The American Egg Board's Web site features product, safety, and industry information for foodservice professionals as well as recipes and other information for the general public.

American Meat Institute

http://www.meatami.org

Provides member information, as well as current and past news and information regarding the meat industry.

Atkins

http://www.atkinstech.com

Atkins offers a variety of tools for reading food temperature. Their Web site includes ordering and operation instructions.

Cooper Instrument

http://www.cooperinstrument.com

Cooper is a supplier of instruments for measuring time, temperature, and humidity. Their Web site features product and ordering information as well as tips and trade show information.

Daydots

http://www.daydots.com

Daydots makes stickers that facilitate efficient food rotation. Their Web site offers safety tips, ordering information, and industry news.

FDA Center for Food Safety and Applied Nutrition (CFSAN)

http://vm.cfsan.fda.gov/list.html

Comprehensive site from CFSAN offers a wealth of food-safety information, from foodborne illness to food labeling. CFSAN strives to be a leader in food safety, and to protect consumers from economic fraud, promote sound nutrition, and encourage innovation.

FDA Seafood and Information Resources

http://vm.cfsan.fda.gov/seafood1.html

The Web site for FDA's Center for Food Safety and Applied Nutrition includes seafood information and resources, including seafood pathogens and contaminants.

Food Marketing Institute

http://www.fmi.org

This nonprofit association conducts programs in research, education, industry relations, and public affairs for its membership of food retailers and wholesalers. Their Web site offers a wealth of food-safety information.

Food Online News

http://www.foodonline.com

This site provides a wealth of information for professionals and vendors in the foodservice equipment industry, including latest news and a searchable database of suppliers and products.

Institute of Food Technologists (IFT)

http://www.ift.org

IFT is a nonprofit scientific society with 28,000 members working in food science, food technology, and related professions in industry, academia, and government. This site includes information on its policies and publications, education and industry news, and meeting and convention locations.

International Association of Milk, Food, and Environmental Sanitarians (IAMFES)

http://www.iamfes.org

IAMFES keeps members informed of the latest scientific, technical, and practical developments in food safety and sanitation. Their Web site includes booklets and links to other food-safety sites.

International Dairy Foods Association (IDFA)

http://www.idfa.org

This site is the link to information on IDFA and its constituent organizations: the Milk Industry Foundation (MIF), the National Cheese Institute (NCI), and the International Ice Cream Association (IICA); as well as information on the entire dairy industry. The site also includes information on current industry events, membership information, and publications.

International Food & Information Council (IFIC)

http://ificinfo.health.org

Web Site for the International Food and Information Council offers information and resources about food safety and food allergies for educators and the public.

National Cattlemen's Beef Association

http://www.beef.org

This site includes up-to-date information from the National Cattlemen's Beef Association, as well as a reference library, information on nutrition and food safety, and links to related sites.

National Food Processors Association (NFPA)

http://www.nfpa-food.org

NFPA is a scientific and technical trade association for the food industry. Their site provides industry news and educational materials on issues of food safety, research, and food science.

National Institute of Standards and Technology

http://www.nist.gov/public_affairs/welcome.htm

The National Institute of Standards and Technology is the government agency that works with industry to develop and apply technology, measurements, and standards. The site includes links to nontechnical information as well.

National Pork Producers Council

http://www.nppc.org

The National Pork Producers Council site provides general information about food and nutrition, as well as information specifically for pork producers. A special section is also provided just for children.

National Restaurant Association

http://www.restaurant.org

The National Restaurant Association site provides information on government agencies affecting the restaurant industry, the latest training and certification updates, and links to state restaurant associations and hospitality schools and universities.

National Shellfish Sanitation Program Manual of Operation

http://vm.cfsan.fda.gov/~ear/nsspman.html

An outline of the FDA's Center for Food Safety and Applied Nutrition National Shellfish Sanitation Program Manual of Operation, which ensures safe molluscan shellfish.

NSF International

http://www.nsf.org

This site offers product information, resources, and publications regarding public health safety.

Produce Marketing Association

http://www.pma.com

The members of this nonprofit trade association market fresh fruits and vegetables to the foodservice industry. Purchasing guidelines are available.

Tyson

http://www.tyson.com

Tyson's Web site offers a consumer kitchen page with recipes and instruction on preparing chicken products. An entire page is also devoted to foodservice professionals and offers information on the industry, Tyson products, and menu planning.

United States Department of Agriculture (USDA)

http://www.usda.gov

The United States Department of Agriculture's Web site features information, publications, and other educational materials about the nation's agriculture.

United States Department of Agriculture Food Safety and Inspection Service

http://www.fsis.usda.gov

The USDA's Food Safety and Inspection Web site offers the latest food-safety news, educational materials, and HACCP-implementation materials.

Chapter 6
Keeping Food Safe in Storage

Knowledge

TEST YOUR FOOD-SAFETY KNOWLEDGE

1. **True or False:** Refrigerated, ready-to-eat foods stored at 41°F (5°C) must be eaten or thrown out within three days. *(See Shelf Life of Food, page 6-13.)*

2. **True or False:** Foodservice chemicals may be transferred to sturdy, properly labeled containers. *(See Cleaning Supplies, page 6-12.)*

3. **True or False:** Freezing destroys all harmful microorganisms in food. *(See Frozen Storage, page 6-4.)*

4. **True or False:** Melissa placed the ready-to-eat pumpkin pie directly below the raw chicken breasts. The raw chicken should have been stored on the bottom shelf below the pie. *(See Refrigerators, page 6-6.)*

5. **True or False:** If stored food has passed its expiration date, you should cook and serve the food at once. *(See Storage Guidelines, page 6-2.)*

Table of Contents

Learning Objectives

After completing this chapter, you should be able to:

○ Label and store specific types of refrigerated food.

○ Label and store frozen foods.

○ Properly store dry and canned foods.

○ Apply first in, first out (FIFO) practices.

○ Properly store raw foods to prevent cross-contamination.

○ Properly transfer food from original containers to storage containers.

○ Follow the proper procedures for using refrigerators, freezers, and dry storage areas.

Key Terms

FIFO
Refrigerated storage
Frozen storage
Deep-chill storage
Shelf life
Dry storage
Hygrometer

Every establishment needs to store food and supplies. How and where they are stored affects food quality and safety. When foods are stored improperly and not used in a timely manner, quality and safety will suffer. Poor storage practices can cause food to spoil quickly, with potentially serious results.

STORAGE GUIDELINES

Every facility has a wide variety of products that need to be stored. Some may be stored for only a few hours. Others may be in storage for several weeks.

A few general rules can be applied to most storage situations.

○ **Use the first in, first out (FIFO) method.** Write the date on each product when it is received or prepared *(see Exhibits 6a and 6b on the next page.)* Train employees to store products with the earliest use-by or expiration date in front of products with later dates. Once items have been properly shelved, items in front of the shelves should be used first. Make sure employees always check the use-by or expiration date. Discard products if the use-by or expiration date has passed.

○ **Keep potentially hazardous foods out of the temperature danger zone of 41°F to 140°F (5°C to 60°C).** Store deliveries as soon as they have been inspected. Take out only as much food as you can prepare at one time. Put prepared foods away until needed. Properly cool and store cooked foods as soon as they are no longer needed. *(See Chapter 7 for more information on chilling cooked foods.)*

○ **Check temperatures of stored foods and storage areas.** Temperatures should be checked at the beginning of the shift. Many establishments use a pre-shift checklist to guide employees through this process.

○ **All foods should be tightly wrapped in clean and moisture-proof materials.** Dirty packaging can contaminate food as it is being opened. It can attract pests too. Always label food you have wrapped, clearly noting the contents of the package and the date it was stored. If you take food out of its original package, put it in a clean, sanitized

Exhibit 6a   **Labeling System**
There are several products on the market that help simplify the labeling process.

food container with a tight-fitting lid. (Eggs and produce should be stored in their original containers.) You should also label the container with the date the package was opened, its contents, and the expiration date of the product.

○ **Keep all storage areas clean and dry.** Floors, walls, and shelving in refrigerators, freezers, dry storerooms, and heated holding cabinets should be properly cleaned on a regular schedule. Clean up spills and leaks right away to keep them from contaminating other foods. Clean, dry floors also prevent employees from slipping.

○ **Clean dollies, carts, transporters, and trays often.**

○ **Store foods only in areas designed for them.** Do not store food products near chemicals or cleaning supplies; in restrooms, locker rooms, janitor closets, furnace rooms, vestibules; or under stairways or pipes of any kind. Foods can be contaminated easily in any of these areas *(see Exhibit 6c).*

TYPES OF STORAGE

An ideal facility has two types of storage areas: food storage and chemical storage. Food storage areas include refrigerators, freezers, deep-chill units, and dry storerooms. Chemical storage areas are used to store cleaning supplies and other chemicals.

Each storage area has its own purpose. Operators should not consider design and management of storage space a low priority because storage areas that aren't

Exhibit 6b

Dating Received Products
Use the FIFO method when storing food and date the product when it is received.

Courtesy of Daydots Foodservice Products, Fort Worth, TX. 800-321-3687.

Exhibit 6c

Poor Dry Storage
Do not store cleaning supplies with food.

Key Point

Most perishable foods should be stored in refrigerators at 41°F (5°C) or below.

used properly or become crowded can cause problems. For example, a refrigerator that is overstocked may not be able to hold temperatures properly. Stock may not get rotated correctly, and there is a greater chance of cross-contamination.

Storage areas also should be located to prevent any food contamination. Foods should be stored away from warewashing areas and garbage rooms. To ease the flow of foods through the operation, storage areas must be accessible to receiving, food prep, and cooking areas. Some facilities have a walk-in refrigerator near the receiving dock and another between the prep area and the cooking line. While space might be at a premium, carefully consider your storage needs. Adding more storage space, or relocating it, could help keep food safer.

Here are the types of storage each facility should have.

○ **Refrigerated storage** holds potentially hazardous foods for short periods of time. Most perishable foods, both raw and cooked, should be stored in refrigerators at 41°F (5°C) or below. Refrigeration slows the growth of microorganisms and helps keep them from multiplying to levels high enough to cause illness.

○ **Frozen storage** is typically used to hold frozen foods at temperatures of 0°F (-18°C) or below. Foods in frozen storage can be held for periods of a few weeks up to several months. Freezing does not kill all microorganisms, but it does slow their growth substantially. Frozen foods should be used as quickly as possible. Freezers should not be used to freeze refrigerated foods, because slow freezing damages the quality of most raw foods.

○ **Deep-chill storage** holds foods at temperatures between 26°F and 32°F (-3°C to 0°C) for short periods of time. For some foods, these temperatures limit growth of microorganisms without damaging food quality. Foods such as poultry, meat, seafood, and *sous vide* products can be stored a little below freezing without forming ice crystals. When stored this way, they have a longer **shelf life,** which is a recommended period of time during which a material may be stored and remain suitable for use.

○ **Dry storage** areas can hold dry foods and canned foods for long periods of time. To maintain the quality of dry and canned foods, dry storage areas should be kept dry and cool. Storerooms should be clean, well ventilated, and well lighted.

○ Cleaning supplies and chemicals should be kept in their own storage area. Store these supplies away from food in clean, dry rooms or cabinets that can be locked. Always keep them in original packages or in sturdy containers that are clearly labeled.

STORAGE TECHNIQUES

A few commonsense rules apply to each of these storage areas. Make sure that all employees follow these rules to keep foods safe.

Refrigerators

In general, the colder food is, the safer it is. Keeping food as cold as possible without freezing it also extends its shelf life. Ideal storage temperatures will vary depending on the food. Fruits and vegetables will freeze if stored at temperatures ideal for fish. Meat and poultry will have a shorter shelf life if stored at temperatures better suited for produce. If possible, store foods such as meat and poultry in separate refrigerators to hold them at optimal temperatures. If this is impractical, store meat, poultry, fish, and dairy products in the coldest part of the unit, away from the door.

While there are many types of refrigeration equipment available to operators, from walk-in refrigerators to refrigerated drawers, some general guidelines apply when using all of them.

○ **To hold food at a certain temperature, refrigerator air temperature usually must be about 2°F (1°C) or lower.** For example, to hold poultry at an internal temperature of 41°F (5°C), the air temperature in the refrigerator should be 38°F (3°C).

○ **Monitor food temperatures regularly.** Use hanging thermometers in the coldest area in the back and the warmest area near the door. Some units have a readout panel on the outside so you can check the temperature without opening the door. These should be checked for accuracy, too. At least once during each shift, check the temperature of the unit, and take the temperature of random food samples with a calibrated probe thermometer.

○ **Use caution when chilling large quantities of hot food in the refrigerator.** This can warm up the interior enough to put other foods into the temperature danger zone.

○ **Do not overload the unit.** Too many products prevent good airflow and make the unit work harder to stay cold *(see Exhibit 6d)*.

○ **Use open shelving.** Lining shelves with aluminum foil or paper to keep them neat restricts circulation of cold air.

Key Point

Generally, the colder food is, the safer it is.

Cross-Contamination

Store raw meat, poultry, and fish separately from cooked and ready-to-eat foods whenever possible.

Exhibit 6d

Overloading a Refrigerator
Overloading a refrigerator will prevent good airflow and make the unit work harder to stay cold.

○ **Keep the door closed as much as possible.** Open it only for short time periods. Opening it too often lets warm kitchen air inside, which can affect food safety and also make the refrigerator work harder. Consider using plastic insulating strips or cooler curtains to help maintain temperatures in walk-ins.

○ **Wrap all foods properly.** If taken out of their packages, products should be stored in clean, covered food containers with clearly marked labels. Uncovered foods can absorb odors from other foods. Leaving food uncovered also can cause cross-contamination.

○ **Store raw meat, poultry, and fish separately from cooked and ready-to-eat foods whenever possible.**

○ **If not, always store prepared or ready-to-eat foods above raw meat, poultry, and fish.** This will prevent raw product juices from dripping onto the prepared foods, which may result in a possible foodborne illness for those who eat it *(see Exhibits 6e and 6f)*. Store raw meat, poultry, and fish in the following order: from top to bottom: fish; whole cuts of beef; pork; ham, bacon, and sausage; ground beef and ground pork; poultry. This order is based on the minimum internal cooking temperature of each food.

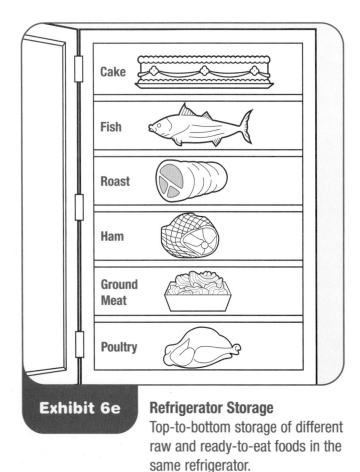

Exhibit 6e **Refrigerator Storage**
Top-to-bottom storage of different raw and ready-to-eat foods in the same refrigerator.

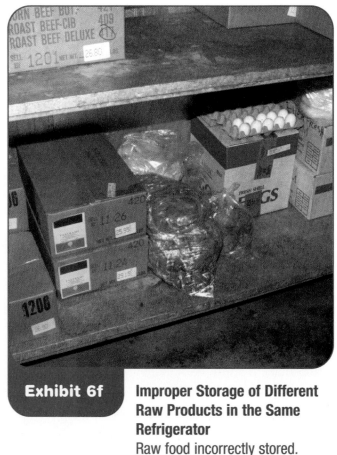

Exhibit 6f **Improper Storage of Different Raw Products in the Same Refrigerator**
Raw food incorrectly stored.

Deep-Chilling Units

Some general guidelines for using deep chillers include:

○ **Hold only foods such as poultry, meat, seafood, and *sous vide* products in deep-chill storage.** The low temperatures of 26°F to 32°F (-3°C to 0°C) will freeze or form ice crystals in most other products and cause them to lose quality.

○ **Use specially designed equipment for deep-chill storage.** Some refrigerators can be used for this purpose if they have the compressor capacity. Check with the manufacturer.

Freezers

Some general guidelines for using freezers include:

○ **Check unit and food temperatures often.** Use a calibrated thermometer to check accuracy of hanging thermometers or unit readout.

○ **Keep freezer temperatures at 0°F (-18°C) or below unless different foods require different temperatures.**

○ **Place frozen food deliveries in freezers as soon as they have been inspected.** Clearly label the contents, date of delivery and use-by date if there is one. If products are going to be used immediately, they can be stored in a refrigerator. Never hold frozen foods at room temperature.

○ **Use caution when placing foods into a freezer.** Warm foods may raise the temperature inside the unit and partially thaw the freezer contents. Preparing several smaller batches of food is a better idea than freezing leftovers. Store foods to allow good air circulation. Overloading the freezer and lining the shelves make it work harder. Overloading also makes it more difficult to find and rotate foods properly.

○ **Defrost freezer units on a regular basis.** They will operate more efficiently when free of frost. Move contents to another freezer when defrosting, or use them immediately.

○ **Never refreeze thawed food unless it has been thoroughly cooked.** Refreezing damages food quality. Thawed food also is more likely to support growth of microorganisms, which refreezing may not kill.

○ **Rotate frozen foods using the FIFO method.** Check use-by dates to be sure they haven't expired. Foods that do not stand up well to long-term frozen storage include ground meat, mackerel, salmon, bluegill, turkey, pork, creamed foods, sauces, custards, gravies, and puddings.

○ **Store foods in their original containers or wrap them tightly in moisture-proof material or containers.**

○ **Open the unit as infrequently as possible.** Use cold curtains to help maintain temperatures.

Dry Storerooms

Storerooms for dry foods should be clean and dry. Dry foods have a long shelf life if held in the right conditions. Moisture and heat are the biggest problems. The temperature of the storeroom should be between 50°F and 70°F (10°C to 21°C). Keep relative humidity around 50 to 60 percent. If high humidity is a problem, consider using a dehumidifier. Some general guidelines for dry storage include:

○ **Monitor temperature and humidity regularly.** Use a hanging thermometer to record temperatures and a **hygrometer** to measure relative humidity. Units are available that combine both instruments.

○ **Store dry foods at least six inches off the floor, away from walls, and out of direct sunlight.**

○ **Store foods in their original packages whenever possible.** Once the packages are opened, store product in tightly covered containers. This will prevent insects, rodents, and microorganisms from contaminating the food.

○ **Clean up any spills immediately.**

○ **Make sure storerooms are well ventilated.** This will help keep temperature and humidity more even and constant throughout the storage area.

STORING SPECIFIC FOODS

The general storage guidelines discussed here apply to most foods. However, certain foods have special requirements.

Meat

○ Store meat immediately after delivery in its own storage unit or in the coldest part of the refrigerator. Fresh meats must be held at a temperature of 41°F (5°C) or below. Frozen meat should be stored at a temperature that will keep it frozen.

○ Wrap raw cuts of meat, especially ground beef, airtight. Meat will turn brown when exposed to air. Frozen meats should be wrapped in airtight, moisture-proof material or containers to prevent freezer burn.

○ Primal cuts, quarters or sides of raw meat, and slab bacon can be hung on clean, sanitized hooks or placed on sanitized racks. To prevent cross-contamination, do not store meats above any other foods.

○ Throw out any meat that shows signs of spoilage.

Key Point

Fresh raw meat, poultry, and fish must be stored at refrigerator temperatures of 41°F (5°C) or below.

Poultry

○ Store raw fresh poultry at 41°F (5°C) or below. Frozen poultry should be stored at temperatures that will keep it frozen. If it is not in its original packaging, wrap it in airtight material or containers.

○ Ice-packed poultry can be stored in a refrigerator as is. Containers must be self-draining. Change the ice and sanitize the container often.

○ Whole fresh birds can be loosely wrapped and should be used within three or four days. Use fresh refrigerated parts and cooked poultry within one to two days.

Fish

○ Store fresh fish at 41°F (5°C) or lower, and use it within forty-eight hours. Keep fillets and steaks in original packaging or in tightly wrapped, moisture-proof wrappings. Fresh whole fish packed in flaked or crushed ice can be stored for up to three days. Ice beds must be self-draining. Change the ice and sanitize the container regularly.

○ Store frozen fish at temperatures that will keep it frozen. Wrap fish in moisture-proof wrappings.

○ Fish meant to be eaten raw must be delivered frozen or be frozen before being served, to kill any parasites. The only exceptions are shellfish and certain species of tuna not susceptible to parasites. Fish meant to be eaten raw can be frozen in the following ways:

   ● at -4°F (-20°C) or lower for 168 hours (seven days) in a storage freezer; or

   ● at -31°F (-35°C) or lower for fifteen hours in a blast freezer.

Shellfish

○ Store live shellfish in their containers at 45°F (7°C) or as low as 35°F (2°C). You can store molluscan shellfish (clams, oysters, mussels, scallops) in a display tank under one of two conditions:

   ● the tanks must carry a sign stating that the shellfish are for display only; or

   ● you must obtain a variance from the health department to serve the shellfish on display.

○ To obtain a variance, you must submit a HACCP-based plan that shows the following:

Time & Temperature

Fish meant to be eaten raw must be delivered frozen or be frozen before being served, to kill any parasites.

Key Point

Store live shellfish in their containers at 45°F (7°C) or as low as 35°F (2°C).

- Water in the tank will not come into contact with any other fish.
- Using the tank won't affect product quality or safety.
- You retain shellstock identification tags as required.

Eggs

○ Eggs received at an air temperature of 45°F (7°C), in compliance with laws governing their shipment from suppliers, must be placed immediately upon their receipt in refrigeration equipment that is capable of maintaining food at 41°F (5°C) or less. Maintain constant temperature and humidity in refrigerators where eggs are stored. Do not wash eggs before storing them. They are washed and sanitized at the packing facility. Eggs should be stored with small end down.

Key Point

Eggs received at an air temperature of 45°F (7°C) must be immediately refrigerated at a temperature of 41°F (5°C) or less.

○ Use the FIFO method of storage. Plan to use all eggs within a few weeks of purchase.

○ Keep shell eggs in cold storage until immediately before use. Take out only as many eggs as are needed at one time.

○ Do not combine cracked eggs in a bowl unless you intend to use them right away.

○ Store frozen eggs at temperatures that will keep them frozen.

○ Liquid egg products should be stored in their original containers at 41°F (5°C) or lower. Do not freeze.

○ Dried egg products can be stored in a cool, dry storeroom away from light. Putting them in a refrigerator is best. Once they are opened, always store them in the refrigerator at 41°F (5°C) or below.

Dairy Products

○ Store dairy products at temperatures of 41°F (5°C) or below.

○ Frozen dairy products such as ice cream and frozen yogurt can be stored at 6°F to 10°F (-14°C to -12°C).

○ Dairy products absorb odors easily. Keep them tightly covered, and store them away from foods with strong odors, such as onions, fish, and cabbage.

○ Dairy products used for cooking should not be held at room temperature for longer than two hours. They can sour quickly. Like all potentially hazardous foods, dairy products must be thrown away if they have been in the temperature danger zone for more than four hours. Do not put products that have been held at room temperature back into refrigerated cartons.

○ Always use the FIFO method of storage. Discard products if the use-by or expiration date has passed.

Fresh Produce

○ Various fruits and vegetables have different temperature requirements for storage. While many whole raw fruits and vegetables can be stored at temperatures of 41°F (5°C) or below, not all will be stored at these temperatures. *(See Appendixes A and B* for recommended storage temperatures for specific produce items.) Whole raw produce and raw cut vegetables (such as celery, carrots, and radishes) that are delivered packed in ice can be stored that way. The containers must be self-draining, and ice should be changed regularly.

○ Fruits and vegetables kept in the refrigerator can dry out quickly. Keep the relative humidity at 85 to 95 percent.

○ Though most fruits can be stored in the refrigerator, avocados, bananas, and pears ripen best at room temperature. Tomatoes ripen best at room temperature as well.

○ Produce should not be washed before storage. Moisture promotes the growth of mold in many instances. Instead, wash produce before preparation or serving.

○ Store whole citrus fruits, hard-rind squash, eggplant, and root vegetables such as potatoes, sweet potatoes, rutabagas, and onions in a cool, dry storeroom. Temperatures of 60°F to 70°F (16°C to 21°C) are best. Make sure containers are well ventilated. Store onions away from other vegetables that might absorb odor.

MAP, Vacuum-Packed, and Sous Vide Foods

○ Always store MAP, vacuum-packed, and *sous vide* foods at temperatures recommended by the manufacturer. Most of these foods should be stored at 41°F (5°C) or below. Frozen MAP, vacuum-packed, and *sous vide* foods should be stored at temperatures that will keep them frozen. Store and handle these products carefully.

○ Check packages for signs of contamination before using the product. Vacuum packaging will not stop the growth of anaerobic microorganisms. MAP fish, for example, is especially susceptible to growth of *Clostridium botulinum.* Do not use any packages that have slime on the outside or bubbles or too much liquid inside. Discard product in torn packaging.

○ Always check the expiration date before using MAP, vacuum-packed, and *sous vide* products. Labels should clearly list contents, storage temperature, preparation instructions, and a use-by date.

○ Operators who produce MAP foods on site must follow specific labeling rules. The FDA rules for processing MAP foods on site are very strict.

Key Point

Store dairy products at temperatures of 41°F (5°C) or below.

Key Point

While many whole raw fruits and vegetables can be stored at temperatures of 41°F (5°C) or below, not all will be stored at these temperatures.

Aseptic and UHT Foods

○ Foods that have been pasteurized at ultra-high temperatures and aseptically packaged (the packaging is free of microorganisms) can be stored at room temperature. Since many of these foods, such as milk and pudding, are served cold, you may want to store them in the refrigerator.

○ Once they are opened, store all UHT and aseptically packaged foods in the refrigerator at 41°F (5°C) or below.

○ UHT products that have not been aseptically packaged must be stored at 41°F (5°C) or below.

Canned Goods

○ Store canned goods and other dry foods at a temperature between 50°F and 70°F (10°C to 21°C). Even canned foods spoil over time. Higher storage temperatures may shorten shelf life. Acidic foods such as canned tomatoes do not last as long as foods low in acid. The acid also can form pinholes in the metal over time.

○ Keep storerooms dry and relative humidity low. Too much moisture will cause cans to rust.

○ Wipe cans clean with a sanitized cloth towel before opening them. That will help prevent dirt from falling into the contents of the can.

Dry Foods

○ Keep flour, cereal, and grain products such as pasta or crackers in airtight containers. They can quickly become stale in a humid room and can become moldy if there is too much moisture.

○ If dry foods are removed from their original packages and stored in containers, clearly label the containers. Many foods, such as salt and sugar, or flour and baking mixes, often look alike.

○ Before using dry foods, check containers or packages for damage by insects or rodents. Cereal and grain products are favorite targets of these pests.

○ Salt and sugar, if stored in the right conditions, can be held almost indefinitely.

Cleaning Supplies

○ Store chemicals and cleaning supplies away from food storage and preparation areas. Keep the area locked, if possible.

○ Never use empty food containers to store chemicals, and never put food in empty chemical containers. Keep chemicals in their original containers. If it is necessary to transfer chemicals, store them in sturdy containers clearly labeled with the contents and their hazards.

Key Point

Storage temperatures higher than 70°F (21°C) may shorten the shelf life of canned goods.

Shelf Life of Food

Different foods have different shelf lives. *Appendix C* lists recommended storage times for refrigerated foods at ideal holding temperatures. *Appendix D* lists recommended maximum storage times for frozen foods. *Appendix E* lists recommended maximum storage times for dry foods. The shelf lives listed in these appendixes serve as a general guide for the maximum storage time for specific foods. Discard any food that has exceeded this time period. Pay close attention to each food's use-by date.

All potentially hazardous ready-to-eat foods stored in refrigeration should be discarded if not consumed within seven days of preparation. Potentially hazardous ready-to-eat foods that have been frozen should be discarded if not consumed within twenty-four hours of being thawed.

SUMMARY

To keep foods safe once they have been received, store them properly. Foods can be exposed to the same dangers in your establishment that they face on their journey to your back door. When stored improperly, foods can quickly spoil or become contaminated and cause a foodborne illness. Though different foods have different storage needs, some commonsense rules apply to all foods. Foods should be stored in clean, dry areas used only for food. Heat and moisture can cause foods to become unsafe. Food's other enemy is time. Use all foods by their use-by date. Also, rotate all food stocks with the first in, first out (FIFO) method.

Refrigerators are used to store potentially hazardous and perishable foods for short periods of time. Two basic types are walk-in and reach-in units. Ideally, two separate units should be used to keep products at their optimum storage temperatures. Use one for meats, poultry, fish, dairy products, and eggs, and one for fresh produce. Other types of refrigerated equipment, such as refrigerated drawers and prep rails in the kitchen, also help keep food safe until it is cooked.

Frequently check the internal temperature of refrigerated foods and the air inside the unit itself. The ambient air temperature of the refrigerator should be 2°F (1°C) lower than the desired food temperature. Wrap foods carefully before storing them, to prevent spoilage. Covering food also prevents food from absorbing odors and becoming contaminated. Always store raw foods on lower shelves, never above cooked or ready-to-eat foods.

Freezers are used to store many foods for longer periods of time. Some foods don't freeze well. Know which foods deteriorate quickly when frozen.

Dry foods can last a long time when stored properly. To maintain quality and safety, you must control temperature, moisture, and ventilation in dry storage areas. Keep storerooms clean and free of insects and rodents.

Health Alert

Store chemicals and cleaning supplies away from food storage and preparation areas.

Remember,
when in
doubt, throw it out.

Store chemicals and cleaning supplies in locked rooms or cabinets away from food storage and prep areas. Keep storage areas clean and dry.

All foods lose quality over time. They may contain high levels of illness-causing microorganisms long before they appear unfit to eat. Proper storage techniques can help preserve food quality and safety. Remember, when in doubt, throw it out.

A CASE IN POINT I

Case Study

On Monday afternoon, the kitchen staff at the Sunnydale nursing home was busy cleaning up from lunch and preparing dinner. Pete, a kitchen assistant, put a large stockpot of hot, leftover vegetable soup in the refrigerator to chill. Angie, a cook, began deboning the chicken breasts that Ed had stored earlier. When she was finished, she put the chicken on an uncovered sheet pan and stored it in the refrigerator. She carefully placed the raw chicken on the top shelf away from the hot soup. Next, Angie iced a carrot cake she had baked that morning. She put the carrot cake in the refrigerator on the shelf directly below the chicken breasts.

What storage errors were made? What food items are at risk?

A CASE IN POINT II

Case Study

Anticipating a slow dinner, Ed, the owner of Big City Diner, gave his kitchen manager the afternoon off. Some time later, Acme Distributors delivered an order of canned and dry foods and supplies, including canned tomato sauce, canned soup, crackers, pasta, paper napkins, and cleaning supplies such as warewashing detergent and sanitizing solution.

After checking the order, Ed stacked the canned and dry foods onto a dolly and wheeled them into the storeroom. There was an open case of tomato sauce on the shelf with one can left in it and a full case behind it. Ed removed the open case and replaced it with the case on the dolly. He put the single can in an open spot on another shelf. By shifting some boxes around, Ed was able to put the cases of soup, napkins, and pasta on shelves in front of cases already stored there. While moving cases, he accidentally knocked over a container of flour, spilling some on the floor. There was no room on the shelves for the crackers, so he left the case on the floor. He was sweating from exertion and wondered if the storeroom was too hot. He checked the hanging thermometer, which read 70°F (21°C).

Ed went back to the dock to get the cleaning supplies. He wheeled those into the dish room and slid the cases under the warewashing machine. When he returned the dolly to the back door, a truck from Mike's Produce pulled up to deliver a case of lettuce, two cases of tomatoes, a case of onions, and a case of cabbage. Ed inspected the delivery and signed for it. He stacked it all on the dolly and wheeled it toward the kitchen. Suddenly remembering some paperwork he had to finish, he left the produce in the hallway outside the restroom. He knew the dinner cook would be coming in soon and would put it away.

What did Ed do wrong? What are the possible consequences?

TRAINING TIPS

Training Tips for the Classroom

1. "Rooms for Rent" Group Activity

Objective: *After completing this activity,, class participants will be able to identify the most effective storage procedures for freezers, refrigerators, deep-chill units, dry storage rooms, and chemical supply rooms. From this activity they will also be able to identify general principles for proper storage. Note: Class participants will meet this objective through group involvement and discussion activity, as opposed to instructor lecture.*

Directions: Create five groups in the classroom, and assign each group one of the following five storage areas:

○ freezer   ○ deep-chill unit

○ refrigerator   ○ dry storage room

○ chemical and cleaning supplies storage

Give each group five minutes to come up with as many storage procedures as possible for its assigned area.

When the time has expired, have each group present its list, soliciting feedback or comments from the entire class on each storage principle or guideline as it is listed.

All principles or guidelines that the class agrees are valid are recorded by a team member on a flipchart or blackboard in the front of the classroom. After a group presents its principles, solicit additional principles or guidelines from the class.

After all five groups have presented their principles, summarize the activity by making a list of general storage principles from guidelines that were common to all five storage lists.

2. "Where Do I Go?" Team Contest

Objective: *After completing this activity,, class participants will be able to identify proper storage areas and handling guidelines for specific types of food. Note: This is a team-building exercise.*

Directions: Divide a sheet of paper into three columns with the following headings:

Item: **Location:**   **Special instructions:**

In the *item* column, list twenty-five to thirty specific food products that may be received into a foodservice facility. Be sure to include a wide variety of foods, including both fresh and frozen meats, poultry, fish, seafood, shellfish, dairy products, eggs, MAP and *sous vide* foods, fresh produce, UHT products, dry foods, and canned or bottled foods.

Assign teams of two to four people per team, and give each team a copy of the worksheet. Give the teams five minutes to determine the proper storage area for each item, as well as at least one specific storage instruction for each item.

Then, allowing each team to score its own worksheet, go down the list of foods, coming to agreement among the class as to where the product should be stored and what specific storage instructions should be given for each item.

The team with the most correct locations and instructions wins.

3. Food Products in Storage Discussion

Objective: *After completing this activity, class participants will be able to develop standard operating procedures (SOPs) for storage of various types of food. Note: This higher-level activity may not be suitable for all classes.*

Directions: Give your class a challenge: develop SOPs for storage of the following items:

- live oysters
- MAP chicken breasts
- prepared shrimp salad
- whole red snapper
- fresh hamburger patties
- fresh shell eggs
- *sous vide* chicken cordon bleu

Divide the class into groups, and ask them to address the following questions:

1. What is the ideal temperature in storage for each product?
2. Who stores the food, where is it to be stored, and how should it be stored?
3. What documentation should you keep?

Allow ample time for each group to develop SOPs for these products in storage. Then review with the class.

Training Tips on the Job

1. "What's This?"

Purpose: *To illustrate the importance of keeping products in their original containers, labeling products in secondary containers, and storing cleaning supplies and chemicals away from food. Note: This is an easy, yet memorable, demonstration.*

Directions: Place a number of food and nonfood products in unlabeled clear containers, or otherwise conceal their identity. Products should include items that are difficult to identify or distinguish from other products, such as:

- sugar
- salt
- flour
- cleanser
- baking soda
- bleach
- vinegar
- oil
- caustic solvent
- window cleaner
- pan spray
- stainless-steel cleaner

Have kitchen employees try to identify each product without smelling, touching, and certainly without tasting!

2. "What's Wrong with This Picture?" Contest

Purpose: *To involve your kitchen staff in the assessment of storage practices in your establishment and to achieve their buy-in when implementing corrective actions that result from their assessment. Note: The competitive nature of this activity should make it fun for your staff.*

Directions: Have each member of your kitchen staff conduct an inspection of the establishment's food storage areas to find mistakes in your freezers, refrigerators, and dry-storage areas, as well as in any temporary storage units for food. The inspector who finds the most mistakes wins a prize.

Equip your employees with clipboards, pencils, blank paper, and thermometers, and have them make note of everything they find to be in violation of safe food-storage principles. They should report on the condition of the storage areas and of the food itself.

When your kitchen staff has finished inspecting, collect and make copies of the reports.

Hold a group meeting, and encourage employees to share their observations, area by area. Have someone record the collective observations. Employees can give themselves a point for each noted violation. (Let them know you have copies of their reports, lest they be tempted to inflate their grades!) The staff member with the most points wins.

Let your staff know that the real winner is the entire operation, if everyone learns from the process. Let them know that the inspection is just the assessment step. Corrective actions need to be taken, and better storage procedures need to be in place. These corrective actions will maximize food safety and quality. They also help control food cost by minimizing waste. Be sure to set an action plan in place, getting everyone involved. Follow-up is essential. Let your staff know that you will be doing the next inspection. If you give storage an A+, you'll treat them all to a party!

3. Developing Charts and Checklists

Purpose: *To involve your kitchen management team in the development of storage job aids that can be used as easy reference tools for employees and management. Note: These job aids will help support your HACCP systems and will lead to management buy-in because the managers were given the opportunity to create them.*

Directions: Get your kitchen management involved in developing a set of colorful, readable, usable charts and checklists that establish your storage guidelines and procedures.

These reference tools may include:

○ A storage temperature list for foods in three areas: freezer, refrigerator, and dry storage.

○ Storage guidelines for each storage area.

○ Time-temperature logs for checking food temperatures during each shift (with corrective action listed).

○ Use-by or discard dates for food products.

Important: Be sure that the standards meet or exceed codes of your regulatory agency, and any recommendations from manufacturers or processors.

Here are some general guidelines for kitchen charts or logs.

○ Keep them simple. Less is more.

○ Make them colorful. Better yet, color code for storage area. Use red for CCPs.

○ Print them in several languages, if necessary.

○ Laminate all materials.

○ Post them in the appropriate areas.

○ Use pictures or graphics, and fewer words, whenever possible.

DISCUSSION QUESTIONS

1. In top-to-bottom order, how would you store the following foods in a refrigerator: raw trout, raw pork chops, an uncooked beef roast, raw chicken, and raw ground beef?

2. What are the proper procedures for storing fish meant to be eaten raw?

3. What is the difference between refrigerated and deep-chill storage?

4. What should you do before storing food that has been taken out of its original package and put into a container?

5. What is the proper storage temperature for frozen dairy products?

6. What temperature should the air inside a walk-in refrigerator be to hold products at an internal temperature of 41°F (5°C)?

7. Ground beef is received at 41°F (5°C) at 9:00 a.m. and stored in a refrigerator. At 1:30 p.m., the temperature of the ground beef is rechecked and is now 60°F (16°C). What corrective action should be taken?

8. How would you verify that a refrigerator is maintaining the ambient air temperature at 38°F (3°C)?

MULTIPLE-CHOICE STUDY QUESTIONS

1. At what storage temperature would ground beef most likely become unsafe to use?

   A. 0°F (-17°C)   C. 41°F (5°C)

   B. 30°F (-1°C)   D. 60°F (16°C)

2. Which storeroom condition is most likely to cause a sack of flour to become contaminated?

   A. A damp floor   C. An air temperature of 50°F (10°C)

   B. Fluorescent lighting   D. Full shelves

3. Under which condition could you display to customers the live mussels that you will be cooking and serving to them?

   A. You have special permission from the health department.

   B. You have a self-cleaning filtration system for the display tank.

   C. You will also cook and serve other mussels that have not been on display.

   D. You have removed the shellstock identification tags as required by law.

4. It is important that storage areas be

   A. close to where cleaning supplies are accessible.

   B. kept safe from all sources of contamination.

   C. kept warm and dry.

   D. opened only by the manager.

5. There is a date written on each of the single-serve boxes of ready-to-eat breakfast cereals in your storage room. What does this date tell you?

   A. You should manage your stock to serve the boxes within a week of the dates.

   B. The contents of the boxes will be unsafe to eat one week before the dates.

   C. You should rotate your stock to serve the boxes with the older dates first.

   D. The manufacturer will take boxes back for credit after the dates.

6. A major problem with the storage of vacuum-packaged foods is that
   A. loss of moisture can lessen food quality.
   B. heat can cause vacuum-packaged cans to explode.
   C. air leakage can contaminate neighboring foods.
   D. bacteria that cause botulism can develop.

7. If you see that a stored *sous vide* package is torn, you should immediately
   A. cook and serve the contents.
   B. throw away the contents.
   C. mend the tear and freeze the package.
   D. measure the internal temperature of the package.

8. In order to guarantee that foods in a refrigerator are being kept at 41°F (5°C) or lower, maintain the air temperature in the refrigerator at
   A. 26°F (-3°C).
   B. 32°F (0°C).
   C. 38°F (3°C).
   D. 0°F (-18°C).

9. The hanging thermometer in your refrigerator has fallen on the floor. What action should you take?
   A. Purchase a new thermometer.
   B. Rehang the thermometer in a safer place in the refrigerator.
   C. Replace the thermometer with a bi-metallic probe thermometer.
   D. Check the accuracy of the thermometer and recalibrate or replace if necessary.

10. Live oysters that will be served raw to customers must be
    A. kept at temperatures between 35°F (2°C) and 45°F (7°C) before serving.
    B. shown live to customers before serving.
    C. cooled to 32°F (0°C) right before serving.
    D. warmed to 41°F (5°C) right before serving.

ADDITIONAL RESOURCES

Resources

Books and Periodicals

Cichy, R. F. (1993). *Sanitation management.* East Lansing, MI: Educational Institute of the American Hotel & Motel Association.

Cross out cross-contamination. (1998). *Best Practices.* 2(1), 6-9.

Equipped for food safety. (1998). *Best Practices,* 2(4), 6-10.

Kotschevar, L. H. (1994). *Quantity food purchasing* (4th ed.). New York: Macmillan.

Loken, J. K. (1995). *The HACCP food-safety manual.* New York: Wiley.

Longree, K. (1996). *Quantity food sanitation* (5th ed.). New York: Wiley.

National Restaurant Association Educational Foundation. (1998). *A Practical Approach to HACCP: Coursebook.* Chicago: Author.

The right way to…prepare beef. (1998). *Best Practices,* 2(3), 12-13.

The right way to…prepare chicken. (1998). *Best Practices,* 2(1), 10.

The right way to…prepare eggs. (1998). *Best Practices,* 2(2), 10-11.

The right way to…preparing fruits and vegetables. (1998). *Best Practices,* 2(4), 12-13.

The right way to…prepare pork. (1997). *Best Practices, 1*(3), 10-11.

The right way to…seafood safety. (1999). *Best Practices,* 3(1), 16.

Secrets of self-inspection. (1998). *Best Practices.* 2(2), 6-9.

Stefanelli, J. M. (1997). *Purchasing: Selection and procurement for the hospitality industry.* New York: Wiley.

United States Dept. of Health and Human Services, Public Health Service, & Food and Drug Administration. (1997). *Food Code: Recommendations of the United Public Health Service Food and Drug Administration.* Springfield, VA: United States Dept. of Commerce. National Technical Information Service. (NTIS No. PB97-133656)

Web Sites

Alaska Seafood Marketing Institute

http://www.alaskaseafood.org

The Alaska Seafood Marketing Institute works to improve seafood quality by teaching fishermen, processors, retailers, and restaurateurs about proper handling of Alaska seafood products. Their Web site also includes industry news, suppliers, and recipes.

American Egg Board

http://www.aeb.org

The American Egg Board's Web site features product, safety, and industry information for foodservice professionals as well as recipes and other information for the general public.

American Meat Institute

http://www.meatami.org

Provides member information, as well as current and past news and information regarding the meat industry.

Daydots

http://www.daydots.com

Daydots makes stickers that facilitate efficient food rotation. Their Web site offers safety tips, ordering information, and industry news.

FDA Center for Food Safety and Applied Nutrition (CFSAN)

http://vm.cfsan.fda.gov/list.html

Comprehensive site from CFSAN offers a wealth of food-safety information, from foodborne illness to food labeling. CFSAN strives to be a leader in food safety, and to protect consumers from economic fraud, promote sound nutrition, and encourage innovation.

Food Marketing Institute

http://www.fmi.org

This nonprofit association conducts programs in research, education, industry relations, and public affairs for its membership of food retailers and wholesalers. Their Web site offers a wealth of food-safety information.

Institute of Food Technologists (IFT)

http://www.ift.org

IFT is a nonprofit scientific society with 28,000 members working in food science, food technology, and related professions in industry, academia, and government. This site includes information on its policies and publications, education and industry news, and meeting and convention locations.

International Association of Milk, Food, and Environmental Sanitarians (IAMFES)

http://www.iamfes.org

IAMFES keeps members informed of the latest scientific, technical, and practical developments in food safety and sanitation. Their Web site includes booklets and links to other food-safety sites.

International Dairy Foods Association (IDFA)

http://www.idfa.org

This site is the link to information on IDFA and its constituent organizations: the Milk Industry Foundation (MIF), the National Cheese Institute (NCI), and the International Ice Cream Association (IICA), as well as information on the entire dairy industry. The site also includes information on current industry events, membership information, and publications.

International Food and Information Council (IFIC)

http://ificinfo.health.org

Web site for the International Food and Information Council offers information and resources about food safety and food allergies for educators and the public.

KatchAll Industries International, Inc.

http://www.katchall.com

KatchAll manufactures food-safety/preparatory kitchenware. Their Web site provides ordering information and links to other food-safety Web sites.

National Cattlemen's Beef Association

http://www.beef.org

This site includes up-to-date information from the National Cattlemen's Beef Association, as well as a reference library, information on nutrition and food safety, and links to related sites.

National Food Processors Association (NFPA)

http://www.nfpa-food.org

NFPA is a scientific and technical trade association for the food industry. Their site provides industry news and educational materials on issues of food safety, research, and food science.

National Frozen Food Association (NFFA)

http://www.nffa.org

NFFA is the voice of the frozen foods industry. Their Web site provides information to consumers as well as foodservice professionals.

National Pork Producers Council

http://www.nppc.org

The National Pork Producers Council site provides general information about food and nutrition, as well as information specifically for pork producers. A special section is also provided just for children.

National Restaurant Association

http://www.restaurant.org

The National Restaurant Association site provides information on government agencies affecting the restaurant industry, the latest training and certification updates, and links to state restaurant associations and hospitality schools and universities.

NSF International

http://www.nsf.org

This site offers product information, resources, and publications regarding public health safety.

Produce Marketing Association

http://www.pma.com

The members of this nonprofit trade association market fresh fruits and vegetables to the foodservice industry. Purchasing guidelines are available.

Tyson

http://www.tyson.com

Tyson's Web site offers a consumer kitchen page with recipes and instruction on preparing chicken products. An entire page is also devoted to foodservice professionals and offers information on the industry, Tyson products, and menu planning.

United States Department of Agriculture Food Safety and Inspection Service

http://www.fsis.usda.gov

The USDA's Food Safety and Inspection Web site offers the latest food-safety news, educational materials, and HACCP-implementation materials.

Chapter 7
Protecting Food During Preparation

TEST YOUR FOOD-SAFETY KNOWLEDGE

1. **True or False:** After the loose dirt has been brushed off the outside of a melon, it can be safely cut open. *(See Fruits and Vegetables, page 7-11.)*

2. **True or False:** A fruit salad containing watermelon, grapes, and strawberries does not need to be maintained under 41°F (5°C) or below to be safe to eat. *(See Fruits and Vegetables, page 7-12.)*

3. **True or False:** Two-stage cooling involves first refrigerating and then freezing foods. *(See Cooling Food, page 7-18.)*

4. **True or False:** A casserole has been removed from the oven before it has reached a safe minimum internal temperature. Before resuming cooking, the casserole should first be cooled to below 41°F (5°C). *(See Poultry, Stuffing, and Stuffed Meats and Casseroles, page 7-14.)*

5. **True or False:** The minimum safe internal temperatures for a beef roast and a pork roast are the same. *(See Pork and Beef, pages 7-14 and 7-15.)*

Table of Contents

Learning Objectives

After completing this chapter, you should be able to:

○ Use time and temperature controls in each step of the flow of food.

○ Properly thaw frozen foods.

○ Safely handle and prepare raw food.

○ Prevent contamination and microbial growth in the flow of food.

○ Identify the minimum internal cooking temperatures of foods.

○ Cool, store, and reheat cooked foods properly.

○ Make corrections if time and temperature standards are not met.

Key Terms

Temperature abuse
Four-hour rule
Cross-contamination
Slacking
Pooled eggs
Highly susceptible populations
Minimum internal temperature
Two-stage cooling method
Ice-water bath
Cold paddle

Serving safe food requires constant attention to the factors that can cause foodborne illness. Even food that is safe when it is delivered, and properly stored, can quickly become unsafe. The extent to which food stays safe once it is in your operation depends on how safely you handle it. From the moment food arrives to the time it is presented to a customer, it is up to you to take precautions that will keep food safe.

You must handle and prepare food with care to make sure it is safe when you serve it. Employees should understand the importance of food safety and how it can be affected during preparation and cooking.

The three most critical factors in preventing the growth and spread of microorganisms that cause foodborne illness are time, temperature, and sanitation. You've already seen how these elements can affect food quality while it's in storage. Once you take food out of storage for use in a recipe, controlling time, temperature, and sanitation becomes even more critical.

SAFE FOODHANDLING

The key to serving safe food is to handle it safely. The causes of most foodborne illness can be grouped into two categories: temperature abuse and cross-contamination. Improper holding temperature was a major factor in all cases of foodborne illness reported to the U.S. Centers for Disease Control and Prevention (CDC). Inadequate cooking temperatures, poor personal hygiene, and cross-contamination were consistently significant as well.

Teach employees how important it is to control sanitation, time, and temperature. Then put practices and procedures in place to help them follow commonsense rules of safe food handling.

Time and Temperature Control

Probably the biggest factor in outbreaks of foodborne illness is **temperature abuse.** Disease-causing microorganisms grow and multiply at temperatures between 41°F and 140°F (5°C and 60°C), which is why this range is known as the temperature danger zone. Microorganisms grow much faster in the middle of the zone, at temperatures between 70°F and 120°F (21°C and 49°C) *(see Exhibit 7a)*. Whenever food is held in the temperature danger zone, it is being abused.

As you learned in Chapter 2, time also plays a critical role in food safety. Microorganisms need both time and temperature to reproduce. The longer food stays in the temperature danger zone, the more time microorganisms have to multiply. When heating or cooling foods, it is important to pass them through the danger zone as quickly and as few times as possible.

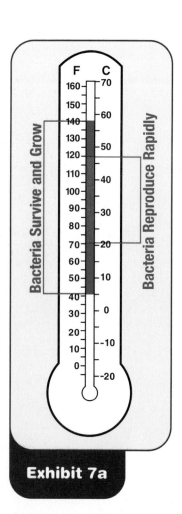

Exhibit 7a

Temperature and Bacterial Growth

Most microorganisms grow rapidly at temperatures between 70°F and 120°F (21°C and 49°C).

To keep food safe during preparation and cooking, you must follow the **four-hour rule.** Never let food remain in the temperature danger zone for more than four hours.

The exposure time adds up during each stage of handling, from the time the food arrives at the receiving dock to the time it is cooked and again when it is held for service, cooled, and reheated. For example, beef used to make a stew could be exposed to dangerous temperatures several times before it is served. Here's what could go wrong:

○ Left on the receiving dock too long while other foods are inspected and put away.

○ Sitting out on the counter while other stew ingredients are cut and prepared.

○ Not heating to minimum cooking temperatures within 4 hours.

Once the food is properly cooked the number of microorganisms in the food will have been reduced. However, you must continue to handle the food safely. There are further opportunities for temperature abuse, such as:

○ Cooling the food at room temperature before storing it in the refrigerator.

○ Failing to reheat the food to 165°F (74°C) for fifteen seconds within two hours.

○ Holding on a steam table that doesn't maintain proper temperature at or above 140°F (60°C).

You can see how four hours in the danger zone adds up easily if you don't handle food with care at each stage of preparation. The best way to avoid a problem is to establish procedures employees must follow and make sure they are always followed. Build time and temperature control into your establishment's HACCP plan and standard operating procedures. Make time and temperature control part of every employee's job. Some suggestions include:

○ **Make sure that the establishment has enough of the right kind of thermometers available in the right places.** Give all employees their own bi-metallic stemmed thermometers. Have employees use timers in prep areas to remind them how long foods are being kept in the temperature danger zone.

Time & Temperature

The time that food spends in the temperature danger zone accumulates at each step in the flow of food. This accumulated time cannot exceed four hours.

Reprinted with permission from Roger Bonifield and Dingbats.

Exhibit 7b **Employees Using Cooling Charts**
Print up simple forms that employees can use to check and record temperatures.

○ **Regularly record temperatures and the times they are taken.** Print up simple forms that employees can use to record temperatures and times throughout the shift. Post these forms on clipboards outside of refrigerators and freezers, near prep tables, and next to cooking and holding equipment.

○ **Establish clear standard operating procedures for employees.** These may include procedures such as:

  ● Do not take more food out of the refrigerator than can be prepared in twenty minutes.

  ● Refrigerate ingredients and utensils before preparing certain recipes such as tuna or chicken salad.

  ● Cook hamburger patties to an internal temperature of 155°F (68°C) for fifteen seconds.

○ **Decide the best way to monitor time and temperature in your establishment.** Determine which foods should be monitored, how often, and who should check them. Then assign responsibilities to employees in each area. Make sure they understand exactly what you want them to do, how to do it, and why it is important.

○ **Develop a set of corrective actions.** Decide what action should be taken if time and temperature standards are not met. For example, an establishment might often hold egg rolls in a steam table at 140°F (60°C) until served. Corrective actions for this product might be to throw out the product if the temperature has fallen below 140°F (60°C) for more than four hours, or to reheat the product to 165°F (74°C) for at least fifteen seconds if the temperature of the product falls below 140°F (60°C) for less than four hours.

Preventing Cross-Contamination

The other major hazard food faces in your operation is **cross-contamination.** Microorganisms move around easily in a kitchen. They can attach themselves to almost anything they contact, including prep tables, equipment, utensils, cutting boards, dish towels, sponges, hands, or other foods. *Exhibit 7c* shows microorganisms living on a cutting board and a sponge. Cross-contamination occurs when microorganisms from one source are spread to other foods.

Cross-contamination can start at almost any point in an operation. When you know what it is and how it happens, cross-contamination is fairly simple to prevent. Prevention starts with the basic principle of creating barriers between food products. Barriers can be physical or procedural. In Chapter 6, for example, you learned to store food in covered containers and to make sure that raw meat or poultry is not stored above cooked food in a refrigerator. These are examples of physical and procedural barriers to cross-contamination.

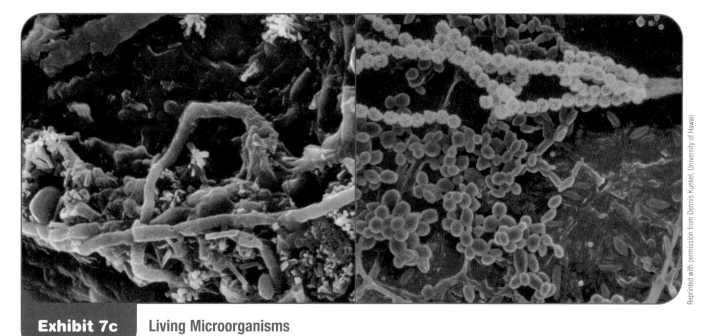

Reprinted with permission from Dennis Kunkel, University of Hawaii

Exhibit 7c **Living Microorganisms**
Left: Kitchen sponge surface with microorganisms.
Right: Kitchen cutting board with bacteria and fungi.

Prevention is a matter of common sense. Here are some simple preventive practices you should consider using in your establishment.

○ **Prepare raw meats, fish, and poultry in separate areas from produce or cooked and ready-to-eat foods.** This can be as easy as using separate prep tables in the same area. If you don't have sufficient space for separate tables, prepare these items at different times, so they don't cross paths.

○ **Assign specific equipment to each type of food product.** For example, use one set of cutting boards, utensils, and containers just for poultry, another set for meat, and a third set for produce. Some manufacturers make colored cutting boards and utensils with colored handles. Color-coding can tell employees which equipment to use with which products, such as green for produce, yellow for chicken, and red for meat.

○ **Use specific containers for each type of food product.** Clearly label containers with their contents, such as "Raw Chicken" or "Tuna Salad," so the containers are less likely to be used for something else. Make sure you have an adequate supply of containers on hand, and keep them clean and sanitized between uses.

○ **Clean and sanitize all work surfaces, equipment and utensils after each task.** After cutting up raw chicken, for example, it is not enough to simply wipe down the counter underneath the cutting board. Run the

Cross-Contamination

Using color-coded cutting boards can prevent cross-contamination by designating a different color for each product, such as yellow for chicken, green for produce, and red for meat.

Manufactured by KatchAll Industries

If gloves
are used,
they should be changed
before starting a new
task, and hands must
be washed before
putting on a new pair.

cutting board and utensils through a warewashing machine, and clean and sanitize the countertop. Employees must know which cleaners and sanitizers to use for each job. Sanitizers used on food-contact surfaces must meet local or state department codes which conform with the Code of Federal Regulations (21CFR178.1010). *(See Chapter 11 for more information on cleaners and sanitizing solutions.)*

○ **Cloths or towels used for wiping food spills must not be used for any other purpose.** There are a few ways to prevent the spread of microorganisms from cleaning cloths to equipment and utensils. One way is to use disposable towels. Another is to use color-coded cleaning cloths that match a specific food preparation area or task. After each use, cloths or towels must be rinsed and stored in a clean sanitizing solution. *(See Chapter 11 for more information on how to make sanitizing solutions.)*

○ **Consider requiring all employees to wear single-use gloves while preparing and serving food in your establishment.** If you adopt this policy, teach employees to use gloves properly. Employees must wash their hands before putting on gloves. Gloves should be used only for the job at hand, and changed each time a new task is started. Hands must be washed before putting on the new pair of gloves. If punctured or ripped, gloves must be changed.

○ **Employees should watch what they touch after handling raw foods and should practice good personal hygiene.** Hands are the most common carriers of microorganisms in the kitchen. Employees should be trained to wash their hands at the beginning of their shift, between tasks, and especially after using the restroom.

PREPARING FOOD

Following the basic principles of time and temperature control and putting up barriers to cross-contamination will help to prevent most potential cases of foodborne illness. However, some food items and methods of preparation require a bit more care.

Thawing Foods Properly

Often, the first step in food preparation is thawing what you intend to cook. If frozen foods are thawed improperly they can become a breeding ground for microorganisms.

Suppose a cook wants to prepare a twenty-pound frozen turkey. The cook is in a hurry, so instead of putting the turkey in the refrigerator, he puts it in a pan on a worktable to thaw overnight. The air temperature in the kitchen is about 72°F (22°C).

There are two problems with this method of thawing. First, when the bird begins to thaw, the skin and outer layers are exposed to the temperature danger zone even though the bird's core is still frozen. Microorganisms may multiply to a point where cooking the turkey may not be sufficient to kill them. The cook may end up serving a potentially dangerous dish to the establishment's customers.

Second, the microorganisms on the turkey may contaminate everything in the area: the cook's hands, the worktable, the pan, the cutting board, knives, and other utensils. If the area and equipment aren't cleaned and sanitized completely before the cook starts another task, microorganisms could spread throughout the entire kitchen.

There are only four acceptable ways to thaw frozen foods (see Exhibit 7d).

○ **Thaw food in the refrigerator at temperatures of 41°F (5°C) or less.** This method requires advance planning. A twenty-pound turkey can take from two to four days to thaw completely in the refrigerator. Employees need to take food out of frozen storage far enough ahead to make sure there is enough thawed product on hand when it's needed.

○ **Submerge the frozen product in running potable water at a temperature of 70°F (21°C) or below.** The water has to flow fast enough to wash loose food particles into the overflow drain. Make sure that the thawed product does not drip water onto other products or food-contact surfaces. Clean and sanitize the sink and work area before and after thawing food this way. Remember, the product cannot remain in the temperature danger zone for more than four hours, which includes the time the food is thawed under the running water and the time needed for preparation for cooking.

○ **Food may be thawed in a microwave oven, only if it will be cooked immediately afterward.** Microwave thawing can actually start cooking the product, so don't use this method unless you intend to continue cooking the food immediately. Large items such as roasts or turkeys don't thaw well in the microwave.

○ **Food may be thawed as part of any cooking procedure as long as the product reaches the required minimum internal cooking temperature.**

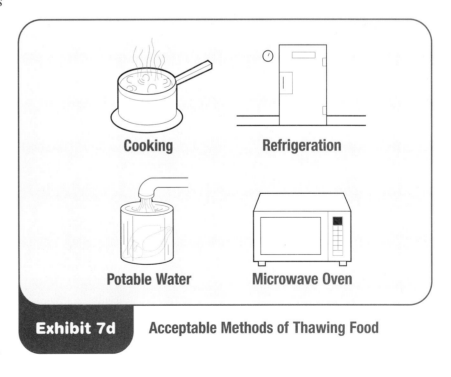

Cooking   Refrigeration

Potable Water   Microwave Oven

Exhibit 7d   Acceptable Methods of Thawing Food

Frozen hamburger patties, for example, can go straight from the freezer onto a grill without being thawed first. Frozen chicken can go straight into a deep fat fryer. These products cook quickly enough from the frozen state to pass through the temperature danger zone without harm. However, always make sure you verify the final cooking temperature with a calibrated thermometer.

Some frozen foods may be *slacked* before cooking. **Slacking** is the process of gradually thawing frozen food in preparation for deep fat frying, or allowing even heating during cooking. For example, you might allow a large block of frozen spinach to warm from -10°F (-23°C) to 25°F (-4°C). Since foods should never be refrozen once they begin to thaw, slack foods just before you cook them. Don't let foods get any warmer than 40°F (4°C). If slacking at room temperature is permitted in your jurisdiction, a system must be in place to ensure that the product does not exceed 40°F (4°C).

PREPARING SPECIFIC FOODS

Meat, Fish, and Poultry

The source of most cross-contamination in an operation is raw meat, poultry, and seafood. Your staff should follow safe procedures when handling these products.

The source of most cross-contamination in an establishment is raw meat, poultry, and seafood.

○ **Use clean and sanitized work areas, cutting boards, knives, and utensils.** Prepare raw meat separately from fresh produce or at a different time.

○ **Be sure to wash your hands.** If you use gloves, change them before you start each new task. Wash your hands again before putting on a new pair of gloves.

○ **Take out of the refrigerator only as much product as you can prepare at one time.** When cubing beef for stew, for example, take out and cube one roast, and refrigerate it before taking out another roast.

○ **Put raw prepared meats away or cook them as quickly as possible.** If you put them back into storage, wrap them first, or put them in covered containers. Store them on bottom shelves in the refrigerator. Never store them above cooked or ready-to-eat food such as fresh produce, because raw meat juices may drip and cause cross-contamination.

Protein-Based Salads

Chicken, tuna, egg, pasta, and potato salads all have been known to cause outbreaks of foodborne illness. These salads are typically made from foods that can easily support the rapid growth of microorganisms. Since they will not typically be cooked after preparation, there is no chance to kill microorganisms that could be introduced to these salads during the preparation. Therefore, extreme care must be taken when preparing protein-based salads.

When preparing salads, handle all ingredients with care and follow these precautions.

○ Make sure that proteins are thoroughly cooked and properly cooled and stored. Cooks often use leftover cooked meats or poultry to make salads. Make sure leftover meats and poultry have been properly cooked, held, cooled, and stored. The recommended storage time for cooked leftover meats is only one to two days. (They *must* be discarded after seven days if stored in the refrigerator.)

○ Make sure foodhandlers wash their hands before preparing food, and, if required by local health codes, have them wear clean, single-use gloves.

○ Prepare other ingredients away from raw meats and poultry. Always prepare vegetables such as celery and carrots using clean, sanitary equipment and utensils.

○ Leave food in the refrigerator until all ingredients are ready to be mixed. Don't take out chicken and leave it on the counter while you chop vegetables, for example.

○ Consider chilling all ingredients before making salads. Put unopened cans of tuna and jars of mayonnaise in the walk-in the day before you make tuna salad, for example. Consider chilling mixing bowls and utensils, too. Taking these steps will provide barriers that may prevent the growth of harmful microorganisms that could be present.

○ Prepare small batches at one time, so large amounts of food don't sit out at room temperature for long periods of time. Set a working time limit of twenty minutes and make sure you take out only enough ingredients to use within that time period. When time is up, refrigerate what you've made, and then get ingredients to make more *(see Exhibit 7e)*.

Exhibit 7e

Preparing Protein Salads
Prepare salads in small batches so large amounts of food don't sit out at room temperature for too long.
Courtesy of Cooper Instrument Corporation, Middlefield, CT

Eggs and Egg Mixtures

Historically, the contents of whole, clean, uncracked shell eggs were considered free of bacteria. Now it is known that a certain bacteria, called *Salmonella* Enteritidis, can be found inside eggs. *Salmonella* Enteritidis can live inside a laying hen and can be deposited in the white of an egg before the shell is formed. Although only a small number of eggs produced in the United States are likely to carry this bacteria, all eggs are considered to be a potentially hazardous food because they are able to support the rapid growth of microorganisms. As a consequence, eggs should be purchased when fresh and stored and handled properly. Eggs should be kept at refrigeration temperatures of 41°F (5°C) or below until immediately before use. Special handling precautions are described below.

○ Use clean bowls, whisks, blenders, and other equipment when working with eggs. To prevent cross-contamination, promptly clean and sanitize all equipment and utensils that were used with eggs. Place freshly made batters or sauces in clean containers. Never reuse containers that held raw eggs, even to replenish the same recipe.

○ Foods containing eggs that receive little or no cooking, such as Caesar salad dressing, meringue, mayonnaise, hollandaise sauce, and eggnog, should be prepared with great care. Follow the directions precisely, and monitor temperatures using a thermometer. Many jurisdictions now require operators to post a notice to consumers when raw eggs are used as an ingredient in these types of food. The use of pasteurized shell eggs may be a safer alternative when preparing these types of dishes.

○ Eggs that are cracked open and combined in a container **(pooled eggs)** must be handled with special care because bacteria present in one egg can be spread to the rest. Cross-contamination from one pooled egg mixture to another may also be a hazard if the food is not prepared or handled properly. Shell eggs must be kept cold until immediately before use. Pooled eggs, if used, must be cooked promptly after mixing or kept at 40°F (4°C) or less for no more than two hours. Containers that held pooled eggs must be washed and sanitized before being used again. Each batch of pooled eggs should be prepared in a container that has been washed and sanitized. Utensils used to prepare them must also be clean and sanitized. There should be no carryover of egg residue from one food batch to another even if it is the same recipe.

○ Federal public health officials recommend that operations that serve **highly susceptible populations,** such as those in nursing homes and hospitals, use only pasteurized shell eggs or pasteurized egg products. This is recommended for all types of eggs and egg dishes served to these populations and is required when liquid egg mixtures are prepared, when eggs are cooked and held for service, and when egg-containing foods will receive little or no cooking (e.g. Caesar salad dressing, meringue, mayonnaise, hollandaise sauce, and eggnog).

Batters and Breading

Batters and breading can be hazardous if made with eggs or milk. They face the same dangers from temperature abuse and cross-contamination as other foods, so they should be handled with care. In some cases, it may be better to buy frozen breaded items that can be taken directly from the freezer and then cooked thoroughly in the oven or fryer. If you prefer to make breaded or battered foods from scratch, follow some simple rules.

Key Point

Batters and breading can be potentially hazardous and face the same risks of temperature abuse and cross-contamination as other foods.

○ Make batters with pasteurized egg products instead of raw shell eggs whenever possible.

○ Prepare batters in small batches and keep what you don't need in a covered container in the refrigerator. Using small amounts prevents temperature abuse of both the batter and the foods being coated.

○ When you make up a batch of batter or breading for later use, store it in the refrigerator at 41°F (5°C) or lower. Don't combine it with other batches already stored in the refrigerator.

○ If you are breading foods to cook later, put the breaded food in the refrigerator as quickly as possible.

○ Cook battered and breaded foods thoroughly. The coating acts as an insulator, which can prevent the food from being cooked thoroughly. When deep-frying foods, make sure the temperature of the oil recovers before loading each batch. Overloading the basket also slows cooking time, which means employees may take product out of the fryer before it is done. Be sure to monitor oil temperature, food temperature, and cooking time.

○ Throw out any used batter or breading left over after each shift. Whenever a potentially hazardous food is dipped in batter or dredged in breading, consider the batter or breading contaminated. Never use a batter or breading for more than one product. For example, if you dip raw chicken into batter, then use the same batter for onion rings, you may contaminate the onion rings and possibly make someone sick.

Fruits and Vegetables

For the most part, fresh fruits and vegetables are not as likely to carry pathogens as are some other foods. But viruses such as Hepatitis A, bacteria such as *Listeria,* and parasites such as *Cryptosporidium* can survive in or on produce, especially cut produce. The risk from such microorganisms can be minimized or eliminated by simple preparation safeguards.

○ Start with a clean, sanitized work space. Prepare vegetables away from raw meats, poultry, and eggs, as well as from cooked and ready-to-eat foods, to prevent cross-contamination.

○ Surfaces that come in contact with raw meat, such as hands, gloves, knives, prep tables, or cutting boards, should never have direct or indirect contact with foods to be eaten raw, such as salad or fruit, unless they have been cleaned and sanitized after use.

○ Wash fruits and vegetables thoroughly under running water to remove dirt or other contaminants just before cutting, combining with other

Preparing Raw Vegetables:

Start with a clean, sanitized work space. Prepare vegetables away from raw meat, poultry, and eggs and from cooked and ready-to-eat foods.

Reprinted with permission from Tony Soluri and Charlie Trotter

ingredients, or cooking. Failure to wash fruit before cutting it may result in contamination of the interior of the fruit.

○ Pay particular attention to leafy items such as lettuce and spinach because dirt and microorganisms can get into inner leaves. Remove the outer leaves, and pull lettuce and spinach completely apart and rinse thoroughly.

○ Cut away bruised or damaged areas when preparing fruits and vegetables.

○ Cooked vegetables must never be left out or held at room temperature.

○ Hold cooked vegetable dishes at 140°F (60°C), or above. If dishes are not held at or above 140°F (60°C) discard them after four hours.

○ Hold cut produce at 41°F (5°C) or below. If cut produce is not held at or below 41°F (5°C), discard it after four hours.

○ Establishments may not add sulfites to any products that are served raw.

○ Any chemical used to wash produce must meet requirements listed in the Code of Federal Regulations (21CFR178.1010). Contact your regulatory agency if you have any questions about the approval or use of a particular chemical.

Ice

People often forget that ice is a food, too. It's used to chill beverages and to chill or dilute foods. In Chapter 6, you learned how to store products on ice, such as poultry, fish, fruits, and vegetables. Many establishments also use ice to chill foods on display, such as canned beverages or fruit. Any time ice has been used to cool foods in this way, you may not reuse it as a food.

Ice used as a food, or used to chill other foods, must always be made with potable water (water that is safe to drink). Remember that ice can become contaminated just as easily as other foods. When transferring ice from an ice machine to an ice bin or to a display, use a clean, sanitized scoop and container. Store ice scoops outside of the ice machine in a clean, protected location. Never hold ice in containers that have been used to store raw meat, poultry, fish, or chemicals.

COOKING FOOD

Cooking is a Critical Control Point for most foods. Whatever cooking method you choose, temperature is key to killing microorganisms that can cause illness. The internal temperature of the product has to reach a certain level for a specific amount of time before you can be sure all microorganisms in the food have been sufficiently reduced in number. While time alone is a good indicator of when a product should be properly cooked, the only way to tell for sure is to measure the internal temperature of the product using a probe thermometer.

Cross-Contamination

When transferring ice from an ice machine, use a clean, sanitized scoop and container.

HACCP Principle

Cooking is a Critical Control Point for most foods.

The **minimum internal temperature** at which microorganisms are destroyed varies from product to product. In general, foods should be cooked to an internal temperature of at least 145°F (63°C) or higher, for at least fifteen seconds. Minimum standards have been developed for most cooked foods, and are included in local and state health codes *(see Exhibit 7f)*.

Here are general guidelines to follow when cooking.

○ **Specify cooking times and proper internal product temperature on all recipes.**

○ **Use properly calibrated thermometers, accurate to within 2°F or 1°C, to measure food temperatures.** Measure internal temperatures in several places, in the thickest part of the food. Clean and sanitize the thermometer after each use.

○ **Avoid overloading ovens, fryers, and other cooking equipment.** The equipment temperature may drop, and the foods may not cook properly. Crowding foods also may cause cross-contamination.

○ **Let the temperature of the cooking equipment recover between batches.**

○ **Use utensils or gloves to handle food after cooking, and taste foods correctly, to avoid cross-contamination.** The safest and most sanitary way to taste food is to ladle a small amount into a small dish. Taste the food in the dish with a clean spoon. When finished, remove the tasting dish and spoon from the area and have them cleaned and sanitized.

While cooking can kill microorganisms, it cannot destroy the spores or toxins they may create. Handling food safely before and after it is cooked will prevent microorganisms from growing or producing spores or toxins.

Exhibit 7f	Minimum Safe Internal Cooking Temperatures

Reprinted from the FDA Model Food Code.

Product	Temperature
Poultry, stuffing, stuffed meats, stuffed pasta, casseroles, field-dressed game	165°F (74°C) for 15 seconds
Pork, ham, bacon, injected meats	145°F (63°C) for 15 seconds
Ground or flaked meats including hamburger, ground pork, flaked fish, ground game animals, sausage, gyros	155°F (69°C) for 15 seconds
Beef and pork roasts (rare)	145°F (63°C) for 3 minutes
Beef steaks, veal, lamb, commercially raised game animals	145°F (63°C) for 15 seconds
Fish	145°F (63°C) for 15 seconds
Shell eggs for immediate service	145°F (63°C) for 15 seconds
Any potentially hazardous food cooked in a microwave oven	165°F (74°C); let food stand for 2 minutes after cooking

Exhibit 7g

Checking the Temperature of Casseroles
Casseroles and previously cooked foods must reach an internal temperature of 165°F (74°C) or above for fifteen seconds.

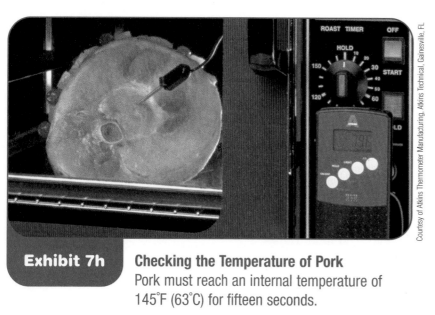

Exhibit 7h

Checking the Temperature of Pork
Pork must reach an internal temperature of 145°F (63°C) for fifteen seconds.

Courtesy of Atkins Thermometer Manufacturing, Atkins Technical, Gainesville, FL

COOKING REQUIREMENTS FOR SPECIFIC FOODS

Poultry, Stuffing and Stuffed Meats, and Casseroles

Poultry, stuffing, and stuffed meats all should be cooked to an internal temperature of 165°F (74°C) or above for fifteen seconds. Poultry tends to have more types and higher counts of micro-organisms than other meats and therefore should be cooked more thoroughly.

Stuffing also poses a hazard for several reasons. Stuffing may be made with potentially hazardous foods, such as eggs, which need to be fully cooked. Stuffing also acts as insulation, preventing heat from reaching the center of the bird. When poultry is stuffed, the thighs, wings, and even the breast can reach temperatures of 165°F (74°C) while the stuffing is still at an unsafe temperature. If the bird is taken out of the oven too soon, microorganisms in the stuffing will not be killed, creating a potential hazard.

Stuffing should be cooked separately, particularly when cooking whole, large birds. Smaller cuts of meat such as pork tenderloins or veal chops may be stuffed before cooking, but verify that internal temperature of both meat and stuffing reaches 165°F (74°C) for at least fifteen seconds.

Casseroles and dishes combining raw and cooked foods, such as soups, also should be cooked to 165°F (74°C) *(see Exhibit 7g)*.

Pork

Cook pork to an internal temperature of 145°F (63°C) or above for fifteen seconds. This includes ham, bacon, and other products containing pork. An internal temperature of 145°F (63°C) for fifteen seconds is enough to destroy any *Trichinella* larvae that may have infested the pork *(see Exhibits 7h and 7i)*.

Beef

Steaks cooked rare must reach and hold an internal temperature of at least 145°F (63°C) for fifteen seconds. Roasts, on the other hand, must hold the same internal temperature for at least three minutes *(see Exhibit 7i).* Depending on the type of roast and the oven used, however, beef can be cooked at different internal temperatures *(see Exhibit 7j).*

Ground Meats

Ground beef, pork, poultry, and other meat or fish must be cooked to an internal temperature of 155°F (68°C) for at least fifteen seconds. Most whole muscle cuts of meat are likely to have microorganisms only on their surface. When beef, pork, poultry, or any meat or seafood is ground, microorganisms on the surface are mixed throughout the product. To make sure all microorganisms are destroyed, thorough cooking is a must. *(See Exhibits 7i and 7j).* Check with your local or state regulatory agency for additional requirements.

Game and Ratites

Commercially raised and inspected game animals, such as elk, deer, bison, and rabbit, can be cooked in much the same way as beef. Small cuts such as steaks can be cooked to a minimum internal temperature of 145°F (63°C) or above for fifteen seconds. Ground meat must be cooked to an internal temperature of 155°F (68°C) or above for fifteen seconds. Stuffed meats should be cooked to 165°F (74°C) or above for fifteen seconds. Roasts can be cooked in the same way and to the same temperatures as beef.

Exhibit 7i

Exhibit 7j

Alternative Temperatures and Cooking Times for Beef and Pork Roasts

Chart: Adapted from the FDA Model Food Code.

Temperature	Time (in minutes)
130°F (54°C)	121
132°F (57°C)	77
134°F (57°C)	47
136°F (58°C)	32
138°F (59°C)	19
140°F (60°C)	12
142°F (61°C)	8
144°F (62°C)	5
145°F (63°C)	3

Minimum Internal Temperatures for Cooking Ground Meats

Photo: Courtesy of Atkins Thermometer Manufacturing, Atkins Technical, Gainesville, FL

Chart: Adapted from the FDA Model Food Code

Temperature	Time
145°F (63°C)	3 minutes
150°F (66°C)	1 minute
155°F (68°C)	15 seconds

Though they are birds, ratites (ostrich, emu, and rhea) are not cooked the same way as poultry. They will have a metallic taste when cooked to internal temperatures higher than 165°F (74°C). Ratites are fully cooked when their internal temperature reaches 155°F (68°C) for fifteen seconds. If portions or cutlets of these birds are stuffed, however, cook these to an internal temperature of 165°F (74°C).

Fish

Cook fish and foods containing fish to an internal temperature of 145°F (63°C) or above for fifteen seconds. Stuffed fish should be cooked to 165°F (74°C) or above for fifteen seconds. This is also true for stuffing which contains fish. If fish has been ground, chopped, or minced, it should be cooked to an internal temperature of 155°F (68°C) or above for fifteen seconds.

Eggs and Egg Mixtures

In general, shell eggs that are cooked for a customer's order should be cooked to 145°F (63°C) or above for fifteen seconds. When eggs are cooked this way, the white is set and the yolk begins to thicken. Properly cooked scrambled eggs are firm, with no visible liquid. To hold eggs for later service, cook them to 155°F (68°C) or above for fifteen seconds, then hold them at 140°F (60°C).

When cooking egg dishes, cook them precisely as directed and make sure that the cooking is complete before serving the item. Egg dishes, especially those prepared in deep dishes such as casseroles or stuffing, or previously frozen dishes, must be cooked to an internal temperature of 165°F (74°C).

When cooking eggs to order, take out only as many eggs as you need or will use within a short time. Never stack egg trays (flats) near the grill or stove.

Vegetables

Though most vegetables can be eaten raw, many are cooked before they are served. When cooking vegetables or fruits for hot holding, cook them to 140°F (60°C) or above.

Tea

Dry tea leaves contain low levels of bacteria, yeast, and mold (like most plant-derived foods). Using improper brewing temperatures to prepare tea and storing it at room temperature for long periods of time can cause these microorganisms to grow to high levels. Improperly cleaned and sanitized equipment can also promote growth of microorganisms. When handling tea, follow these recommendations.

Key Point

Shell eggs cooked to order should be cooked to 145°F (63°C) for fifteen seconds.

○ Store tea bags in a dark, cool, dry place away from strong odors and moisture. Do not refrigerate.

○ Brew only as much tea as you reasonably expect to sell within a few hours.

○ Never hold finished brewed tea for more than eight hours at room temperature. Discard any unused tea after eight hours.

○ To protect tea flavor and avoid microbial contamination and growth, clean and sanitize tea brewing, storage, and dispensing equipment at least once a day. Equipment should be disassembled, washed, rinsed, and sanitized. Urn faucets should be replaced at the end of each day with a freshly sanitized faucet.

○ For any brewing method, use a thermometer to make sure that brewing water in your equipment meets the temperature specified.

● 195°F (91°C) for automatic iced tea and automatic coffee machine equipment. Tea leaves should remain in contact with the water for a minimum of one minute.

● 175°F (80°C) minimum when using the traditional steeping method. Tea leaves must be exposed to the water for approximately five minutes by this method.

Microwave Cooking

Microwave ovens tend to cook food more unevenly than other methods of cooking. For this reason, there are special rules for using microwave ovens to cook meats, poultry, and fish.

○ Rotate or stir food halfway through the cooking process to distribute heat more evenly.

○ Cover food to prevent the surface from drying out.

○ Let food stand after cooking for at least two minutes to let product temperature equalize.

Meat, poultry, or fish cooked in a microwave oven must be heated to 165°F (74°C) or above. Check the internal temperature of the food in several places to make sure it has cooked through.

COOLING FOOD

You've already seen how important it is to keep foods out of the temperature danger zone. When food that has been cooked isn't going to be served right away, it is essential to cool it as quickly as possible.

The FDA Model Food Code recommends a **two-stage cooling method.** By this method, cooked foods must be cooled from 140°F (60°C) to 70°F (21°C) within two hours and from 70°F (21°C) to below 41°F (5°C) in an additional

---

Key Point

Never hold brewed tea for more than eight hours at room temperature.

Key Point

Meat, poultry, or fish cooked in a microwave oven must be heated to 165°F (74°C) or above.

Time & Temperature

In the two-stage cooling method, food must be cooled from 140°F (60°C) to 70°F (21°C) within two hours and to 41°F (5°C) or lower within four hours.

four hours for a total cooling time of six hours. Some jurisdictions, however, follow the one-stage, four-hour method, by which foods must be cooled to 41°F (4°C) or lower in less than four hours total after cooking or hot holding.

When using the two-stage cooling method, if the food has not reached 70°F (21°C) within two hours, the food must be reheated. Reheat the food to 165°F (74°C) for fifteen seconds within two hours and then cool it properly. When using the one-stage cooling method, if the food has not reached 41°F (5°C) within four hours, reheat it to 165°F (74°C) for fifteen seconds within two hours, or throw it out. Re-evaluate your cooling methods if they are consistently inadequate.

At first glance, the two-stage cooling method seems to be less strict than the four-hour method because it appears you have a total of six hours to cool the food rather than only four. But keep in mind that this is a two-stage process (two hours plus four hours). You have learned that microorganisms grow best in the temperature danger zone, but particularly, the temperature range of 120°F (49°C) to 70°F (21°C) is *ideal* for the growth of pathogenic microorganisms. Foods must pass through this temperature range quickly during cooling (as well as cooking and reheating) to ensure minimal microorganism growth. Because only two hours are allowed to cool foods from 140°F (60°C) to 70°F (21°C), the two-stage cooling method passes potentially hazardous foods through this temperature range more quickly, and therefore more safely, than the four-hour method.

While common sense may suggest that the quickest way to cool something off is to put it in the refrigerator, it isn't. Refrigerators are designed to hold chilled foods at cold temperatures. They usually don't have the capacity to chill hot foods quickly unless the portion is reduced in size.

In general, the thickness of the food, or distance to its center, plays the biggest part in how fast it cools. A large stockpot of beef stew, for example, may take four times as long to cool as a pot that is one-half the size. Twelve gallons of stew in a sixteen-inch diameter stockpot would take more than thirty-six hours to cool from 140°F (60°C) to below 50°F (10°C). That is more than enough time for a tremendous growth of microorganisms.

The more dense a food product is, the slower it will cool. Refried beans will take longer to cool than vegetable broth since the beans are more dense. Refried beans in an eight-inch-deep metal pan, for example, will take more than twenty hours to cool from 140°F (60°C) to 41°F (5°C) or lower.

The container in which food is stored also affects how fast it will cool. Stainless steel transfers heat faster than plastic. Shallow pans will let product disperse heat faster than deep ones.

There are a number of methods you can use to cool food quickly (*see Exhibit 7k*).

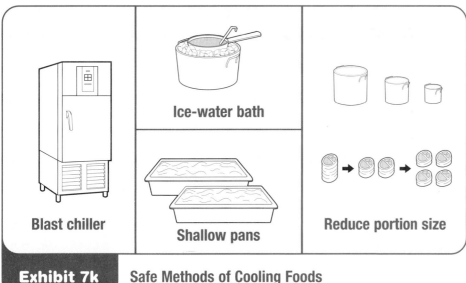

| **Exhibit 7k** | Safe Methods of Cooling Foods |

○ **Reduce the size of the food you're trying to cool.** Divide hot food into smaller quantities, put it into shallow pans, and then refrigerate it. If you pre-chill the pans, the food will cool off even faster.

○ **Use ice-water baths to bring food temperatures down quickly in establishments that don't have quick-chill units.** After dividing food into small quantities, put the pans in ice water in a sink or large pot *(see Exhibit 7l)*.

○ **Use blast chillers to cool foods before placing them into the walk-in for storage.** Small quick-chill units are available for use in most establishments. Larger establishments often have a tumbler chiller that's part of a cook-chill system.

○ **A steam-jacketed kettle also can serve as a cooler.** Simply run cold water through the jacket.

○ **Stir foods as they cool.** This allows the product to cool faster and more evenly. Some manufacturers make plastic wands and paddles that you can fill with water and freeze. Stirring food products with these **cold paddles** chills food very quickly. Make sure all utensils you use to stir product are cleaned and sanitized between each use *(see Exhibit 7m)*.

| **Exhibit 7l** | Ice-Water Bath Method |

Ice-water baths can be used to cool foods quickly.

Exhibit 7m   **Cold Paddle Method**
Stirring products with cold paddles chills food very quickly.

Once a food has been cooled, the refrigerator can handle it more easily. Foods will continue to chill faster if you do the following.

○ **Keep food in shallow stainless-steel pans.** Put dense foods, like chili or stew, in pans that are two inches deep. Thinner liquids can be stored in three-inch pans.

○ **Always place pans on top shelves in the cooler.** Leave them uncovered if they are protected from overhead contaminates. Otherwise, cover them loosely with foil or plastic wrap. Stir the contents of the pans occasionally, using caution while stirring. Make sure products stored on the lower shelves are covered or properly wrapped.

○ **Position pans so that air can circulate around them.** This will help transfer heat from the pan. When food has cooled to 41°F (5°C) or below, cover the pan.

All cooked food should be stored in containers that are labeled with the date and time that the food was prepared and stored.

REHEATING POTENTIALLY HAZARDOUS FOOD

When previously cooked food is reheated for hot holding, take it through the temperature danger zone again as quickly as possible. Food must be reheated to an internal temperature of 165°F (74°C) for fifteen seconds, within two hours. Don't mix leftover foods with freshly prepared foods. When adding previously cooked food as an ingredient to another food, such as adding ground beef to spaghetti sauce, the whole mixture must be reheated to 165°F (74°C).

When using the microwave oven to reheat previously cooked foods, follow the same rules used for microwave cooking. Cover the product. Rotate or stir it midway through cooking. Allow it to stand for two minutes. Check internal temperature in several places to see if it has reached at least 165°F (74°C).

Foods that are reheated for immediate service to a customer, such as a roast beef sandwich, may be served at any temperature, as long as the beef was properly cooked.

If the food has not reached 165°F (74°C) for fifteen seconds within two hours, throw it out.

Key Point

Keep containers labeled with the date and time.

Courtesy Daydots Foodservice Products, Forth Worth, Texas, 800-321-3687.

SUMMARY

To protect food during preparation you must handle it safely at every step of the process. The keys are time and temperature control and a clean, sanitary establishment.

All potentially hazardous foods should be kept out of the temperature danger zone—41°F (5°C) to 140°F (60°C)—as much as possible. The more time foods are exposed to these temperatures, the greater the risk that microorganisms will grow. As soon as deliveries arrive at your back door, food is subject to the four-hour rule: potentially hazardous foods that are exposed to temperatures between 41°F (5°C) and 140°F (60°C) for more than four hours, *including the time it takes to prepare and cook them,* must be discarded.

The two main causes of foodborne illness are temperature abuse and cross-contamination. You can prevent both by putting up physical, chemical, and procedural barriers throughout the establishment.

Barriers to temperature abuse start with proper thawing of frozen foods. Thaw frozen foods in the refrigerator or under cool running water, never at room temperature. Have employees prepare food in small batches, use chilled utensils and bowls, and record product temperatures and preparation times. Use timers to remind employees how long products have been in the danger zone.

To erect barriers against contamination, prepare raw foods separately from other ready-to-eat or cooked foods. Use equipment and utensils that are dedicated to each type of product. Color code them so employees know which cutting boards and knives to use with poultry and which to use with produce, for example. Always clean and sanitize equipment, utensils, and containers before and after each use. Always wash your hands and consider using gloves when handling food products, especially ready-to-eat foods that will not be cooked.

Cooking is typically considered a Critical Control Point because it can kill almost all microorganisms present in food. To ensure that most microorganisms are destroyed, food must be cooked to required minimum internal temperatures for a specific time. Different types of food must be cooked to different temperatures that range from 145°F (63°C) to 165°F (74°C). Employees should learn what these times and temperatures are and how to check them. Cooking does not kill the spores or toxins that some microorganisms produce. That's why it is so important to inspect product once it arrives and handle it safely during preparation.

Once food is cooked it should be served as quickly as possible. If it is going to be stored and served later, it must be cooled rapidly. Cooked food must be cooled to 70°F (21°C) or below within two hours and then chilled to 41°F (5°C) or below within four hours, unless otherwise required by your local health code. Placing large containers of hot food in refrigerators may put all the other stored foods in

Time & Temperature

If food has not cooled to 41°F (5°C) or below in four hours, you must take corrective action. Within two hours, reheat the food to 165°F (74°C) for fifteen seconds, or throw the food out.

HACCP Principle

Reheating cooked foods may be considered a Critical Control Point. Take corrective action if food isn't reheated properly. If food hasn't reached 165°F (74°C) within two hours, throw it out.

danger. To cool large quantities of cooked food quickly, divide it into smaller portions, put it in shallow stainless-steel pans, use ice-water baths or quick-chill units, and stir it often. When the food is cold enough, store it properly in the refrigerator.

Previously cooked food that is to be held hot must be reheated to an internal temperature of 165°F (74°C) within two hours before it can be served.

Directions for safe foodhandling practices should be written into all recipes, including what corrective action to take if a standard is not met. Employees should have their own probe thermometers and should be trained to take responsibility for time and temperature management and sanitation in their own areas.

To keep food safe during preparation, keep your establishment in good condition and train your employees. If you control time, temperature and sanitary conditions, microorganisms will have little chance to grow and cause problems.

A CASE IN POINT I

Case Study

On Friday, John went to work at The Fish House knowing he had a lot to do. After changing clothes and punching in, he grabbed a bus tub off the counter in the dish room and emptied it. He brought it into the refrigerator and scooped the ice from the crates of fresh fish into it. He dumped the old ice in the dish room sink and rinsed out the bus tub. Then he used the bus tub to transfer fresh ice into the ice flaker. When the bus tub was filled with flaked ice, he took it into the refrigerator and emptied it into the crates of fresh fish. When he was finished, he left the bus tub next to the ice machine.

Next, John took a case of raw shrimp out of the freezer. To thaw it quickly, he put the block of frozen shrimp into the prep sink and turned on the hot water. While waiting for the shrimp to thaw, John took several fresh whole fish out of the refrigerator. He brought them back to the prep area and began to clean and fillet them. When he finished, he put the fillets in a pan in the refrigerator. He rinsed off the boning knife and cutting board in the sink and wiped off the worktable with a dish towel.

Next, John transferred the shrimp from the sink to the worktable in a large plastic bucket. He peeled, deveined, and butterflied the shrimp with the boning knife. He put the prepared shrimp in a covered container in the refrigerator, then started preparing fresh produce.

What did John do wrong?

A CASE IN POINT II

Case Study

By 7:30 in the evening, all the residents at Sunnydale nursing home had eaten dinner. As she began cleaning up, Angie realized that she had a lot of chicken breasts left over. Betty, the new assistant manager, had forgotten to inform Angie that several residents were going to a local festival and would miss dinner.

"No problem," Angie thought. "We can use the leftover chicken to make chicken salad."

Angie left the chicken breasts in a pan in the steam table while she started putting other foods away and cleaning up the kitchen. At 9:45 p.m. when everything else was clean, she put her hand over the pan of chicken breasts and decided they were cool enough to handle. She took the pan out of the steam table, covered it with plastic wrap, and put it in the refrigerator.

Three days later, when she came in to work on the early shift, Angie decided to make chicken salad from the leftover chicken breasts. After she hung up her coat and put on her apron, Angie took all the ingredients she needed for chicken salad out of the refrigerator and put them on a worktable. Then she started breakfast.

First, she cracked three dozen eggs into a large bowl, added some milk, and set the bowl near the stove. Then she took a slab of bacon out of the refrigerator and put it on the worktable next to the chicken salad ingredients. She peeled off strips of bacon onto a sheet pan and put the pan into the oven. After wiping her hands on her apron, she went back to the stove to whisk the eggs and pour them onto the griddle. When they were almost done, Angie scooped the scrambled eggs into a hotel pan and put it in the steam table.

As soon as breakfast was cooked, Angie went back to the worktable to wash and cut up celery and cut up the chicken for chicken salad.

What did Angie do wrong?

TRAINING TIPS

Training Tips for the Classroom

1.   Group Breakout Activity: SOP for Menu Items

Objective: *After completing this activity, class participants will be able to develop standard operating procedures (SOPs) for preparing food and be able to identify some Critical Control Points for various food groups. Note: this higher-level activity may not be suitable for all classes.*

Directions: Make a list of six to eight different menu items, including at least one item from each major food category discussed in Chapter 7 under "Preparing Food." A representative list might include:

○ Fruits and Vegetables: Chef's Salad

○ Meats, Fish, and Poultry: Fresh Whitefish, Roast Turkey, Loin of Pork

○ Protein Salads and Sandwiches: Chicken Salad

○ Eggs: Denver Omelet

○ Batters and Breading: Hand-Breaded Fried Chicken

Break the class into groups of about three people each. Give each group a sheet of paper with one of these menu items printed at the top. Each group will work on a different menu item.

Ask each group to create a recipe for its assigned menu item. The recipe should include ingredients, preparation steps, and, especially, sanitation instructions. (Note: Accurate weights and measures are not necessary for these recipes, since this activity should focus on food-safety guidelines rather than culinary accuracy.)

Give the groups a time limit to develop recipes for their items. Remind them that these recipes should focus on Critical Control Points and SOPs for safe food handling.

When time is up, have each group present its recipe. Allow time for discussion from the class. Make necessary changes and additions. Distribute copies of the recipes as a handout at the end of the day.

2.   "Chill Out!" Team Contest

Objective: *After completing this activity, class participants will be able to identify the various methods of cooling food, discuss advantages and disadvantages, and identify foods best suited for each cooling method.*

Note: Before this activity, remind students that failure to cool foods properly is the number-one cause of outbreaks of foodborne illness. Present this exercise after reviewing the critical limits for cooling foods but before a discussion of cooling methods.

Directions: Print up a form that asks students to list as many cooling methods as they can think of, along with a list of food types best suited for each method, and advantages and disadvantages of each method.

Divide the class into teams of two and give each team one form. Allow about seven minutes (closed book) for the teams to list cooling methods, food types, and advantages and disadvantages.

After time is up, let the teams grade their sheets as follows:

○ For each cooling method (recognized as safe)   5 points
○ For each type of food to be cooled by this method   2 points
○ Each advantage of the cooling method   3 points
○ Each disadvantage of the cooling method   3 points

The team with the most point wins.

Allow time for discussion, and let students make additions and correct errors on their papers. Create a master list of cooling methods from this activity as a reference handout.

3. "Cook-Kill" Temperature Quiz

Objective: *After completing this activity, class participants will be able to identify safe internal cooking temperatures for different food groups.*

Directions: As a pop quiz, give students a printed list of twelve to fifteen varied menu items with a blank space next to each item. Ask students to fill in the minimum safe internal temperature for cooking each item. Put in some trick menu items, such as rare roast beef, stuffed ostrich cutlets, and buffalo burger.

Limit this activity to two minutes. Review the different temperatures with the class. See if anyone in the class got a perfect score.

Use this quiz as a springboard to a group discussion on safe cooking temperatures for different food groups. Discuss how to monitor cooking temperatures (visual check versus thermometers or timers) and how often such monitoring is necessary. Discuss how one might serve food that does not reach these critical limits, due to customer requests (rare hamburger) or type of item (sushi, raw oysters).

Training Tips on the Job

1. "Cool It" Cooling Activity

Purpose: *To examine the methods used in your operation to cool food and to identify any problems with your current practices. When your kitchen staff is involved in this analysis, they can be the ones to identify any problems and will buy into the decision to change methods if necessary.*

Directions: Invite the kitchen staff to participate in a competition.

Select a regular menu item that your establishment prepares in advance and then cools down for future use. Chili is a good example. The next time you're scheduled to make chili, have the cook take the temperature of the chili just before taking it off the stove. Record this temperature, then tell the cook to cool the chili as she normally does.

Ask each staff member to guess what the temperature of the chili will be two hours after removing it from the stove, and again four hours later. Record their guesses on index cards.

Have someone take and record the temperature of the chili every hour. After the first two hours, and after six hours, verify the final temperature yourself, and then check the index cards. List the different temperatures that people guessed, and announce a winner. (If the chili does not reach 70°F [21°C] within two hours and 41°F [5°C] in less than six hours, ask if they want to eat the chili for lunch!)

Discuss with your staff the importance of cooling foods. Solicit their input to develop a system for cooling chili and other food products to 41°F (5°C) within two plus four hours. Discuss and brainstorm different methods, and then implement the best method as your standard cooling process.

2. "Cooking Pays" Temperature Quiz

Purpose: *To reinforce the importance of monitoring temperature when handling food in your operation.*

Directions: Every day, carry a number of tokens in your pocket, along with your calibrated thermometer or thermocouple. Whenever you walk through the kitchen and see an employee using a thermometer to check the temperature of the food, give him or her a token. If employees can tell you the temperature of the food product and are accurate within 2°F (1°C), give them two more tokens. During your regular staff meetings, give out prizes based on the number of tokens collected. This activity verifies that employees are using thermometers.

As you conduct this activity, remind the cooks of the importance of monitoring both time and temperature when handling potentially hazardous food, and solicit their cooperation in minimizing the time these foods are in the temperature danger zone.

3. Work Station Safety Checklists

Purpose: *To involve kitchen staff members in the creation of food-safety checklists for specific work areas in your establishment.*

Directions: Create a food-safety team from a group of volunteers from the kitchen staff. Ideally all different shifts and work areas should be represented.

Assign members of the team to study different work areas in your operation (cold prep, grill, stove, oven, bakery, cleanup, storage, and so on) and ask them to develop food-safety checklists for those specific work areas. They should keep in mind the three major food-safety concerns: monitoring time and temperature, personal hygiene, and cross-contamination.

After receiving the checklists, edit them to make them consistent with one another and easy to read. Suggest ways to make them colorful or color-coded, and translate them into several languages if necessary. Review your work with the team. Have the checklists professionally printed and laminated, and post them in the appropriate work areas for easy reference.

DISCUSSION QUESTIONS

1. What are the minimum required internal temperatures of cooked poultry, stuffing, pork, and ground beef?

2. What is the four-hour rule?

3. Discuss methods for cooling cooked foods.

4. What are the four rules for using a microwave oven to cook potentially hazardous foods?

MULTIPLE-CHOICE STUDY QUESTIONS

1. You have only one ceramic cutting board available for food preparation. You have just sliced some chicken breasts for cooking and now need to prepare a green salad. What should you do to the cutting board before you use it for preparing the salad?

   A. Scrub it using hot, potable water and a detergent, then sanitize.
   B. Dry it with a paper towel.
   C. Rinse it under very hot water.
   D. Turn it over and use the reverse side.

2. Which of the following is not a safe method for thawing a frozen brisket of beef?

   A. Let it sit at room temperature for five hours.
   B. Put it in a microwave set on automatic defrost.
   C. Immerse it in room-temperature running water for one hour.
   D. Let it sit in the refrigerator overnight.

3. With a probe thermometer, you measure the temperature of the breast meat and the stuffing of a stuffed chicken. What would tell you that the chicken has been safely cooked?

   A. Both temperatures read 155°F (68°C).
   B. The stuffing temperature reads 160°F (71°C).
   C. Both temperatures read 165°F (74°C).
   D. The meat temperature reads 170°F (77°C).

4. You are making omelets to order for the Sunday brunch at the hotel where you work. How should you handle the eggs to ensure that you are serving safe omelets?

   A. Keep all of the eggs in their shells. Take out only the number of eggs you expect to use and store them away from, but near, the stove.
   B. Keep all of the eggs in their shells next to your cooking station.
   C. Reserve one special bowl for cracking the eggs into before cooking.
   D. Crack all of the eggs at once into a large container.

5. You are making a mixed vegetable tray for the salad bar. Which of the following is a proper procedure for preparing the vegetables to ensure food safety?

    A. The tomatoes should be scrubbed with a bristle vegetable brush.

    B. Each leaf of the romaine lettuce should be washed separately.

    C. The carrots should be soaked in ice water.

    D. The mushrooms should be dry brushed to remove any field soil.

6. Your ice maker produces ice for chilling the plates for the salad bar and for beverage service. Which is a proper procedure for dispensing the ice from the ice maker storage bin?

    A. Ice for both uses should be dispensed with a handled scoop stored properly outside of the ice bin.

    B. Ice for beverages must be taken from the back of the salad bar.

    C. Ice for the salad bar should be dispensed in large plastic containers using a clean sanitized glass.

    D. Ice for both uses should be made with carbon-filtered water.

7. You need to cool a large stockpot of clam chowder for use the following day. What is the first thing you should do?

    A. Put the large pot containing soup in the refrigerator.

    B. Transfer the soup to a different large pot.

    C. Transfer the soup from the large pot to a shallow stainless-steel pan.

    D. Put some ice cubes into the soup to help the cooling.

8. Your soup and salad bar is serving minestrone soup that was prepared the day before and refrigerated overnight. The serving table has a temperature-controlled heater that can maintain an equipment temperature of 165°F (74°C). What do you need to do to the soup to prepare it for serving?

    A. Reheat the cold soup on the serving table and stir it to speed the reheating.

    B. Reheat only small amounts of the soup at a time on the serving table.

    C. Microwave the soup to 75°F (24°C) and then put it on the serving table to finish reheating to 165°F (74°C).

    D. Reheat the soup quickly to 165°F (74°C) on the stove before putting it on the serving table.

9. Which of the following foods has been safely cooked?

    A. A rare beef roast that has been cooked to an internal temperature of 135°F (57°C)

    B. A pork roast that has been cooked to an internal temperature of 135°F (57°C)

    C. A whole turkey that has been cooked to an internal temperature of 155°F (68°C)

    D. A tuna casserole that has been cooked to an internal temperature of 165°F (74°C)

10. How should a batch of beef stew that has been cooled to 70°F (21°C) be stored for service the following day?

    A. In a shallow, loosely covered pan on the top shelf of the refrigerator

    B. In a plastic jar at the back of the refrigerator

    C. In a pot in the storage freezer

    D. In a warming oven set to 140°F (60°C)

Resources

ADDITIONAL RESOURCES

Books and Periodicals

Bax, B. (1997). *Handbook for safe food-service management.*
    Upper Saddle River, NJ: Prentice-Hall.

Cross out cross-contamination. (1998). *Best Practices, 2*(1), 6-9.

Equipped for food safety. (1998). *Best Practices, 2*(4), 6-10.

Laconi, D. (1995). *Fundamentals of professional food preparation:*
    A laboratory text workbook. New York: Wiley.

Loken, J. (1995). *The HACCP food-safety manual.* New York: Wiley.

Longree, K & Armbruster, G. (1996). *Quantity food sanitation.*
    New York: Wiley.

The right way to…prepare beef. (1998). *Best Practices, 2*(3), 12-13.

The right way to…prepare chicken. (1998). *Best Practices, 2*(1), 10.

The right way to…prepare eggs. (1998). *Best Practices, 2*(2), 10-11.

The right way to…prepare pork. (1997). *Best Practices, 1*(3), 10-11.

The right way to…preparing fruits and vegetables. (1998). *Best Practices, 2*(4), 12-13.

The right way to…seafood safety. (1999). *Best Practices, 3*(1), 16.

Secrets of self-inspection. (1998). *Best Practices, 2*(2), 6-9.

Web Sites
Alaska Seafood Marketing Institute
http://www.alaskaseafood.org
The Alaska Seafood Marketing Institute works to improve seafood quality by teaching fishermen, processors, retailers, and restaurateurs about proper handling of Alaska seafood products. Their Web site also includes industry news, suppliers, and recipes.

American Egg Board
http://www.aeb.org
The American Egg Board's Web site features product, safety, and industry information for foodservice professionals as well as recipes and other information for the general public.

American Meat Institute
http://www.meatami.org
Provides member information, as well as current and past news and information regarding the meat industry.

Daydots
http://www.daydots.com
Daydots makes stickers that facilitate efficient food rotation. Their Web site offers safety tips, ordering information, and industry news.

FDA Center for Food Safety and Applied Nutrition (CFSAN)

http://vm.cfsan.fda.gov/list.html

Comprehensive site from CFSAN offers a wealth of food-safety information, from foodborne illness to food labeling. CFSAN strives to be a leader in food safety, and to protect consumers from economic fraud, promote sound nutrition, and encourage innovation.

Food Marketing Institute

http://www.fmi.org

This nonprofit association conducts programs in research, education, industry relations, and public affairs for its membership of food retailers and wholesalers. Their Web site offers a wealth of food-safety information.

Institute of Food Technologists (IFT)

http://www.ift.org

IFT us a nonprofit scientific society with 28,000 members working in food science, food technology, and related professions in industry, academia, and government. This site includes information on its policies and publications, education and industry news, and meeting and convention locations.

International Association of Milk, Food and Environmental Sanitarians (IAMFES)

http://www.iamfes.org

IAMFES keeps members informed of the latest scientific, technical, and practical developments in food safety and sanitation. Their Web site includes booklets and links to other food-safety sites.

International Dairy Foods Association (IDFA)

http://www.idfa.org

This site is the link to information on IDFA and its constituent organizations: the Milk Industry Foundation (MIF), the National Cheese Institute (NCI), and the International Ice Cream Association (IICA); as well as information on the entire dairy industry. The site also includes information on current industry events, membership information, and publications.

National Association of College and University Food Services

http://www.nacufs.org

NACUFS is a trade association for foodservice professionals at institutions of higher learning in the United States and Canada. Their Web site includes educational materials as well as tours of members' foodservice operations.

National Cattlemen's Beef Association

http://www.beef.org

This site includes up-to-date information from the National Cattlemen's Beef Association, as well as a reference library, information on nutrition and food safety, and links to related sites.

National Food Processors Association (NFPA)

http://www.nfpa-food.org

NFPA is a scientific and technical trade association for the food industry. Their site provides industry news and educational materials on issues of food safety, research, and food science.

National Frozen Food Association (NFFA)

http://www.nffa.org

NFFA is the voice of the frozen foods industry. Their Web site provides information to consumers as well as foodservice professionals.

National Pork Producers Council

http://www.nppc.org

The National Pork Producers Council site provides general information about food and nutrition, as well as information specifically for pork producers. A special section is also provided just for children.

National Restaurant Association

http://www.restaurant.org

The National Restaurant Association site provides information on government agencies affecting the restaurant industry, the latest training and certification updates, and links to state restaurant associations and hospitality schools and universities.

Produce Marketing Association

http://www.pma.com

The members of this nonprofit trade association market fresh fruits and vegetables to the foodservice industry. Purchasing guidelines are available.

Tyson

http://www.tyson.com

Tyson's Web site offers a consumer kitchen page with recipes and instruction on preparing chicken products. An entire page is also devoted to foodservice professionals and offers information on the industry, Tyson products, and menu planning.

United States National Dairy Database

http://www.inform.umd.edu/EdRes/Topic/AgrEnv/ndd

The National Dairy Database is a large collection of the best and most up-to-date publications on dairy production available from across the United States. All publications are written and peer-reviewed by top dairy experts.

United States Department of Agriculture (USDA)

http://www.usda.gov

The United States Department of Agriculture's Web site features information, publications, and other educational materials about the nation's of agriculture.

United States Department of Agriculture Food Safety and Inspection Service

http://www.fsis.usda.gov

The USDA's Food Safety and Inspection Web site offers the latest food-safety news, educational materials, and HACCP-implementation materials.

Chapter 8
Protecting Food During Service

Knowledge

TEST YOUR FOOD-SAFETY KNOWLEDGE

1. **True or False:** A hot-holding device can be used for reheating foods, provided it is capable of reaching a temperature of 140°F (60°C). *(See Hot Foods, page 8-2.)*

2. **True or False:** Raw chicken can be served next to ready-to-eat foods on a self-service buffet if care is taken during setup. *(See Self-Service Areas, page 8-8.)*

3. **True or False:** A sneeze guard protects the foods on a salad bar from cross-contamination. *(See Self-Service Areas, page 8-8.)*

4. **True or False:** Catered foods should be delivered in insulated food containers. *(See Delivery, page 8-10.)*

5. **True or False:** Foods that have been prepared safely and cooked properly in the kitchen may not necessarily be safe to eat when they reach the table. *(See Serving Foods Safely, page 8-4.)*

Key Terms

Hot-holding equipment
Personal hygiene
Food bar
Sneeze guard
(food shield)

Off-site services
Single-use items
Mobile unit
Temporary unit
Vending machine

Table of Contents

Learning Objectives

After completing this chapter, you should be able to:

○ Identify proper hot-holding and cold-holding temperatures.

○ Prevent cross-contamination when serving foods.

○ Safely set up buffets and food bars and replenish them.

○ Identify foods that can be re-served.

○ Explain the proper procedures for serving food off-site.

○ List and apply the eight rules of safe food handling.

Key Point

When holding foods for service, keep cold foods cold and hot foods hot.

Key Point

Any conflict between food quality and food safety must always be decided in favor of food safety.

Time & Temperature

Never use hot-holding equipment to reheat foods.

Time & Temperature

Discard hot foods after four hours if they have not been held at or above 140°F (60°C).

The job of protecting food continues even after it has been properly prepared and cooked, since microorganisms still have many chances to contaminate food before it is eaten.

The key to serving safe food is to prevent temperature abuse and cross-contamination. Again, commonsense rules apply. Food must be held, displayed, and served at the right temperature and handled in a sanitary manner. People do many things without thinking that can lead to contamination. Train employees to serve food properly, and make sure rules are followed.

Customers who serve themselves at buffets and self-service displays may be at even greater risk. If you don't set up procedures to minimize hazards, one customer's bad habits can easily cause an outbreak of foodborne illness.

There are a number of ways to serve food, ranging from self-service to tableside preparation. Whichever service method is used, the key is to provide service without compromising food safety.

HOLDING FOODS FOR SERVICE

In many establishments, foods are cooked to order. When foods are stored, prepared, and cooked properly, then served immediately, there's little chance that anyone will get sick. Even in facilities that cook food to order, however, many menu items are cooked and held for service. A coffee shop might hold soup in a warming kettle. A family steak house might keep baked potatoes and prime rib warm on a steam table and keep salad ingredients cold on a refrigerated prep table. Many establishments, such as cafeterias and buffets, hold almost all of the foods they serve.

Kitchen staff may be tempted to hold hot foods at lower temperatures that will keep food warm but will not affect quality. For example, a chef may want to hold prime rib at 120°F (49°C) to keep it rare, or to turn down the heat on the warming kettle so the soup doesn't skin over or become thick. However, all employees must remember that microorganisms can grow at temperatures between 41°F (5°C) and 140°F (60°C).

To ensure the safety of foods that are held hot or cold, specific procedures must be followed.

Hot Foods

○ **Never use hot-holding equipment to reheat foods.** Reheat foods first to 165°F (74°C), then transfer to holding equipment. Most hot-holding equipment is incapable of passing food through the temperature danger zone quickly enough to prevent the growth of microorganisms.

○ **Use only hot-holding equipment that can keep foods at 140°F (60°C) or higher.** Hot-holding equipment includes steam tables, *bains maries,* chafing dishes, double boilers, and heated cabinets *(see Exhibit 8a).*

○ **Stir foods at regular intervals.**
Stirring will distribute heat evenly
throughout the food.

○ **Keep food covered.** Covers retain
heat and keep contaminants from
falling into the food.

○ **Measure internal food temperature
at least every two hours.** Use a
probe thermometer. The thermostat
of the holding equipment measures
the temperature of the equipment,
not of the food. Record temperatures
in a log.

○ **Discard hot foods after four hours
if they have not been held at or
above 140°F (60°C).**

○ **Never mix freshly prepared food
with foods being held for service.**
Mixing foods may cause cross-
contamination *(see Exhibit 8b).*

○ **Prepare food in small batches so it
will be used faster.** Don't prepare hot
foods any farther ahead than necessary.

Cold Foods

○ **Use only cold-holding equipment
that can keep foods at 41°F (5°C)
or lower** *(see Exhibit 8c on page 8-4).*

○ **Most foods should not be stored
directly on ice.** Place foods in pans
or on plates first. Whole fruits and
vegetables and raw cut vegetables are
the only exceptions. Ice used on a
display should be self-draining. Wash
and sanitize drip pans after each use.

○ **Measure internal food temperatures
at least every two hours.** If above
41°F (5°C), take corrective action.
Record temperatures (and corrective
actions) in a log.

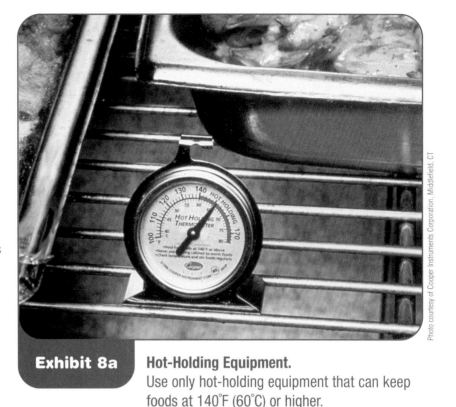

Photo courtesy of Cooper Instruments Corporation, Middlefield, CT

Exhibit 8a — **Hot-Holding Equipment.**
Use only hot-holding equipment that can keep
foods at 140°F (60°C) or higher.

Exhibit 8b — **Refilling Food in Holding Equipment**
Never mix freshly prepared food with foods being
held for service.

Exhibit 8c

Cold-Holding Equipment
Cold-holding equipment must keep foods
at 41°F (5°C) or lower.

○ **Protect food from contaminants with covers or food shields.**

SERVING FOODS SAFELY

If you have handled foods safely and cooked them properly, you don't want to risk contamination when you serve them. Microorganisms can easily spread from the front of the house to the kitchen, as well as from the kitchen to the customer. Make sure your efforts to keep food safe don't end once food comes off the stove or out of the oven.

Kitchen Staff

Train your kitchen staff to follow these procedures to serve food safely.

○ **Store utensils properly.** Serving utensils can be stored in the food, with the handle extended above the rim of the container *(see Exhibit 8d);* or they can be placed on a clean, sanitized food-contact surface. Spoons or scoops used to serve foods such as ice cream or mashed potatoes may be stored under running water.

○ **Serving utensils should have long handles.** A long-handled ladle will keep a server's hands away from soup in a kettle; a cup will not.

○ **Use clean and sanitized utensils for serving.** Utensils should be used for only one food and must be properly cleaned and sanitized after each task. Utensils should be cleaned and sanitized at least once every four hours during continuous use.

○ **Never touch cooked or ready-to-eat foods with bare hands.** Always use gloves or utensils to handle these foods *(see Exhibit 8e).*

○ **Practice good personal hygiene.** Wash hands after using the restroom, after hands have come into contact with food, and after handling dirty equipment or utensils.

Servers

Food servers need to be just as careful as kitchen staff. Servers can contaminate food simply by handling the food-contact area of a plate. Train servers in these safe serving procedures *(see Exhibit 8f on page 8-6).*

Time & Temperature

When holding food, measure its internal temperature at least every two hours.

Cross-Contamination

Always use gloves or utensils to handle food that is cooked or ready-to-eat.

- **Handle glassware and dishes properly.** Never touch the food-contact area of plates, bowls, glasses, or cups. Hold dishes by the bottom or the edge. Hold cups by their handles. Hold glassware by the middle, bottom, or stem.

- **Never stack glassware or dishes when serving.** The rim or surface of one can be contaminated by the one above it. Stacking china and glass also can cause it to chip or break.

- **Hold flatware and utensils by the handles.** Store flatware so servers will grasp the handles, not the food-contact surfaces.

- **Never touch food with bare hands.** Serve with tongs or gloves, for example.

- **Serve milk from refrigerated bulk dispensers or in single-serve cartons.** Serve cream and half-and-half in single-serve containers or in covered pitchers.

- **Use plastic or metal scoops or tongs to get ice.** Never use a glass to scoop ice, because it might chip or break in the ice. Don't use bare hands to scoop ice. Store the ice scoop in a secure spot, not in the ice bin. Remember that ice is also a food and can be contaminated.

- **Practice good personal hygiene.** Be neat and clean. Keep hair pulled back and covered. Avoid touching hair or face when serving food. Eliminate bad habits such as chewing fingernails, licking fingers, rubbing or touching your nose and face.

- **Cloths used for cleaning food spills shouldn't be used for anything else.** When cleaning tables between guest seatings, wipe up spills with a

Exhibit 8d **Properly Stored Utensils**
If stored in food, utensils should be stored with the handle extended above the rim of the container.

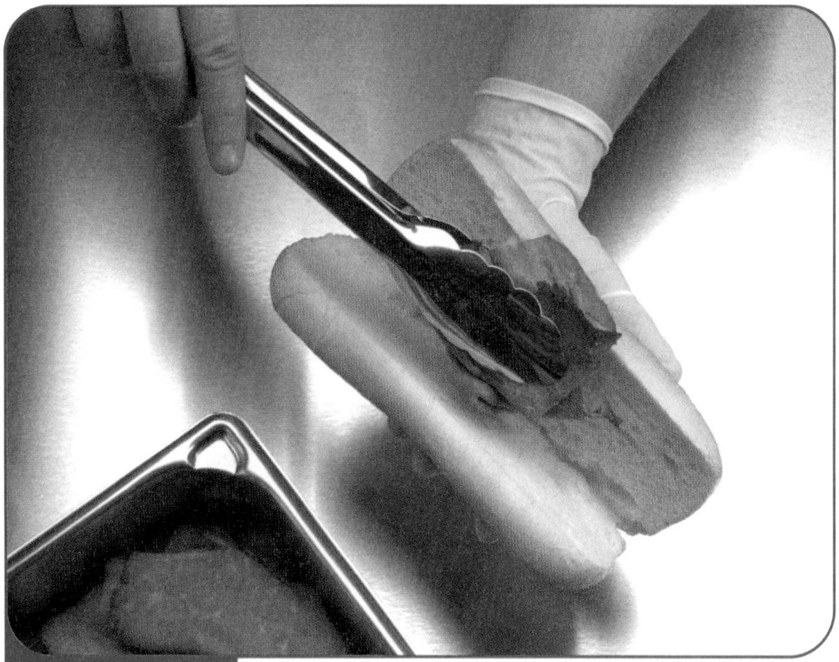

Exhibit 8e **Handling Cooked Food**
Never touch cooked or ready-to-eat foods with bare hands.

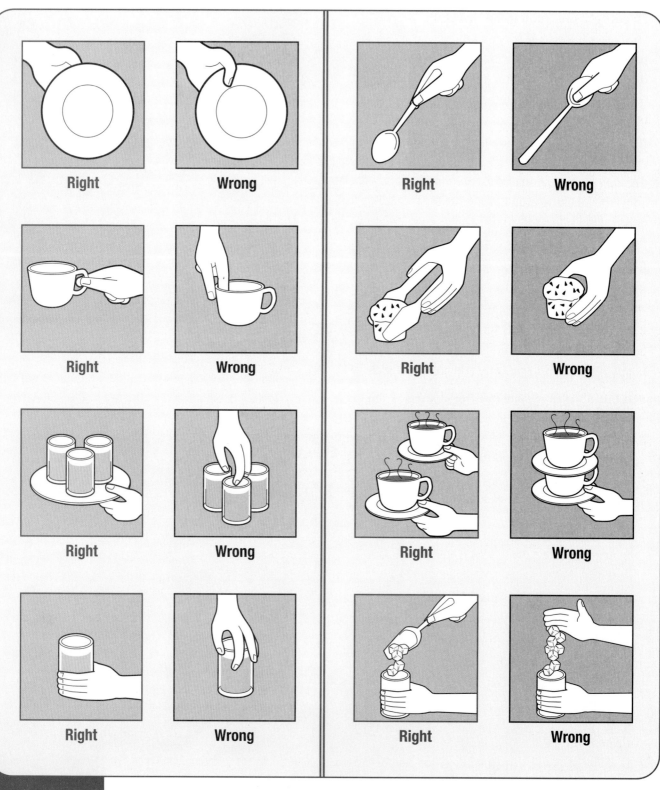

Right   Wrong   Right   Wrong

Right   Wrong   Right   Wrong

Right   Wrong   Right   Wrong

Right   Wrong   Right   Wrong

Exhibit 8f   **Carrying Utensils and Serving Food**
Proper and improper ways to carry utensils and serve food.

disposable dry cloth. Then clean the table with a moist cloth that has been stored in a fresh sanitizing solution.

Division of Labor

It's a good idea to schedule staff so that they are not assigned to do more than one job during a shift. Serving food, setting tables, and busing dirty dishes are separate tasks, with different responsibilities. Since this division of labor is difficult to manage in most establishments, it's important for servers and busers who do double duty to wash their hands often and handle food safely. After wiping tables or busing dirty dishes, servers must wash their hands before handling place settings or food.

Re-Serving Food

Servers and kitchen staff also should know rules about re-serving foods that have been previously served to a customer. In general, the only foods that can be re-served are unopened, prepackaged foods. Condiment packets, wrapped crackers or breadsticks, and other sealed foods generally offer adequate protection from contamination.

Never re-serve plate garnishes such as fruit or pickles to another customer. Served but unused garnish must be discarded. Never re-serve uncovered condiments. Don't combine leftovers with fresh food. Opened portions of salsa, mayonnaise, mustard, or butter, for example, should be thrown away.

Uneaten bread and rolls may not be re-served to other customers. Linens used to line bread baskets must be changed each time a customer is served.

Self-Service Areas

Buffets and **food bars** give patrons the chance to choose what they want to eat. Unless these self-service areas are protected and closely monitored microorganisms can spread.

Customers at food bars often unknowingly serve themselves in ways that can put themselves and other customers in danger. A customer may eat from her plate or nibble from the food bar while moving through the line; may pick up carrot sticks, pickles, and olives with her fingers, or dip a finger into salad dressing to taste it. Another customer may return unwanted food items to the food bar, use a soiled plate for a second helping, or put his head under the sneeze guard to reach items in the back of the display.

Buffets and food bars should be monitored by employees trained in food-safety procedures. Assign a staff member to replenish food-bar items and to hand out fresh plates for return visits. Post signs with polite tips about food-bar etiquette. These types of practices will do a lot to keep self-service areas more sanitary.

Cross-Contamination

Never re-serve plate garnishes, uncovered condiments, or uneaten bread.

Cross-Contamination

Assign a trained staff member to monitor food bars.

Cross-Contamination

Never mix fresh food with food being replaced.

Here are more rules for food bars:

○ **Protect food on display with sneeze guards or food shields** *(see Exhibit 8g)*. These must be between fourteen and forty-eight inches above the food, in a direct line between the food and the mouth or nose of an average customer.

○ **Identify all food items.** Label containers on the food bar (but don't put probe-type tags into the food itself). Write the names of salad dressings on ladle handles *(see Exhibit 8h)*.

○ **Maintain proper food temperatures.** Keep hot food hot—140°F (60°C), and cold food cold—41°F (5°C). Measure and record food temperatures in a log at least every two hours.

○ **Replenish foods on a timely basis.** Practice the FIFO method of product rotation. Prepare and replenish small amounts at a time so food is fresher and has less chance of being exposed to contamination. Never mix fresh food with food being replaced.

○ **Keep raw foods separate from cooked and ready-to-eat foods.** Customers can easily spill foods when serving themselves. Use separate displays or food bars for raw foods and cooked foods so there's less chance of cross-contamination.

○ **Do not let customers use soiled plates or silverware for refills.** Encourage all customers to take a clean plate for return trips to the food bar. Customers can use glassware for refills as long as beverage dispensing equipment doesn't come in contact with the rim or interior of the glass.

OFF-SITE SERVICE

Establishments use many different ways to deliver food to people. Restaurants offer home delivery. Grocery stores deliver deli trays to private parties and company picnics. Hospitals cater off-site meetings and corporate functions. Nursing homes deliver meals to the elderly. Community groups run refreshment

Exhibit 8g

Protecting Food on Display
Sneeze guards must be between fourteen and forty-eight inches above the food, in a direct line between the food and the mouth or nose of an average customer.

stands at local sporting events. On-site food services vend food from machines in college dorms.

In all cases, operators must follow the same rules of food safety. No matter how food is prepared and delivered, it must be protected from contamination and temperature abuse. Facilities and equipment used to prepare food must be clean and sanitary. Menu items must lend themselves to safe service. Food must be handled safely. **Off-site services,** including delivery, mobile and temporary kitchens, and vending all present special challenges.

Delivery

Many establishments such as schools, hospitals, caterers, and even restaurants use a central commissary to prepare food. Food from the central kitchen is then delivered to remote locations for service. The greater the time and distance from the point of preparation to the point of consumption, the greater the risk that food will be exposed to contamination or temperature abuse. Equipment used to transport food—both food containers and vehicles—must be designed to maintain safe food temperatures and be easy to keep clean.

Exhibit 8h | **Identifying Food Items**
Label all items on a food bar.

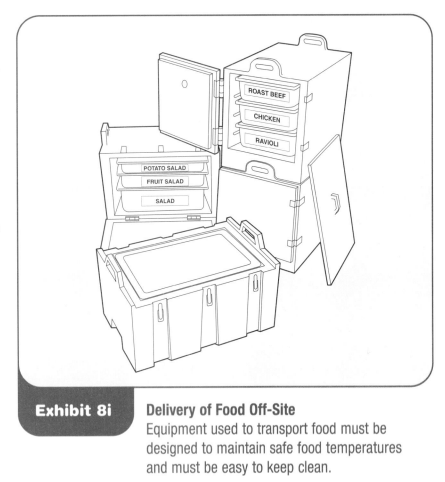

Exhibit 8i | **Delivery of Food Off-Site**
Equipment used to transport food must be designed to maintain safe food temperatures and must be easy to keep clean.

Here are some safe procedures to follow when transporting food from a central kitchen.

○ **Use rigid, insulated food containers capable of maintaining food temperatures above 140°F (60°C) or below 41°F (5°C).** Containers should be sectioned so foods don't mix, leak, or spill and must also allow air circulation to keep temperatures even. Keep containers clean and sanitized *(see Exhibit 8i).*

○ **Clean and sanitize the inside of delivery vehicles regularly.**

○ **Make sure employees practice good personal hygiene when distributing food.**

○ **Check internal food temperatures regularly.** Take corrective action if food isn't at the proper temperature. Within two hours, reheat hot foods to 165°F (74°C) for fifteen seconds, before serving. Re-chill cold foods to 41°F (5°C) or below within four hours. If containers or delivery vehicles aren't maintaining proper food temperatures at the end of each route, reevaluate the length of delivery routes or the efficiency of the equipment you use.

○ **Label foods with storage, shelf life, and reheating instructions for employees at off-site locations.**

○ **Provide food-safety guidelines for consumers.** If you or your employees won't be serving the foods you deliver, provide customers with information on which items should be eaten immediately, which may be saved for later, and how to serve them.

Catering

Caterers provide food for private parties and events, and public and corporate functions. They may bring in prepared food, or cook food on-site in a mobile unit, a temporary unit, or in the customer's own kitchen.

Caterers must follow the same rules of food safety as other establishments. Food must be protected from contamination and temperature abuse. Facilities must be clean and sanitary. Food must be prepared and served safely.

In addition, caterers must meet the special challenges of off-site food handling. They must make sure there is safe drinking water for cooking, warewashing, and handwashing, as well as adequate power for holding and cooking equipment; and proper arrangements for garbage disposal.

Outdoor catering for barbecues and cookouts may require special arrangements. When power or running water aren't available, caterers may have to change their procedures and improvise.

○ **Deliver raw meats frozen and wrapped, on ice.** Deliver milk and dairy products in a refrigerated vehicle or on ice. Use ice chests or insulated

Time & Temperature

When delivering food off-site, check temperatures regularly and take corrective action if the food isn't the proper temperature.

containers for all potentially hazardous foods. Keep them cold with ice or frozen gel-ice packs.

○ **Serve cold foods in containers on ice.** If that isn't possible, record the time when the potentially hazardous food was first taken out of cold storage, and then discard the food after four hours. Note: This method is not permitted in some jurisdictions.

○ **Keep raw and ready-to-eat products separate.** Store raw chicken separately from ready-to-eat salads, for example.

○ **Use only single-use items.** Make sure customers get a new set of disposable tableware for refills. Arrange proper garbage disposal away from food prep and serving areas.

○ **Provide instructions for proper storage, shelf life, and reheating if food is left with the customer after the event.**

Mobile Units

These portable facilities range from concession vans to elaborate field kitchens. **Mobile units** that serve only frozen novelties, candy, packaged snacks, and soft drinks have to meet basic sanitation requirements. Mobile kitchens that prepare and serve potentially hazardous foods, however, must follow the same rules required of permanent foodservice kitchens. Both might be required to apply for a special permit or license from the local regulatory agency.

Like permanent establishments, mobile kitchens have to be outfitted with adequate cooking equipment, mechanical cooling units and hot-holding units, and warewashing and handwashing sinks with hot and cold drinking water under pressure. Mobile kitchens must provide adequate ventilation, garbage storage and disposal facilities, and pest control. Operators should clean and maintain mobile units as needed every day.

Temporary Units

Temporary foodservice units typically operate in one location for less than fourteen days. Tents or kiosks set up for food fairs, special celebrations, or sporting events may be temporary units *(see Exhibit 8j on the next page)*. In some areas, the definition also extends to units set up for longer periods of time. **Temporary units** usually serve prepackaged food or food that requires limited preparation, such as hot dogs. It is best to keep the menu simple to limit the amount of on-site food preparation. Check with regulatory agencies for operating requirements.

Key Point

Mobile units that serve frozen novelties, candy, packaged snacks, and soft drinks have to meet basic sanitation requirements.

Courtesy of the Illinois Restaurant Association

Exhibit 8j

Temporary Unit
If food is prepared on-site, the unit must have adequate cold storage, cooking, and hot-holding equipment.

Temporary units should be constructed so that dirt and pests are kept out. If floors are dirt or gravel, cover them with mats or platforms to control dust and mud. Construct walls and ceiling with materials that will protect food from weather and windblown dust.

In addition, the same safe-handling rules discussed above apply to food preparation in temporary units. If food is prepared on-site, the unit must have adequate cold storage, cooking, and hot-holding equipment. If food is prepared off-site, it must be transported and held at proper temperatures below 41°F (5°C) or above 140°F (60°C).

Safe drinking water must be available for cleaning and sanitizing, and handwashing. Since warewashing facilities will most likely be limited, use disposable single-service items.

Vending Machines

Food prepared and packaged for **vending machines** has to be handled with the same care as any other food served to a customer. Vending operators also have to protect vended foods from contamination and temperature abuse during transport, delivery, and service.

Keep foods at the right temperature. Machines that contain potentially hazardous foods must be able to maintain proper food temperatures of 41°F (5°C) or below and 140°F (60°C) or above (see Exhibit 8k). They must also have automatic cutoff controls that prevent foods from being dispensed if the temperature stays in the danger zone for a certain amount of time.

Check product shelf life daily. Replace foods with expired code dates. If refrigerated foods aren't used within seven days after preparation they must be thrown out. Dispense potentially hazardous foods, such as milk, in their original containers. Fresh fruits with an edible peel should be washed and wrapped before being put in a machine.

Place machines in appropriate locations. Keep machines away from garbage containers, sewage drains, and overhead pipes. Make sure the vending area is clean and well lighted. Supply safe drinking water for beverage machines. Vending areas often provide a microwave oven, refrigerator, ice machine, or other equipment for customer convenience. Make sure that there is adequate power for such equipment.

Clean and service vending machines regularly. Sanitize food-contact surfaces in machines each time food is replenished. Train employees to use good personal hygiene. Employees should wash their hands before and after servicing or refilling machines.

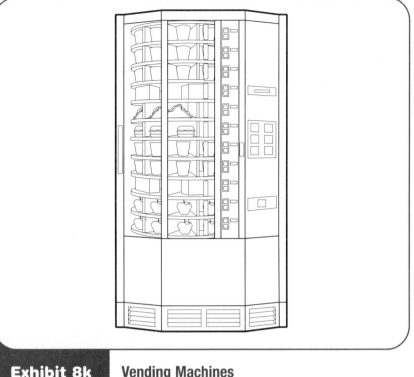

Exhibit 8k   **Vending Machines**
Replace foods with expired code dates and throw out foods if they aren't used within seven days of preparation.

EIGHT RULES OF SAFE FOOD HANDLING

All employees should know the eight basic rules of safe food handling and should be responsible for safe food practices in their assigned areas.

1. Practice strict personal hygiene. Employees should be clean and neatly groomed, should wash their hands regularly and never touch ready-to-eat food with their bare hands. Employees who are ill should not report to work or should be assigned to nonfoodhandling tasks (check local regulations).

2. Monitor time and temperature and prevent cross-contamination when storing and handling food during preparation.

3. Make sure raw products are kept separate from ready-to-eat foods.

4. Avoid cross-contamination by cleaning and sanitizing food-contact surfaces, equipment, and utensils before and after every use, and at least once every four hours during continuous use.

5. Cook foods to their required minimum internal cooking temperature or higher.

6. Hold hot foods at 140°F (60°C) or above and cold foods at 41°F (5°C) or below.

7. Chill cooked food to 41°F (5°C) within four hours. (Alternatively, cool cooked food to 70°F [21°C] within two hours, and then chill to 41°F [5°C] within four hours.)

8. Reheat foods for service to an internal temperature of 165°F (74°C) for fifteen seconds within two hours.

SUMMARY

Safe foodhandling practices don't stop once food is properly prepared and cooked. To make sure the food you serve is safe, you must continue to protect it from temperature abuse and contamination until it is eaten. Reheat cooked foods to 165°F (74°C) for fifteen seconds within two hours. When holding foods for service, keep hot foods hot, above 140°F (60°C), and cold foods cold, below 41°F (5°C). Use clean, sanitized utensils to serve foods. Never touch ready-to-eat food with bare hands.

Make sure all employees, including servers, practice good personal hygiene. Train employees to avoid cross-contamination when handling service items and tableware. Teach them the potential hazards posed by contamination from dirty tables and tableware and from re-serving plate garnishes, breads, or open dishes of condiments.

Customers can unknowingly contaminate foods in self-service areas such as food bars and buffets. Post signs to communicate self-service rules, and station employees in these areas to make sure that customers follow the rules. Protect foods in food bars and buffets with food shields and make sure equipment can hold food at the proper temperature.

Take special precautions when preparing, delivering, or serving food off-site. Catering, mobile kitchens, temporary units, and vending machines all pose unique challenges to food safety. Learn and follow all regulations in your jurisdiction. Be prepared with backup plans if off-site facilities aren't adequate or equipment breaks down.

Finally, make sure all employees know and understand the eight rules of safe food handling.

A CASE IN POINT I

Case Study

Jill, a line cook on the morning shift at Memorial Hospital, was busy helping the kitchen staff put food out on display for lunch in the hospital cafeteria. Ann, the kitchen manager who usually supervised lunch in the cafeteria, was at an all-day seminar on food safety. Jill was responsible for making sure that meals were trayed and put into food carts for transport to the patients' rooms. The staff also packed two dozen meals each day for a neighborhood group that delivered them to homebound elderly people.

Jill first looked for insulated food containers for the delivery meals. When she couldn't find them, she loaded the meals into cardboard boxes she found near the back door, knowing the driver would be there soon to pick them up. To help the cafeteria staff, Jill filled a warming kettle insert with soup by dipping a two-quart measuring cup into the stockpot and pouring it into the insert. She carried the insert out to the cafeteria, put it into the warming kettle, and turned the kettle on low.

The lunch hour was hectic. The cafeteria was busy, and the staff had a lot of patient meals to tray and deliver. Halfway through lunch, a cashier came back to the kitchen to tell Jill the salad bar needed replenishing. Since she was busy, Jill asked a kitchen employee to take pans of prepared ingredients out of the refrigerator and put them on the salad bar. When she looked up a few moments later, she saw the kitchen employee shoo away two children who were eating carrot sticks from the salad bar. The employee then dumped the pans of lettuce and vegetables into the salad bar.

With lunch almost over, Jill breathed a sigh of relief. She moved down the cafeteria serving line, taking food temperatures. One of the casseroles was about 130°F (54°C). Jill checked the water level in the steam table and turned up the thermostat, then went to clean up the kitchen and finish her shift.

What did Jill do wrong? What should have been done?

A CASE IN POINT II

Case Study

Megan, a new server at The Fish House, reported for work ten minutes early on Thursday. Excited about her new job, she had made sure to shower and wash her hair before going to work. When she arrived, she changed into a clean uniform, pulled her hair back tightly into a ponytail, and checked her appearance. She had used makeup sparingly and wore only small hoop earrings and a short chain necklace for jewelry.

Her shift started off well. One of her customers ordered a menu item Megan hadn't tried yet. When the order came up, Megan dipped her finger into the sauce at the edge of the plate for a taste. As the shift progressed, Megan's station got busier. When the hostess came to ask how soon one of Megan's tables could be cleared, Megan decided to do it herself. She took the dirty dishes to a bus station, then wiped down the table with a serving cloth she kept in her apron.

The buser, John, finished another task and came to help her set the table. While he put out silverware and linens, Megan filled water glasses with ice by scooping the glasses into the ice bin in the bus station. Only one slice of bread and one pat of butter were missing from the basket that had been on the table, so she put that on a tray with the water glasses and took it to the table.

The table was reset in record time, and Megan soon had more guests. While taking their orders, Megan reached up to scratch a mosquito bite on her neck. Then she went to the kitchen to turn in the order and pick up a dessert order for another table.

What errors did Megan make? What should she have done?

TRAINING TIPS

Training Tips for the Classroom

1. Group Activity: Food Contamination During Service

Objective: *After completing this activity, class participants will be able to identify ways food can be contaminated during service; standard operating procedures (SOPs) and Critical Control Points (CCPs) that can minimize contamination; and corrective actions that should be taken for various modes of food service.*

Directions: Break the group into teams, and assign one team to each of the following topics about safe food handling that were discussed in Chapter 8.

○ hot holding
○ cold holding
○ table service and re-service of food
○ food bars, self service, buffets
○ off-site delivery
○ carryout
○ catering
○ mobile units
○ temporary units
○ vending machines

Have the teams address the following questions about the assigned topics.

○ How can food become contaminated at this stage or under these conditions?
○ What standard operating procedures (SOPs) must be in place to minimize food contamination?
○ When you look at the flow of a particular food, what is the last point where a hazard can be eliminated, prevented, or reduced?
○ How do you verify that food-safety procedures have been followed and standards have been met?
○ What corrective actions and documentation must be performed, and who will perform them?

Allow about ten minutes for the teams to answer the questions, and allow time for each team to present its answers to the class. Then ask the entire class the following questions, to answer in group discussion.

○ Which team's information applies to your establishment?
○ Does your establishment have these systems and procedures in place?
○ If not, how might they be put in place?

2. The Eight Rules of Safe Food Handling: A Self-Assessment Test

Objective: *After completing this activity, class participants will be able to assess foodhandling practices in their establishments, discover any weaknesses, and identify systems and procedures that could be put in place to improve foodhandling practices.*

Directions: Ask students to honestly assess how well their establishments follow the Eight Rules of Safe Food Handling listed at the end of Chapter 8.

Print a rating sheet with a list of the Eight Rules of Safe Food Handling. Next to each rule, create a rating scale from one to four. Ask students to rate foodhandling practices at their own establishments on a scale of one to four, one (1) being "perfect" and four (4) being "poor" or "needs work." The sheet should look like this.

Perfect	Good	Average	Poor	
1	2	3	4	
—	×	—	—	Keep raw products separate from ready-to-eat foods.
—	—	×	—	Chill cooked food to 41°F (5°C) within four hours.

Allow ten or fifteen minutes for this self-assessment test, then solicit responses from the class for each of the rules. Ask for a show of hands: how many students reported a one, two, three, or four on each safe foodhandling rule? Record the results on the worksheet, and report the results orally, or write the tally on a blackboard or flipchart if one is available.

Then allow time for in-depth discussion of each rule. If any students indicated a score of "average" or "poor" on one or more of the rules, ask them the following questions.

○ What food-safety system and procedures do you currently have in place regarding the foodhandling rule?

○ What additional food-safety systems and procedures could you put in place to improve food handling?

Solicit class input to help improve foodhandling practices at the establishment in question.

For students who scored "good" or "perfect" on any rule, ask if their establishment ever falls short of the rule, and ask if there is anything they could do to further improve current foodhandling practices.

Some students may be reluctant to report anything negative about their establishment, so you should ask for volunteers to answer some of these questions. The idea is to have a lively class discussion about each of the rules, not to make anyone uncomfortable.

Remind the class that while they may find their establishments to be safe in general, there is always room for improvement. Managers should take every opportunity to improve systems and procedures to minimize food contamination.

Training Tips on the Job

1. Front-of-the-House Food-Safety Task Force

Objective: *To involve service staff in an assessment of front-of-the-house foodhandling practices and to solicit recommendations for improving current practices.*

Directions: Develop a food-safety task force made up of the front-of-the-house personnel. Include at least one person from every front-of-the-house position in the task force. This should include servers, busers, cashiers, bartenders, hosts and hostesses, catering personnel, delivery drivers, shift supervisors, and managers.

Plan to hold three or more meetings with your food-safety task force. Before your first meeting, do some research. Get information from state and local health departments about outbreaks of foodborne illness that have occurred in recent years. Ask which were directly caused by breakdowns in temperature control, personal hygiene, or cross-contamination, or due to mistakes during holding, service, or transportation.

First Meeting: Start your first meeting by presenting the information about recent outbreaks of foodborne illness. Then ask the members of your food-safety team to look for and document hazards and food-safety violations in their particular areas of the establishment. Schedule another meeting to hear their reports.

Second Meeting: Your next meeting should allow time for each person to report on the hazards or safety violations found in his or her work area. Encourage and allow time for group feedback. The next assignment for the task force is to research possible solutions to problem areas. Suggestions for possible solutions can come from fellow employees and other establishments.

Third Meeting: Receive recommendations from members of the task force. Allow time for group feedback. Discuss how these recommended changes can be implemented. Remember to reward your task force for their work!

Follow-Up: Implement the solutions, evaluate, monitor, and modify where necessary.

2. Food-Safety Training Materials

Objective: *To assess current food-safety training practices for front-of-the-house personnel in your establishment.*

Directions: Most front-of-the-house personnel have received some orientation and training materials, but these materials may lack up-to-date information about adequate food safety and standards. Consider taking the following steps:

○ Assess all current training materials (videos, manuals, checklists, etc.) for the various front-of-the-house personnel. Are the appropriate food-safety issues addressed?

○ What additional safety issues and standards need to be addressed? How can these be integrated into existing materials?

○ How can you effectively communicate to training personnel the need for additional training in food-safety issues and standards? How can you be certain this information is reinforced during training?

○ Are employees properly tested about food-safety issues? Is there follow-up after the evaluation?

If you insist that employees follow safe foodhandling standards from the very beginning of their employment, communicate that these standards are to be met, and consistently monitor employee performance, employees will form safe foodhandling habits.

DISCUSSION QUESTIONS

1. What are some ways a customer might contaminate food in self-service areas?

2. List hazardous conditions to food safety that occur during food service and explain how to prevent them.

3. What types of foods should not be prepared in a temporary unit?

4 Explain the procedures for and the importance of separating raw and ready-to-eat foods in self-service facilities.

5. How often should you check the temperature of foods being held for service?

MULTIPLE-CHOICE STUDY QUESTIONS

1. Which serving method is most likely to protect the safety of the food being served?

   A. Using stainless-steel tongs to serve hot rolls directly to customers at the table

   B. Using a clean coffee cup to ladle soup into individual serving bowls

   C. Using a drinking glass to scoop ice cubes out of the icemaker storage compartment

   D. Using a teaspoon to scoop a serving of ice cream out of a five-gallon tub

2. Which procedure is the safest for dispensing ice cubes for beverage service?

   A. From an ice bucket with clean hands

   B. From the ice maker storage bin with a clean, sanitary metal or plastic scoop

   C. From the ice maker storage bin with a clean drinking glass

   D. From an ice bucket with a handled cup

3. Which bread product may be re-served to customers?

   A. Hard-crust rolls that were not eaten and are less than one day old

   B. Bread slices that were not eaten by the previous customer

   C. Bread sticks that are wrapped in cellophane

   D. Oyster crackers that were served in a bowl to the previous customer

4. Which of the following does not represent a potential food-safety problem for a self-service lunch buffet?

   A. A bowl of macaroni salad that has been sitting on a table at room temperature from 11:00 a.m. to 3:30 p.m.

   B. A chafing dish of chicken à la king served at a temperature of 120°F (49°C)

   C. Vegetable soup served from a large bowl

   D. A platter of fresh vegetables presented directly on ice

5. You are in charge of maintaining your restaurant's salad bar. Which of the following procedures is not necessary to prevent the customers from accidentally contaminating the food?

A. Provide a supply of clean drinking glasses every two hours at a minimum.

B. Give each customer a clean plate when he or she comes back for a second helping of food.

C. Locate the salad greens at the opposite end of the table from the roast ham.

D. Provide a proper serving utensil for every dish.

6. Which buffet setup practice demonstrates the FIFO principle?

A. Bowls of washed salad greens are put out for consumption one at a time.

B. Dated single-serve containers of milk are offered for consumption in order by date.

C. A bowl of carrot and raisin salad is refilled as needed throughout the day.

D. A basket of sliced bread is replaced every hour throughout the day.

7. You supply home-delivery boxed meals to elderly people in your community. The meals, one hot and one cold meal daily, are prepared in a hospital kitchen. The van driver takes you to the clients' homes in a specially equipped van. Which of the following conditions might give you reason to question the safety of the meals?

A. The temperature of the first cold meal delivered registered at 35°F (2°C).

B. The total time for the delivery route was more than two hours.

C. The driver didn't wash her hands before starting on the route.

D. The temperature of the last hot meal delivered registered at 135°F (57°C).

8. Hash brown potatoes are a popular item on your hotel's breakfast buffet. They are served in a chafing dish which has its own heat source and an external dial thermometer. How can you ensure that the potatoes are safe to eat?

A. Every two hours, measure the internal temperature of the potatoes with a thermometer.

B. Every two hours, record the temperature reading on the chafing dish's external thermometer.

C. Every hour, add fresh hot potatoes to the potatoes already in the chafing dish.

D. Every hour, turn up the heat on the chafing dish to 165°F (74°C).

9. Which of the following is an appropriate service of an uncooked food?

A. Raw fish in sushi, held for several hours and served at room temperature

B. Unshucked oysters on ice, partitioned off from other ready-to-eat foods

C. Raw chicken with fresh bread, served on a salad bar next to cut raw vegetables

D. Raw lamb kibbe, served on an appetizer tray with cold stuffed grape leaves, relishes, salads, and other ready-to-eat foods

Resources

ADDITIONAL RESOURCES

Books and Periodicals

Equipped for food safety. (1998). *Best Practices, 2*(4), 6-10.

Fairbrook, P. (1994). The trouble with steamtables. *Food Management, 29*(3), 48.

Lachney, A. M. (1997). *The HACCP cookbook and manual* (2nd ed.). Washington, DC: Nutrition Development Systems.

The right way to…calibrate equipment. (1997), *Best Practices, 1*(3), 12.

The right way to...chill foods. (1998). *Best Practices, 2*(2), 12.

The right way to...prepare beef. (1998). *Best Practices, 2*(3), 12-13.

The right way to...prepare eggs. (1998). *Best Practices, 2*(2), 10-11.

The right way to...preparing fruits and vegetables. (1998). *Best Practices, 2*(4), 12-13.

The right way to...seafood safety. (1999). *Best Practices, 3*(1), 16.

Secrets of self-inspection. (1998). *Best Practices, 2*(2), 6-9.

Townsend, R. (1990). Proper holding patterns maintain food quality. *Restaurants & Institutions, 100*(9), 125.

Web Sites

1999 FDA Model Food Code

http://vm.cfsan.fda.gov/~dms/fc99-toc.html

Complete outline of the FDA's latest code for regulating operations that provide food directly to consumers. Also includes a quick synopsis of changes from the 1997 Food Code.

Alaska Seafood Marketing Institute

http://www.alaskaseafood.org

The Alaska Seafood Marketing Institute works to improve seafood quality by teaching fishermen, processors, retailers, and restaurateurs about proper handling of Alaska seafood products. Their Web site also includes industry news, suppliers, and recipes.

American Egg Board

http://www.aeb.org

The American Egg Board's Web site features product, safety, and industry information for foodservice professionals as well as recipes and other information for the general public.

American Meat Institute

http://www.meatami.org

Provides member information, as well as current and past news and information regarding the meat industry.

Atkins

http://www.atkinstech.com

Atkins offers a variety of tools for reading food temperature. Their Web site includes ordering and operation instructions.

Cooper Instrument

http://www.cooperinstrument.com

Cooper is a supplier of instruments for measuring time, temperature, and humidity. Their Web site features product and ordering information as well as tips and trade show information.

Daydots

http://www.daydots.com

Daydots makes stickers that facilitate efficient food rotation. Their Web site offers safety tips, ordering information, and industry news.

FDA Center for Food Safety and Applied Nutrition (CFSAN)

http://vm.cfsan.fda.gov/list.html

Comprehensive site from CFSAN offers a wealth of food-safety information, from foodborne illness to food labeling. CFSAN strives to be a leader in food safety, and to protect consumers from economic fraud, promote sound nutrition, and encourage innovation.

FDA Seafood and Information Resources

http://vm.cfsan.fda.gov/seafood1.html

The Web site for FDA's Center for Food Safety and Applied Nutrition includes seafood information and resources, including seafood pathogens and contaminants.

Food Marketing Institute

http://www.fmi.org

This nonprofit association conducts programs in research, education, industry relations, and public affairs for its membership of food retailers and wholesalers. Their Web site offers a wealth of food-safety information.

Hobart

http://www.hobartcorp.com

Hobart designs food equipment such as mixers and slicers, and warewashing, cooking, bakery, refrigeration, weighing, and wrapping equipment. This site includes information on all the equipment Hobart manufactures.

Institute of Food Technologists (IFT)

http://www.ift.org

IFT is a nonprofit scientific society with 28,000 members working in food science, food technology, and related professions in industry, academia, and government. This site includes information on its policies and publications, education and industry news, and meeting and convention locations.

International Association of Milk, Food and Environmental Sanitarians (IAMFES)

http://www.iamfes.org

IAMFES keeps members informed of the latest scientific, technical, and practical developments in food safety and sanitation. Their Web site includes booklets and links to other food-safety sites.

International Dairy Foods Association (IDFA)

http://www.idfa.org

This site is the link to information on IDFA and its constituent organizations: the Milk Industry Foundation (MIF), the National Cheese Institute (NCI), and the International Ice Cream Association (IICA); as well as information on the entire dairy industry. The site also includes information on current industry events, membership information, and publications.

International Food & Information Council

http://ificinfo.health.org

Web site for the International Food and Information Council offers information and resources about food safety and food allergies for educators and the public.

National Cattlemen's Beef Association

http://www.beef.org

This site includes up-to-date information from the National Cattlemen's Beef Association, as well as a reference library, information on nutrition and food safety, and links to related sites.

National Food Processors Association (NFPA)

http://www.nfpa-food.org

NFPA is a scientific and technical trade association for the food industry. Their site provides industry news and educational materials on issues of food safety, research, and food science.

National Institute of Standards and Technology

http://www.nist.gov/public_affairs/welcome.htm

The National Institute of Standards and Technology is the government agency that works with industry to develop and apply technology, measurements, and standards. The site includes links to nontechnical information as well.

National Frozen Food Association (NFFA)

http://www.nffa.org

NFFA is the voice of the frozen foods industry. Their Web site provides information to consumers as well as foodservice professionals.

National Pork Producers Council

http://www.nppc.org

The National Pork Producers Council site provides general information about food and nutrition, as well as information specifically for pork producers. A special section is also provided just for children.

National Restaurant Association

http://www.restaurant.org

The National Restaurant Association site provides information on government agencies affecting the restaurant industry, the latest training and certification updates, and links to state restaurant associations and hospitality schools and universities.

North American Association of Food Equipment Manufacturers (NAFEM)

http://www.nafem.org

NAFEM represents approximately seven hundred companies throughout the United States, Canada and Mexico that manufacture commercial foodservice equipment and supplies. This site allows you to access Web sites for many foodservice equipment manufacturers.

NSF International

http://www.nsf.org

This site offers product information, resources, and publications regarding public health safety.

Produce Marketing Association

http://www.pma.com

The members of this nonprofit trade association market fresh fruits and vegetables to the foodservice industry. Purchasing guidelines are available.

Tyson

http://www.tyson.com

Tyson's Web site offers a consumer kitchen page with recipes and instruction on preparing chicken products. An entire page is also devoted to foodservice professionals and offers information on the industry, Tyson products and menu planning.

United States Department of Agriculture's Food Safety and Inspection Service

http://www.fsis.usda.gov

The USDA's Food Safety and Inspection Web site offers the latest food-safety news, educational materials, and HACCP implementation materials.

Chapter 9
Principles of a HACCP System

TEST YOUR FOOD-SAFETY KNOWLEDGE

1. **True or False:** A control point (CP) is the last point where you can intervene to prevent, control, or eliminate the growth of microorganisms in food. *(See Determine Critical Control Points, page 9-11.)*

2. **True or False:** Cooking chicken to a minimum internal temperature of 165°F (74°C) for fifteen seconds would be an appropriate critical limit if the cooking step is a CCP for chicken. *(See Establish Critical Limits, page 9-12.)*

3. **True or False:** The first step in developing a HACCP system is to identify all Critical Control Points in the flow of food in your establishment. *(See HACCP Principles, page 9-4.)*

4. **True or False:** An establishment's HACCP plan must be reevaluated if there is a menu change. *(See Verify that the System Works, page 9-15.)*

5. **True or False:** The receiving step would be a Critical Control Point when receiving fresh clams for clam chowder in an establishment. *(See Determine Critical Control Points, page 9-11.)*

Learning Objectives

After completing this chapter, you should be able to:

○ Identify the flow of a food through an establishment.

○ Discuss the importance of prerequisite programs for a Hazard Analysis Critical Control Point (HACCP) system.

○ Define HACCP.

○ Identify the HACCP principles for food safety.

○ Discuss how HACCP is important to food safety.

○ Identify Critical Control Points (CCPs) for various foods and processes.

Key Terms

Hazards

Hazard Analysis Critical Control Point (HACCP)

HACCP plan

Prerequisite programs

Hazard analysis

Control point (CP)

Critical Control Point (CCP)

Critical limit

Monitoring

Corrective action

Verification

The last several chapters focused on the flow of food through a typical establishment. In these chapters, you learned how to receive, store, prepare, cook, cool, reheat, and serve food.

In this chapter, we'll discuss a system that will enable you to consistently serve safe food by identifying and controlling possible **hazards** (biological, chemical, or physical agents that may cause illness or injury if not controlled) throughout the flow of food. This system is called **Hazard Analysis Critical Control Point (HACCP).** A HACCP system is a dynamic process that uses a combination of proper foodhandling procedures, monitoring techniques, and record keeping to help ensure that the food you serve is safe. Because HACCP is dynamic, it allows you to continuously improve your food-safety system.

The focus of this chapter is to provide you with a basic understanding of HACCP and its application to the restaurant and foodservice industry; however, if you want to develop and implement a HACCP plan, you will require additional information. A **HACCP Plan** is a written document, based on HACCP principles, which describes the procedures a particular establishment will follow. Refer to the additional resources listed at the end of this chapter to help you develop your plan.

WHAT IS HACCP?

HACCP was developed by the Pillsbury Company in the early 1960s for the National Aeronautics and Space Administration (NASA). The system was designed to make the food served to astronauts in outer space as safe as possible. Since a foodborne illness on a space flight could be disastrous, the system had to ensure safe food. The HAACP system is based on the idea that if biological, chemical, or physical hazards are identified at specific points within the flow of food, they can be prevented, eliminated, or reduced to safe levels.

HACCP was so beneficial that it has been adopted by many segments of the food industry from growers to manufacturers, and from distributors to operators. Over the years the system has been refined and improved, and is now accepted worldwide. The National Restaurant Association and the Food and Drug Administration recommend that all foodservice facilities, no matter how large or small, develop a HACCP system. A HACCP system helps you to do the following.

○ Identify the foods and procedures that are most likely to cause foodborne illness.

○ Develop procedures that will reduce the risk of a foodborne illness outbreak.

○ Monitor procedures to keep food safe.

○ Verify that the food you serve is consistently safe.

HACCP gives operators the tools to identify and prevent potential food-safety problems before they happen. As with any effective food-safety practice, management must be strongly committed to food safety and the HACCP concept, and HACCP must be built on a solid foundation of prerequisite programs. These are proper sanitation practices and procedures that must be in place, also called Standard Operating Procedures (SOPs).

Prerequisite Programs

Prerequisite programs support your HACCP plan and are the basic operating conditions for producing safe food.

Prerequisite programs (SOPs) protect your food from contamination, minimize microbial growth, and ensure the proper functioning of equipment *(see Exhibit 9a)*. They may include the following:

- Proper personal hygiene
- Proper facility design
- Choosing good suppliers
- Creating supplier specifications
- Proper cleaning and sanitation
- Appropriate equipment maintenance

HACCP Principle

Prerequisite programs support your HACCP plan and are the basic operating conditions for producing safe food.

Proper personal hygiene

Proper facility design

Choose good suppliers and develop supplier specifications

Proper cleaning and sanitation

Appropriate equipment maintenance

Courtesy of Hobart Corporation

Exhibit 9a   **HACCP Must Be Built on a Solid Foundation of Prerequisite Programs**

For example, ineffective handwashing practices, improper cooling procedures, and faulty equipment that won't maintain proper temperatures will sabotage even the best HACCP plan.

DEVELOPING A HACCP PLAN

A HACCP plan is a written document that describes the procedures a particular establishment will follow. A HACCP plan is developed using the HACCP principles and is specific to the facility, its menu, its equipment, its processes, and its operations.

> **HACCP Principle**
>
> A HACCP plan is developed using the HACCP principles and is specific to the facility, its menu, its equipment, its processes, and its operations.

The HACCP plan for a product prepared in one facility will be different from the HACCP plan for the same product prepared in another facility. For example, a pot of soup prepared in a deli may be purchased in cans and simply heated to the proper temperature; soup prepared in a restaurant may be made entirely from fresh ingredients on the premises; and soup prepared in a hospital may be made from a combination of *sous vide* and fresh ingredients. In each of these establishments, the soup should be prepared using a customized HACCP plan.

While generic HACCP plans can serve as useful guides, each facility must develop a HACCP plan that addresses its own unique conditions.

HACCP Principles

> **HACCP Principle**
>
> Each HACCP principle builds upon the information gained from the previous principle.

The plan you develop will be based on the seven basic HACCP principles. Principles one, two, and three help you design your system. Principles four and five help you implement it. Principles six and seven help you maintain the system and verify its effectiveness. Each principle builds upon the information gained from the previous principle. For the plan to be complete, you must consider all seven principles in order.

Principle One: Conduct a Hazard Analysis

Identify and assess potential hazards in the food you serve. This means taking a look at the flow of food in your establishment and determining where food-safety hazards (biological, chemical, and physical) are likely to occur. Think about how food comes in contact with the people and equipment in your establishment.

Principle Two: Determine the Critical Control Points (CCPs)

Critical Control Cop

Find those points in the flow of food that are essential to prevent or reduce a food-safety hazard. If this is the last point where this hazard can be controlled, then it is a CCP.

Principle Three: Establish Critical Limits

For each CCP, set standards that food must meet to be considered safe. These are the boundaries for the CCPs. For example, if you have identified cooking as a CCP for pork, the critical limit will be to cook it to a minimum internal temperature of 155°F (68°C) for fifteen seconds.

Principle Four: Establish Monitoring Procedures

Once limits have been established, you must determine ways for checking them and also decide who should check them and how often. For the pork in the previous example, you might establish a procedure requiring kitchen staff to use a thermometer to monitor the final internal temperature every time pork is cooked.

Principle Five: Establish Corrective Actions

Determine what you will do if the critical limit is not met. If the pork in the previous example does not meet the critical limits you established, you may decide to continue cooking it until it does.

Principle Six: Establish Verification Procedures

Determine if the plan is working as intended. Plan to evaluate on a regular basis your monitoring charts, records, how you performed your hazard analysis, etc., and determine if your plan needs modification.

Principle Seven: Establish Record-Keeping and Documentation Procedures

Develop procedures that specify who should perform the documentation, when it should be performed, and how. You should maintain records obtained while performing monitoring activities, whenever a corrective action is taken, when equipment is validated (checked to make sure it is in good working condition), and when dealing with supplier specifications. You also need to keep all documentation created while you were developing the plan.

Now let's look at each HACCP principle in more detail and illustrate each principle, using chicken as an example.

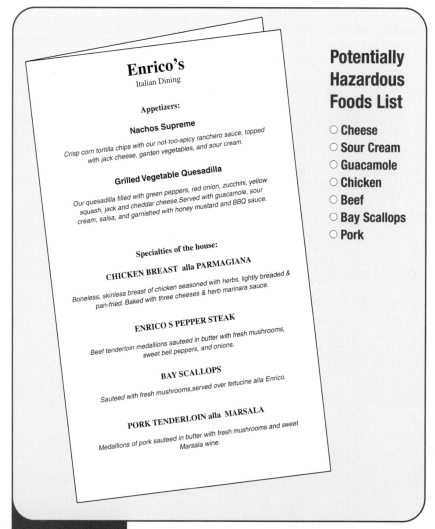

HACCP Principle

The hazard analysis is the key to determining where hazards may occur in the flow of food if care is not taken to prevent or control them.

Hazard Analysis (Principle 1)

Hazard analysis is the process of identifying and evaluating potential hazards associated with foods, in order to decide which must be addressed in a HACCP plan.

The hazard analysis is key to determining where hazards may occur in the flow of food if care is not taken to prevent or control them. To perform a hazard analysis in your establishment, you need to consider your menu item's ingredients, your equipment and processes, and your employees. You also need to consider the customers you serve. It is very important to look closely at each of these areas to make sure you identify all potential hazards. This will determine the success of your plan. Since most foodborne illnesses are caused by biological hazards, we will concentrate more on these. However, you also should be aware of any potential chemical or physical hazards as well.

Here are some key steps to follow to identify all potential hazards in your establishment.

○ **Identify potential food hazards.** Identify any foods that may become contaminated if handled incorrectly at any stage in the flow of food, or that may allow the growth of harmful microorganisms. Because potentially hazardous foods are known to support the rapid growth of pathogens, it is especially important to identify them and determine how they are handled in your establishment. Make a list of these foods. For example, your list may include menu items that contain chicken, seafood, pork, and beef, all of which are potentially hazardous *(see Exhibit 9b)*.

Enrico's
Italian Dining

Appetizers:

Nachos Supreme
Crisp corn tortilla chips with our not-too-spicy ranchero sauce, topped with jack cheese, garden vegetables, and sour cream.

Grilled Vegetable Quesadilla
Our quesadilla filled with green peppers, red onion, zucchini, yellow squash, jack and cheddar cheese. Served with guacamole, sour cream, salsa, and garnished with honey mustard and BBQ sauce.

Specialties of the house:

CHICKEN BREAST alla PARMAGIANA
Boneless, skinless breast of chicken seasoned with herbs, lightly breaded & pan-fried. Baked with three cheeses & herb marinara sauce.

ENRICO S PEPPER STEAK
Beef tenderloin medallions sauteed in butter with fresh mushrooms, sweet bell peppers, and onions.

BAY SCALLOPS
Sauteed with fresh mushrooms, served over fettucine alla Enrico.

PORK TENDERLOIN alla MARSALA
Medallions of pork sauteed in butter with fresh mushrooms and sweet Marsala wine.

Potentially Hazardous Foods List

○ Cheese
○ Sour Cream
○ Guacamole
○ Chicken
○ Beef
○ Bay Scallops
○ Pork

Exhibit 9b   **Identifying Potentially Hazardous Foods on a Menu**

○ **Determine where hazards can occur in the flow of food.** Using your list, write down each step in the flow of food for each particular item. Examine and write down all the potentially adverse conditions the food may be exposed to during receiving, storage, preparation and handling by your employees; holding and service; and cooling and reheating. You must also consider how your operation's equipment may affect these foods. Determine the potential hazards that could occur, such as contamination and time-temperature abuse, and write them down.

In our example, chicken was identified as a potentially hazardous food. We determined and recorded its flow, which includes receiving, storage, preparation, cooking, and service. There are potentially adverse conditions at each step. Those identified at the preparation step include thawing at room temperature (time-temperature abuse); using one prep table for all foods (cross-contamination); and allowing hand contact with product (contamination) *(see Exhibit 9c)*.

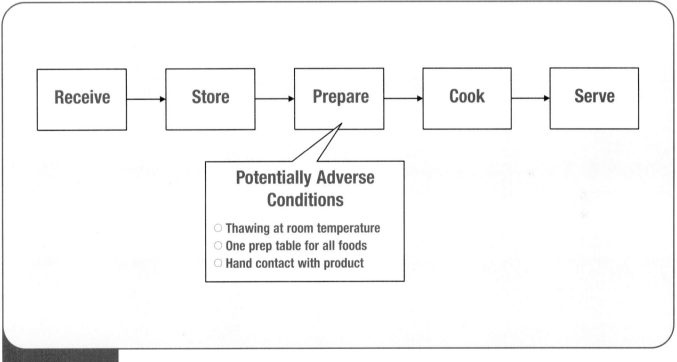

Exhibit 9c **Determining the Steps in the Flow of Chicken Through an Establishment**
The flow of chicken consists of five steps in this establishment. Potentially adverse conditions have been identified for the preparation step in this example, but must be identified for all steps in your establishment.

○ **Group foods by processes.** It may help to group foods that were identified as potentially hazardous according to how they are processed in your establishment *(see Exhibit 9d).* The most common groups include:

● Foods that are prepared and served without cooking (salads, raw oysters, cheeses, and sandwich meats). Hazards involve contamination of the food by employees or equipment.

● Foods that are prepared and cooked for immediate service (hamburgers, scrambled eggs, and hot sandwiches). Hazards involve improper cooking, which may not eliminate biological hazards.

● Foods that will be prepared, cooked, held, cooled, reheated, and served (chili, soups, and sauces). Hazards may occur at many points.

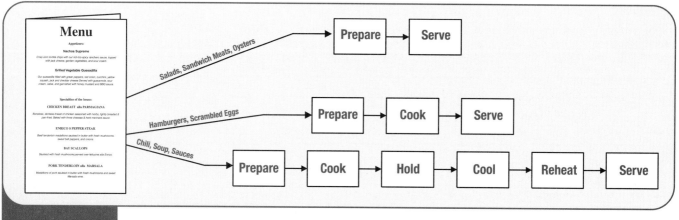

Exhibit 9d   Grouping Foods by Processes
When developing your HACCP plan, you may choose to group foods by how they are processed in your establishment.

The chicken in our example is prepared and cooked for immediate service. Pork, which is also found on the list of potentially hazardous menu items, is processed in this establishment in the same manner, and therefore can be grouped with chicken. Since beef is used only for minestrone soup in this establishment, it falls under the third grouping of processes.

○ **Identify your customers.** This is particularly important if the customers you serve are very young or elderly, or people who are ill or immunocompromised. Because these groups are more susceptible to foodborne pathogens, hospitals, nursing homes, daycare centers, and schools need to be extra careful when preparing food.

In our example, we identified that our customers are elderly, and are therefore more susceptible to foodborne pathogens that may be found in the chicken.

HACCP Principle

A control point (CP) is any step in the flow of food where a physical, chemical, or biological hazard can be controlled.

Determine Critical Control Points (Principle 2)

Once you've identified all potential food hazards and the step or steps at which they occur in your establishment, the next task is determining at which steps you can intervene to control these hazards. A **control point (CP)** is any step in the flow of food where a physical, chemical, or biological hazard can be controlled. While all control points in the flow of food are important in preventing a hazard from occurring, some are critical. To assess whether a control point is critical, you need to determine if it is the last step where you can intervene to prevent, control, or eliminate the growth of microorganisms before the food is served to customers. If it is, the step is called a **Critical Control Point (CCP).**

We know that the raw chicken in our example may be contaminated with the biological hazards *Salmonella* and *Campylobacter.* While care is needed during preparation to prevent cross-contamination, proper cooking is essential to prevent your customers from becoming ill. Therefore, preparation is a control point, while cooking is a Critical Control Point for this menu item prepared with raw chicken *(see Exhibit 9e).*

Typically, steps such as cooking, chilling, or holding food are named CCPs. However, these may not be the CCPs for all foods or all processes in your establishment. Each situation must be evaluated on an individual basis. Different establishments may prepare a similar food quite differently and will therefore have different Critical Control Points for that food.

Establish Critical Limits (Principle 3)

Once you've determined the CCPs for the potential food hazards in the flow of food, you need to establish **critical limits,** minimum and maximum limits that the CCP must meet in order to prevent, eliminate, or reduce a hazard to an acceptable limit. For example, in our establishment, we have determined

HACCP Principle

A Critical Control Point (CCP) is the last step where you can intervene to prevent, control, or eliminate the growth of microorganisms before the food is served to customers.

HACCP Principle

Cooking, cooling, or holding food are typically named CCPs. However, these may not be the CCPs for all foods or all processes in your establishment. Each situation must be evaluated on an individual basis.

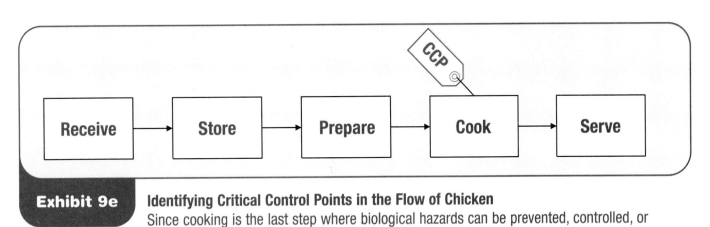

Exhibit 9e **Identifying Critical Control Points in the Flow of Chicken**
Since cooking is the last step where biological hazards can be prevented, controlled, or eliminated for the chicken in this establishment, this step is a Critical Control Point (CCP).

HACCP Principle

Critical limits are minimum and maximum limits that the CCP must meet in order to prevent, eliminate, or reduce a hazard to an acceptable limit.

that cooking is a CCP for chicken, and we have set our critical limit for cooking to a minimum of 165°F (74°C) for fifteen seconds *(see Exhibit 9f)*. We will accomplish this by putting the chicken in our convection oven, which is set to 350°F (177°C), and cooking it for twenty-five minutes. When establishing critical limits, keep the following points in mind. The limit must be:

○ Measurable (such as a time or a temperature)

○ Based on scientific data, food regulations (such as the FDA Model Food Code), and expert advice

○ Appropriate for the food and equipment when prepared under normal conditions, and specific to your establishment

○ Clear and easy to follow

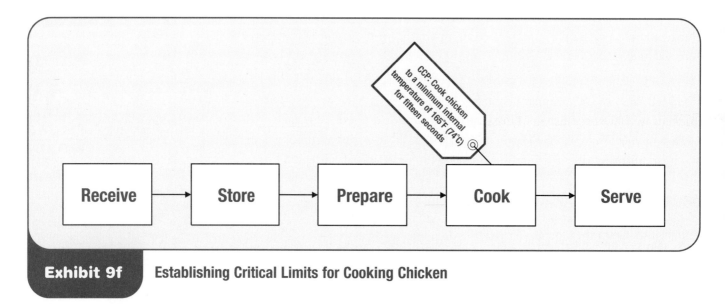

CCP: Cook chicken to a minimum internal temperature of 165°F (74°C) for fifteen seconds

Receive → Store → Prepare → Cook → Serve

Exhibit 9f Establishing Critical Limits for Cooking Chicken

HACCP Principle

Monitoring lets you know that critical limits are being met.

Monitoring Critical Control Points (Principle 4)

By establishing procedures to monitor your Critical Control Points and consistently following them you will be alerted to food-safety problems in the flow of food that need to be corrected. **Monitoring** also lets you know that critical limits are being met, and that you are doing things right.

To develop a successful monitoring program, you need to consider the following. Focus on each CCP. Establish clear directions that will determine the following:

○ **How to monitor the CCP.** This depends on the critical limits you have established, and may include measuring time, temperature, pH, oxygen, water activity, etc.

- ○ **When and how often to monitor the CCP.** Continuous monitoring is preferred but not always possible. Regular monitoring intervals should be determined based on the normal working conditions in your establishment, depending on volume, equipment, and procedures.

- ○ **Who will monitor the CCP.** Assign responsibility to a specific employee or position, and make sure that person is trained to perform the monitoring.

- ○ **Equipment, materials, or tools needed to monitor the CCP.** Provide employees with the proper items to make the task as easy as possible.

Going back to the example of the chicken, we have established (1) that cooking is a CCP; and (2) the critical limits are 165°F (74°C) for fifteen seconds. Because our facility cooks chicken to order, we can take the temperature of each chicken at this point. We determined that Clay, our cook, needs to monitor the chicken has reached the proper temperature for the required amount of time (reached the critical limit) by inserting a thermometer into several places in the thickest part of each chicken *(see Exhibit 9g)*.

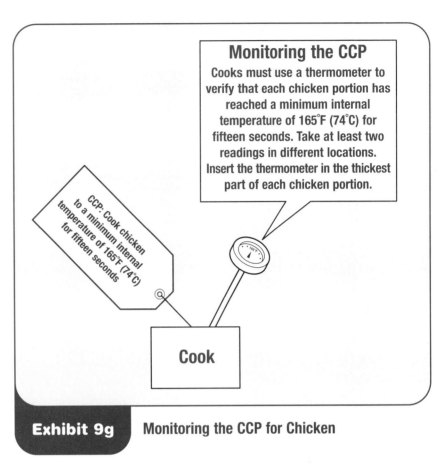

Monitoring the CCP

Cooks must use a thermometer to verify that each chicken portion has reached a minimum internal temperature of 165°F (74°C) for fifteen seconds. Take at least two readings in different locations. Insert the thermometer in the thickest part of each chicken portion.

CCP: Cook chicken to a minimum internal temperature of 165°F (74°C) for fifteen seconds

Cook

Exhibit 9g   **Monitoring the CCP for Chicken**

Taking Corrective Action (Principle 5)

Because you have selected a particular point or step in the flow of food as critical to ensure the safety of the food, specific and immediate actions must be established to correct food-safety errors as they occur. **Corrective actions** are predetermined steps taken when food doesn't meet a critical limit.

Remember, this will be the last opportunity you have to ensure the safety of the food served. A corrective action may be as simple as continuing to cook the food to the correct temperature. Other actions may include throwing food away after a specified amount of time, or rejecting a shipment that is not received at the temperature you specified. When developing corrective actions, be sure that the action you name is specific, in that it states exactly what should be done to correct the situation. Employees should also know

HACCP Principle

Corrective actions are predetermined steps taken when food doesn't meet a critical limit.

who is responsible for taking a corrective action, how to perform the action, and where and how to record the action that was taken.

In our previous example, Kristie, the manager, had determined that if cooked chicken doesn't reach its critical limit, the corrective action is for Clay to continue cooking the chicken until it reaches that limit. Clay is also instructed to record this corrective action in the log *(see Exhibit 9h)*.

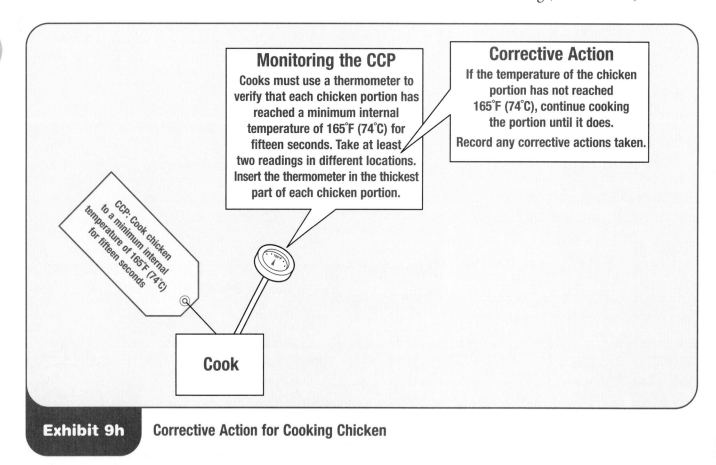

Monitoring the CCP

Cooks must use a thermometer to verify that each chicken portion has reached a minimum internal temperature of 165°F (74°C) for fifteen seconds. Take at least two readings in different locations. Insert the thermometer in the thickest part of each chicken portion.

Corrective Action

If the temperature of the chicken portion has not reached 165°F (74°C), continue cooking the portion until it does.

Record any corrective actions taken.

CCP: Cook chicken to a minimum internal temperature of 165°F (74°C) for fifteen seconds

Cook

Exhibit 9h   **Corrective Action for Cooking Chicken**

Verify that the System Works (Principle 6)

HACCP Principle

Verification confirms that the system you developed works according to the plan.

After you have developed your system, you need to confirm that it works according to the plan. This is called **verification.**

During this step you verify that the CCPs and critical limits you selected are appropriate, that monitoring alerts you to hazards, that corrective actions are adequate to prevent foodborne illness from occurring, and that employees are following established procedures.

Verification of your plan should be performed on a regular basis, but if you notice that critical limits are frequently not being met, or if you receive a foodborne-illness complaint, you need to reevaluate your plan. You also need to reevaluate your plan if there are any changes in your menu, equipment, processes, suppliers, or product.

In our example, the six-month plan evaluation revealed that we frequently failed to meet the critical limits set for our chicken. Upon reevaluating our cooking process, we determined that our supplier had started providing a larger product. This caused the chicken to be undercooked, given the equipment and cooking parameters we had established. We revised our plan by adjusting our cooking process to account for larger weights.

Record Keeping and Documentation (Principle 7)

Recording how food is produced and kept safe is important to the success of a HACCP system. Proper records allow you to document that you are continuously preparing and serving safe food.

Detailed records also serve as an indicator for modifying your processes if any food-safety problems should occur. Examples of records include time-temperature logs, procedures for taking temperatures, SOPs, calibration records, corrective actions, monitoring schedules, product specs, etc.

Training

Training is critical in making a HACCP plan successful. HACCP works best when it is integrated into each employee's job description and duties. Awareness of the HACCP plan depends on your role in the establishment. For example, you may be responsible for training members of your management staff to conduct activities required by the plan. To do this effectively, consider the following:

○ Help management staff understand the importance of food safety and the benefits of HACCP.

○ Train staff (management and employees) to perform specific tasks required by the HACCP plan. A good training program should:

⦿ Explain the importance of what the staff is learning.

⦿ Demonstrate steps and procedures.

⦿ Let employees practice.

⦿ Give feedback on their performance.

⦿ Review materials.

⦿ Test employees on their knowledge.

⦿ Retrain if needed.

⦿ Evaluate employees on job performance as well as their food-safety practices.

⦿ Encourage employee involvement regarding food-safety issues.

HACCP Principle

Proper records allow you to document that you are continuously preparing and serving safe food.

SUMMARY

Hazard Analysis Critical Control Point (HACCP) is a food-safety system designed to keep food safe throughout its flow in an establishment. HAACP is based on the idea that if biological, chemical, or physical hazards are identified at specific points within a food's flow through the operation, then the hazards can be prevented, eliminated, or reduced to safe levels. A successful HACCP system uses a combination of proper foodhandling procedures, monitoring techniques, and record keeping to keep food safe.

HACCP must be built on a solid foundation of prerequisite programs. These programs protect your food from contamination, minimize microbial growth, and ensure the proper functioning of equipment. They include programs for proper personal hygiene, proper cleaning and sanitation, proper facility design, choosing good suppliers, and equipment maintenance.

While generic HACCP plans can serve as useful guides, each facility must develop a plan that addresses its own unique conditions. The plan should be specific to the facility, its menu, its equipment, its processes, and its operations. An effective HACCP plan will be based on the following seven basic HACCP principles.

○ **Conduct a hazard analysis.** Identify and assess potential hazards in the foods you serve.

○ **Determine the Critical Control Points (CCPs).** Find those points in the flow of food that are essential to prevent or reduce a food-safety hazard. If this is the last point where this hazard can be controlled, then it is a CCP.

○ **Establish critical limits.** For each CCP, set standards that food must meet to be considered safe. These are the boundaries for the CCPs.

○ **Establish monitoring procedures.** Once limits have been established, you must determine ways for checking them, and also decide who should check them and how often.

○ **Establish corrective actions.** Determine what you will do if the critical limit is not met.

○ **Establish verification procedures.** Determine if the plan is working as intended.

○ **Establish record-keeping and documentation procedures.** Develop procedures that specify who should perform the documentation, when it should be performed, and how.

Training is critical in making a HACCP plan successful. HACCP works best when it is integrated into each employee's job description and duties.

A CASE IN POINT

Case Study

Jason, the manager at Cal's Catering, received several calls from customers complaining that they had contracted a foodborne illness after attending a picnic catered by Cal's the previous weekend. Jason was concerned and reviewed his menu for the event and noticed that the only potentially hazardous food served was barbecued rib sandwiches. He looked at his HACCP plan and noticed that the barbecued rib sandwiches were received, stored, prepared, and cooked in the establishment. They were put in large, insulated containers and reheated and held for self-service during the picnic.

At what steps in the flow of the barbecued rib sandwiches should Jason have identified Critical Control Points? What critical limits should have been established for each CCP? What records should be available for this plan?

TRAINING TIPS

Training Tips for the Classroom

1. Developing HACCP Systems—A Breakout Exercise

Objective: *After completing this activity, class participants should be able to assess hazards and identify Critical Control Points (CCPs) for various menu items.*

Directions: Make a list of four to eight different menu items. Identify the method of preparation for each item and the type of establishment that might serve it. Print up handouts with one menu item per page. Examples may include:

Menu Item	Method of Preparation	Establishment
Chili	Prepare, cook, hold, cool, reheat	Full service
Potato Salad	Prepare, cook, cool, hold	Vending machine
Baked Chicken	Prepare, cook, hold, serve	Catered event
Cream Pie	Prepare, serve	Retail

When designing this list, include different types of foods (meats, fish, dairy, vegetables, etc.). Include potentially hazardous foods (PHFs) or items that contain PHFs. Also, vary the methods of preparation (cook, cool and reheat; prepare and serve; hot hold), and assign these menu items to different types of establishments (quick service, full service, catering, vending, retail, and so on). When developing this list, keep your audience in mind. If possible, the list should reflect the types of food and the types of establishments that your students represent.

Break your class into four to eight teams, with two to four members each. Assign a captain for each team. (Choose team captains who have some foodservice experience.) Assign each team a menu item. If the captain comes from a specific type of establishment, assign that team a menu item produced in an establishment that is similar.

Give each team a handout with the seven HACCP principles listed on it. Ask each team to develop a HACCP plan for their assigned menu item within the specified type of foodservice operation. It is not as important that each group develop a complete plan as it is that they properly assess the hazards associated with their menu item and identify the Critical Control Points (CCPs).

This exercise will last about an hour. Allow thirty minutes for the teams to develop the outline for the HACCP plan, and thirty minutes for the teams to present their plans to the class. You should also allow adequate time for questions and comments from the class.

2. Control Point (CP) or Critical Control Point (CCP)?

Objective: *After completing this activity, class participants should be able to differentiate between Control Points (CPs) and Critical Control Points (CCPs).*

Directions: Develop a list of ten food items along with a specific step or process related to the preparation and service of that particular item. Here are some examples (with answers in bold):

Item	Step or Process
Fresh chicken	Receive chicken at 41°F (5°C) or below. **(CP)**
Fresh ground beef	Discard ground beef that has been in the temperature danger zone for more than four hours. **(CP)**
Fresh pork	Cook pork to a minimum internal temperature of 155°F (68°C). **(CCP)**
Iceberg lettuce	Wash lettuce prior to making salads. **(CP)**
Chili	Hold cooked chili for service at 140°F (60°C). **(CCP)**
Clam chowder	Cool cooked clam chowder to 70°F (21°C) within two hours and to 41°F (5°C) within an additional four hours. **(CCP)**

Give a copy of the list to each student. Ask them to determine whether the step or process for each food in the list represents a Control Point (CP) or Critical Control Point (CCP).

As a group, discuss each menu item. Discuss rationales for each student's position on CPs and CCPs. If possible, reach an agreement about each item.

3. The Seven Principles Pop Quiz

Objective: *After completing this activity, class participants should be able to explain each of the seven HACCP principles and give an example of each.*

Directions: After discussing Chapter 9, have your students close their books and put away their notes. Give the students a blank sheet of paper and ask them to write down, in order, the seven principles for setting up a HACCP system. They should explain what each principle means and give a specific example.

Allow fifteen minutes for this quiz. Then review the seven principles, and provide students with a completed, one-page synopsis of the seven principles to use as a study aid.

Training Tips on the Job

1. Developing and Implementing a HACCP Plan

Purpose: *To provide some practical tips to help organize the development and implementation of a HACCP plan.*

Directions: Developing a HACCP plan can take a fair amount of time and effort, so it may be more effective if it is developed and implemented in several stages. Here are a few suggestions.

Stage 1. First, make sure that your establishment has strong prerequisite programs in place. These include programs for personal hygiene, cleaning and sanitation, equipment maintenance, supplier selection, and so on. Strong prerequisite programs are the foundation of a strong food-safety system, and are essential to the success of any HACCP plan.

Stage 2. Build a HACCP team. Ideally, the team should represent people from different job positions who handle some aspect of food production and service. For example, the team might include an employee responsible for ordering, receiving, and storing foods; a chef and a prep cook; a server; and a manager, among others. Team members should have food-safety knowledge and strong foodservice experience. They should also be respected by their co-workers, and be committed to the program.

Stage 3. The members of the team should meet to discuss the food served in your establishment, how it is handled, the nature of your customers, and the prerequisite programs currently in place.

Stage 4. After the initial discussion, the team will address the seven HACCP principles. This stage should probably be broken down into several meetings. Keep in mind that all seven principles must be performed in order, because each principle builds on the previous one. The seven principles can be combined into three groups to aid in development: Principles one, two, and three (Stage 4); principles four and five (Stage 5); and principles six and seven. The team must begin by conducting a hazard analysis. A brainstorming session may help the team identify a list of potential hazards likely to occur in the flow of food in the establishment (hazard analysis, Principle 1). Each team member brings to the table potential hazards specific to their area in the establishment. The team should review ingredients used in menu items, processes for preparing the items, equipment used to process it, and how the product will be stored and distributed.

After the list has been completed, the team must then determine which potential hazards will be addressed in the HACCP plan (determining Critical Control Points, Principle 2). The choice should be based on the severity of the hazard and the likelihood that it will occur in the establishment. These are the Critical Control Points (CCPs) that must be identified by the team. Once the CCPs are identified, the team should brainstorm methods for controlling them (setting critical limits, Principle 3). The controls will be critical to preventing, eliminating, or reducing the hazard to an acceptable level. If necessary, the team should look to regulatory standards and guidelines or experts to help set the critical limits.

Stage 5. Next, the team must establish procedures for monitoring the CCPs (Principle 4), and corrective actions to take to keep the CCPs in control (Principle 5). Make sure these procedures are specific so that the employee responsible for performing the monitoring or corrective action knows exactly what is expected. Training is the key to a successful implementation.

Stage 6. A HACCP plan should be implemented gradually over a period of time in order to be successful. Furthermore, you must frequently revisit and reassess your HACCP plan, particularly when changes in your establishment take place (this is Principle 6, verification). Changes that might affect your plan could include employee turnover, a change in menu items and products, changes in equipment, changes in company standards, or new local or state laws.

Be sure your team establishes sound record-keeping procedures. Records will keep an accurate account of the who, what, when, where, and why for monitoring and corrective actions, and will also aid in verifying your plan.

DISCUSSION QUESTIONS

1. What is HACCP?

2. Identify and explain the seven basic HACCP principles.

3. What are three types of hazards? Give an example of each.

4. What is the difference between a control point and a Critical Control Point?

5. Explain what a critical limit is and give an example.

MULTIPLE-CHOICE STUDY QUESTIONS

1. Checking the internal temperature of a pork fillet with a bi-metallic stemmed thermometer is an example of which HACCP principle?

   A. Verification   C. Record Keeping
   B. Monitoring   D. Hazard Analysis

2. You place chicken salad from a buffet bar in an ice-water bath to cool after you find that its temperature is 55°F (13°C) instead of 41°F (5°C) or below. This is an example of

   A. monitoring.   C. a hazard analysis.
   B. corrective action.   D. verification.

3. Your restaurant plans to add linguini in a red clam sauce to the menu. The fresh clams will be cooked before being mixed with the pasta. Should receiving be a CCP for the clams?

   A. Yes, receiving is always a CCP for clams.
   B. No, clams do not pose a health hazard.
   C. No, because the clams will be cooked thoroughly.
   D. Yes, because the clams will be cooked thoroughly.

4. All of the following are methods of monitoring a CCP except

    A. taking the temperature of food during the cooking process.
    B. observing the amount of time food remains in the temperature danger zone.
    C. checking the pH of a food.
    D. analyzing a food's flow through the establishment.

5. Which of the following would not be a corrective action?

    A. Continuing to cook a hamburger that has not reached an internal temperature of 155°F (68°C)
    B. Throwing out potato salad that has remained at room temperature for longer than four hours
    C. Covering a cut with a bandage and finger cot
    D. Rejecting a delivery of fish received at an internal temperature of 60°F (16°C)

6. Which statement best describes the purpose of verification in a HACCP plan?

    A. To determine if the CCPs and critical limits that have been chosen are appropriate.
    B. To determine whether monitoring is alerting you to hazards.
    C. To determine if corrective actions are adequate to prevent foodborne illness.
    D. All of the above

7. Which of the following records would not be useful to your HACCP plan?

    A. Time and temperature logs
    B. Thermometer calibration records
    C. Corrective action logs
    D. Workplace accident records

8. Which of the following prerequisite programs should be included in your HACCP Plan?

    A. A personal hygiene program
    B. An incentive program
    C. Workplace accident prevention program
    D. None of the above

9. Which of the following situations could sabotage even the best HACCP plan at an establishment?

   A. Improper procedures for cooling food
   B. Faulty equipment that won't maintain temperatures
   C. Ineffective handwashing practices
   D. All of the above

10. The purpose of a HACCP system is to

    A. identify and control possible hazards throughout the flow of food.
    B. identify the proper methods for receiving foods.
    C. keep the establishment pest free.
    D. identify faulty equipment within the establishment.

11. Which of the following steps is likely to be a Critical Control Point for oysters that will be eaten raw?

    A. Receiving
    B. Storage
    C. Preparation
    D. All of the above

12. A chef took the temperature of a container of minestrone soup which was being held in a hot-holding unit for service. The temperature of the soup was 120°F (49°C), which he recorded in a temperature log. The chef reheated the soup to 165°F (74°C) for fifteen seconds and placed it in a different holding unit. Which of the following actions was a corrective action?

    A. Taking the temperature of the soup
    B. Reheating the soup
    C. Recording the temperature of the soup in the temperature log
    D. Placing the soup in a different holding unit

13. Which of the following statements is an example of a critical limit?

    A. Cook ground beef to 155°F (68°C).
    B. Store ground beef at 41°F (5°C).
    C. Discard ground beef if it remains at temperatures between 41°F and 140°F (5°C and 60°C) for more than four hours.
    D. All of the above

14. Your deli serves cold sandwiches in a grab-and-go display. Which step in the preparation of these sandwiches may be a CCP?

A. Storage
B. Cooking
C. Reheating
D. Cooling

15. Noting food temperatures in a temperature log is an example of which HACCP principle?

A. Verification
B. Monitoring
C. Record Keeping
D. Hazard Analysis

ADDITIONAL RESOURCES

Books and Periodicals

Bertagnoli, L. A clean start. (1996). *Restaurants & Institutions, 106*(22), 90.

Bolton, L. (1997). Food safety I.Q.: Don't wait for the health inspector! *Restaurant Hospitality, 81*(2), 85.

Cichy, R. F. (1993). *Sanitation management.* East Lansing, MI: Educational Institute of the American Hotel & Motel Association.

Durocher, J. (1996). A clean sweep. *Restaurant Business, 95*(4), 202.

Durocher, J. (1997). Germ warfare. *Restaurant Business, 96*(22), 123.

Equipped for food safety. (1998). *Best Practices, 2*(4), 6-10.

Frable, F., Jr. (1997). 10 common errors in kitchen planning and ways to avoid them. *Nation's Restaurant News, 31*(6), 22.

Hertneky, P.B. (1996). You and your health inspector. *Restaurant Hospitality, 80*(6), 57.

ICMSF (1988). *Microorganisms in foods 4: Application of the Hazard Analysis Critical Control Point (HACCP) system to ensure microbiological safety and quality.* Oxford: Blackwell Scientific Publications.

Kratt, D. T. (1997). Running a clean operation. *Nightclub & Bar, 13*(11), 79.

Marriott, N. G. (1994). *Principles of food sanitation.* New York: Chapman & Hall.

National Advisory Committee on Microbiological Criteria for Foods. (1997). *Hazard Analysis and Critical Control Point principles and application guidelines.* Available on-line at http://www-seafood.ucdavis.edu/Guidelines/nacmcf1.htm

National Restaurant Association Educational Foundation. (1998). *A Practical Approach to HACCP: Coursebook.* Chicago: Author.

National Restaurant Association. (1996). *Sanitation survival kit for restaurant operators.* Washington, DC: Author.

The right way to...organize workspaces. (1998). *Best Practices, 2*(3), 15-16.

Sanson, M. (1996). A blueprint for safety: Restaurant kitchen design. *Restaurant Hospitality, 80* (8), 65.

Secrets of self-inspection. (1998). *Best Practices, 2*(2), 6-9.

Stevenson, K. E. and Bernard, D. T. (Eds.). (1995). *Establishing Hazard Analysis Critical Control Point Programs: A workshop manual.* Washington, DC: The Food Processors Institute.

Vin Quetua, V. (1997). Scratching the surface. *Restaurants & Institutions, 107*(20), 196.

Walczak, D. (1997). The sanitation imperative: Keep people from getting sick in your restaurant. *Cornell Quarterly, 38*(2), 68.

Web Sites

1999 FDA Model Food Code

http://vm.cfsan.fda.gov/~dms/fc99-toc.html

Complete outline of the FDA's latest code for regulating operations that provide food directly to consumers. Also includes a quick synopsis of changes from the 1997 Food Code.

Archives of FDA Publications

http://www.fda.gov/opacom/archives.html

A complete database of FDA materials from the last six years, including press releases, speeches, and consumer publications.

Food and Drug Administration (FDA)

http://www.fda.gov

Web site for the Food and Drug Administration features links to the FDA's Center for Food Safety and Applied Nutrition.

FDA Center for Food Safety and Applied Nutrition (CFSAN)

http://vm.cfsan.fda.gov/list.html

Comprehensive site from CFSAN offers a wealth of food-safety information, from foodborne illness to food labeling. CFSAN strives to be a leader in food safety, and to protect consumers from economic fraud, promote sound nutrition, and encourage innovation.

FDA Seafood and Information Resources

http://vm.cfsan.fda.gov/seafood1.html

The Web site for FDA's Center for Food Safety and Applied Nutrition includes seafood information and resources, including seafood pathogens and contaminants.

Institute of Food Technologists (IFT)

http://www.ift.org

IFT is a nonprofit scientific society with 28,000 members working in food science, food technology, and related professions in industry, academia, and government. This site includes information on its policies and publications, education and industry news, and meeting and convention locations.

International Association of Milk, Food, and Environmental Sanitation (IAMFES)

http://www.iamfes.org

IAMFES keeps members informed of the latest scientific, technical, and practical developments in food safety and sanitation. Their Web site includes booklets and links to other food-safety sites.

International HACCP Alliance

http://ifse.tamu.edu/haccpall.html

The International HACCP Alliance was developed to provide a uniform program to assure safer meat and poultry products. The site offers information about HACCP implementation for the meat and poultry industry as well as a calendar on industry events and food safety.

National Advisory Committee for the Microbiological Criteria for Foods

http://www-seafood.ucdavis.edu/Guidelines/nacmcf.htm

This site provides a downloadable version of the 1997 document, Hazard Analysis and Critical Control Point Principles and Application Guidelines.

National Food Safety Database

http://www.foodsafety.org

A compilation of food-safety database information from government, consumer, and public health organizations, this is a one-stop Web site for food-safety information on the Internet.

National Restaurant Association

http://www.restaurant.org

The National Restaurant Association site provides information on government agencies affecting the restaurant industry, the latest training and certification updates, and links to state restaurant associations and hospitality schools and universities.

National Shellfish Sanitation Program Manual of Operation

http://vm.cfsan.fda.gov/~ear/nsspman.html

An outline of the FDA's Center for Food Safety and Applied Nutrition's National Shellfish Sanitation Program Manual of Operation, which ensures safe molluscan shellfish.

Sanitation Standard Operating Procedures Reference Guide

http://bbq.tamu.edu/USDA/ssop/ssop.html

This USDA site includes a reference guide, additional resources, and flow chart for sanitation compliance.

Seafood HACCP Alliance for Education and Training

http://www-seafood.ucdavis.edu/haccp/ha.htm

Information from the Seafood HACCP Alliance for Education and Training; includes updates and research.

United States Department of Agriculture (USDA)

http://www.usda.gov

The United States Department of Agriculture's Web site features information, publications, and other educational materials about the nation's agriculture.

United States Department of Agriculture Food Safety and Inspection Service

http://www.fsis.usda.gov

The USDA's Food Safety and Inspection Web site offers the latest food-safety news, educational materials, and HACCP implementation materials.

UNIT 3

CLEAN AND SANITARY FACILITIES AND EQUIPMENT

Jack in the Box instituted HACCP in its restaurants in 1993. The program consists of farm to fork procedures from microbial meat testing by our suppliers to in-restaurant grilling procedures for ensuring properly cooked hamburgers. Jack in the Box makes proactive efforts to join forces with state legislators, regulators and advocacy groups like the National Restaurant Association Educational Foundation to help ensure safe food for customers.

David Theno, Ph.D.
Vice President, Quality Assurance, Product Safety, Research & Development
Foodmaker, Inc.

Chapter 10
Sanitary Facilities and Equipment

Knowledge

TEST YOUR FOOD-SAFETY KNOWLEDGE

1. **True or False:** A hose attached to a sink faucet and left sitting in a bucket of dirty water could contaminate the water supply. *(See Cross-Connections, page 10-20.)*

2. **True or False:** It is alright to use nonpotable water for scrubbing pots and pans. *(See Water Supply, page 10-18.)*

3. **True or False:** Food-contact surfaces can be made from galvanized metal. *(See NSF International Standards, page 10-10.)*

4. **True or False:** Any clean sink may be used for scrubbing baking potatoes. *(See Sinks, page 10-9.)*

5. **True or False:** Grease on the ceiling of an establishment can be a sign of inadequate ventilation. *(See Ventilation, page 10-23.)*

Table of Contents

Learning Objectives

After completing this chapter, you should be able to:

○ Respond to an interruption in the internal water supply.

○ Respond to wastewater overflows.

○ Identify methods to prevent backflow problems.

○ Identify potable water sources.

○ Identify uses of nonpotable water.

○ Handle waste properly.

○ Clean and maintain restrooms properly.

○ Identify the requirements of a handwashing station.

○ Position equipment and facilities to make sanitation easier.

Key Terms

ADA
Porosity
Resiliency
Acrylic wood
Coving
Food-contact surface
Service sink
NSF International
UL (Underwriters Laboratories)

Blast chiller
Tumble chiller
Cantilever mounted
Potable water
Single-use item
Booster heater
Cross-connection
Flood rim
Backflow
Vacuum breaker

Air gap
Foot-candle
Environmental Protection Agency
Solid waste
Garbage
Pulper

Equipment must be designed and installed in the establishment so that the building and the equipment can be cleaned easily. Many breakdowns in sanitation are caused by facilities and equipment that are simply too hard to keep clean. Sanitary facilities and equipment are basic parts of a well-designed HACCP program.

DESIGNING A SANITARY ESTABLISHMENT

When designing a sanitary facility you must consider how every area of it will be kept clean. The building and all equipment must be easy for employees to clean. Areas and equipment that are not properly cleaned allow bacteria, viruses, and molds to remain. These can become serious problems when food comes into contact with them. Facilities should be arranged so contact with contaminated sources such as garbage or dirty tableware, utensils, and equipment is unlikely to occur.

This section focuses on four topics related to the sanitary layout and design of equipment and facilities.

○ Materials for walls, floors and ceilings that will make cleaning these surfaces easier.

○ Arrangement and design of equipment and fixtures to comply with sanitation standards.

○ Design of utilities to prevent contamination and to make cleaning easier.

○ Proper solid waste management to avoid contaminating food and attracting pests.

The Plan Review

A sanitary foodservice layout and design should begin in the planning stage of the facility. Prior to starting construction, a manager should consult local regulations. Many jurisdictions require approval of layout and design plans by the health department or local regulatory agency prior to new construction or extensive remodeling. These plans should include the proposed layout, the mechanical plans (and specifications for utilities), construction materials, and the types or models of proposed equipment. Specifications for utilities, plumbing, and ventilation will probably be required.

Some local jurisdictions will require the approval of design plans by building and zoning departments. In addition, the **ADA (Americans with Disabilities Act)** requires reasonable accommodation for access to the building by both patrons and employees with disabilities. These guidelines may be found in the ADA accessibility guide (ADAAG).

Even if local laws do not require it, a manager should have layout and design plans reviewed by local or state regulatory agencies. In addition to ensuring compliance with sanitation requirements, such reviews can save time and money. This is true for remodeling as well as for new construction.

To assist establishments, regulatory agencies may provide checklists of features they consider necessary for good sanitation. Managers should also ask for guides on how to submit plans and specifications.

Once construction is completed, the establishment applies for a permit to operate. Before granting the permit, the regulatory agency may conduct a pre-opening inspection to make certain that all design and installation requirements have been met. Once the establishment has passed the inspection and obtains a certificate of operation, it may open for business.

Materials for Interior Construction

Materials used in the construction of a facility must be selected with several factors in mind. Materials chosen should create an attractive facility, contribute to workplace safety, and be reasonably priced. A mixture of floor, wall, and ceiling surfaces that are sound absorbent, that resist absorption of grease and moisture, and have light-reflective surfaces will probably create an environment that is acceptable to your regulatory agency.

However, the most important consideration when selecting construction materials is how easy the establishment will be to clean and maintain.

There are advantages and disadvantages of different construction materials for floors, walls, and ceilings when constructing a new facility or simply remodeling an existing one.

Flooring

Flooring materials in the kitchen and service areas should meet requirements for health and safety, strength and durability, and appearance. The floor surfaces should be easy to maintain, wear resistant, slip resistant, and nonporous. Flooring should be kept in good repair and replaced if damaged or worn. When selecting flooring materials for your facility, choose for ease of cleaning and resistance to damage or wear.

One of the most important factors to consider when selecting floor covering for an establishment is the **porosity** of the material. Porosity is the extent to which a floor covering can become saturated by liquids. When liquids are absorbed, the flooring can be damaged, and microorganisms and mold may grow. It may also cause a potential slip-and-fall situation. The

Key Point

Even if local laws do not require it, a manager should have plans reviewed by the local regulatory agency.

Key Point

Select nonabsorbent flooring materials for food preparation areas, walk-in refrigerators, warewashing areas, and restrooms.

Food Code recommends the use of nonabsorbent flooring in food preparation areas, walk-in refrigerators, warewashing areas, restrooms, and other areas that are subject to moisture, flushing, or spray cleaning.

Nonporous Resilient Flooring

In most areas of the establishment, nonporous resilient flooring is the best choice. **Resiliency** means a material has the ability to react to a shock without breaking or cracking.

Nonporous resilient materials are relatively inexpensive and are easy to clean and maintain. They are rated for light, moderate, and heavy traffic and resistance to grease and alkalis. Some are easily damaged by cigarette burns or sharp objects, but they are also easy to repair or replace. They tend to be slippery when wet. Types of resilient materials to consider are rubber tile and light and medium weight vinyl. Vinyl tile is not recommended in dining rooms or public areas because it requires a high level of maintenance, such as waxing and frequent machine buffing. Vinyl tile is a suggested choice in employee dressing rooms, employee dining rooms, or foodservice offices. *See Exhibit 10a* for general physical characteristics and recommended uses of nonporous resilient flooring.

Hard-Surface Flooring

The second major type of flooring commonly used in an establishment is hard-surface flooring and includes quarry tile, ceramic tile, brick, terrazzo, marble, and hardwood. These materials are nonporous, but are not resilient. Quarry and ceramic tile are excellent for use in public restrooms or high-soil areas, but unglazed tiles should be selected for these areas due to their slip-resistant qualities.

Exhibit 10a **Characteristics of Resilient Flooring**

Materials	Where to Use	Durability	Advantages	Disadvantages
Rubber Tile	Kitchens; restrooms	Less durable or resistant to grease or alkalis	Anti-slip; resilient	Use in moderate traffic areas
Vinyl Sheet	Offices; back of the house; corridors	Less resistant to grease and alkalis	Very resilient	Use only in light or moderate traffic areas
Vinyl Tile	Offices; employee restrooms	Wears out quickly with high traffic	Very resilient	Requires waxing and machine buffing

Exhibit 10b — Hard-Surface Flooring

Material	Where to Use	Durability	Advantages	Disadvantages
Marble; Terrazzo	Public corridors; back of the house; dining rooms; public restrooms	Wear resistant	Non-porous; good appearance	Non-resilient; expensive; requires special care such as buffing and polishing; heavy and difficult to install
Quarry Tile	Kitchen; dishwashing areas; receiving areas; offices; restrooms; dining rooms; service areas	Wear resistant	Non-porous	Non-resilient; heavy; relatively expensive; slippery when wet unless an abrasive is added
Wood	Offices; dining rooms	Durable in lower traffic areas	Good appearance and sound absorption	Requires frequent polishing and periodic refinishing to maintain surface qualities
Acrylic Wood (plastic absorbed into wood)	Offices; dining rooms	High abrasion resistance	Less vulnerable to stains, scratches, chemical damage; high resistance to bacterial growth	Much like resilient flooring

Hard-surface floors are very durable but may crack or chip if heavy objects are dropped on them. In addition, breakable objects dropped on hard surface flooring will be more likely to shatter than those dropped on resilient flooring. Hard-surface flooring does not absorb sound and is somewhat difficult to clean compared to resilient surfaces. Some types, such as marble, are slippery. In general, these materials are very heavy and more expensive to install and maintain than resilient flooring. *See Exhibit 10b* for general physical characteristics and recommended uses of hard-surface flooring.

Carpeting in dining rooms is popular because it absorbs sound but is not recommended in high-soil areas such as waitstaff service areas, tray and dish drop-off areas, beverage stations, and major traffic aisles. Carpets can be maintained by simple vacuum cleaning. Areas prone to heavy traffic and moisture will require routine cleaning. Where sanitation, soiling, moisture, and fire safety are concerns, special carpet can be purchased.

Special Flooring Needs

Each area of an establishment has its own particular flooring needs. Non-slip surfaces should be used in traffic areas. In fact, non-slip surfaces are best used for the entire kitchen because slips and falls are very common accidents for foodservice employees. Rubber mats are allowed for safety reasons in areas where standing water may occur, such as the dish room. Rubber mats should be picked up and cleaned separately when the floors are scrubbed.

Coving is required in all new or remodeled establishments that use resilient or hard-surface flooring materials. **Coving** is a curved, sealed edge placed between the floor and the wall to eliminate sharp corners or gaps that would be impossible to clean *(see Exhibit 10c)*. The coving tile or strip must adhere tightly to the wall to eliminate hiding places for insects and to prevent moisture from deteriorating the wall.

Finishes for Interior Walls and Ceilings

Interior finishes are the materials used on the surface of walls, partitions, or ceilings of the establishment. As with flooring, the most important criteria when choosing interior finishes are ease of cleaning and porosity.

When selecting finishes for walls and ceilings, consider the location. One type of material may be suitable in one area but may be a poor choice for another area. Walls and ceilings in food preparation areas must be light in color to distribute light and make soil more visible for cleaning. They should be kept in good repair, without cracks, holes, or peeling paint. The best wall finish in cooking areas is ceramic tile; however, it must be monitored for grout loss and re-grouted whenever necessary. Stainless steel is occasionally used because of its durability and resistance to moisture. The most common ceiling materials are acoustic tile, painted drywall, painted plaster, or exposed concrete.

The support structures for walls and ceilings (studs, joists, and rafters) and pipes should not be exposed unless they are finished and sealed for cleaning. Flexible materials such as papers, vinyls, and thin wood veneers are often used for walls and ceilings. Vinyl wall coverings such as FRP are used in many areas of establishments because these materials are attractive, relatively inexpensive, easy to clean, and durable. Vinyl wall coverings are rated for flammability by testing agencies. Plaster or cinder-block walls that are sealed and painted with soil-resistant and easy-to-wash glossy paints are appropriate for dry areas of the facility.

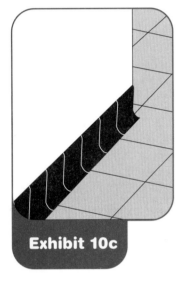

Exhibit 10c

Coving
Coving is a curved, sealed edge placed between the floor and the wall to eliminate sharp corners or gaps.

CONSIDERATIONS FOR OTHER AREAS OF THE FACILITY

Dry Storage

Dry storerooms should be constructed of easy-to-clean materials that allow good air circulation *(see Exhibit 10d)*. Shelving, table tops, and bins for dry ingredients should be made of corrosion-resistant metals or food-grade plastics.

Any windows in the storeroom should have frosted glass or shades. Direct sunlight can increase the temperature of the room and affect food quality.

Steam pipes, heating or ventilation ducts, water lines, and other conduits have no place in a well-designed storeroom. Dripping condensation or leaks in overhead pipes can promote microbial growth in such normally stable items as crackers, flour, and baking powder. Leaking overhead sewer lines are a highly dangerous source of contamination for any food. Hot water heaters or steam pipes can increase the temperature of the storeroom to harmful levels.

Dry foods are especially susceptible to attack by insects and rodents. Cracks and crevices in the floor or walls should be filled. Doorways to the outside should be fitted with solid or screened self-closing doors to keep out flying insects. Screens for windows and doors should be sixteen mesh to the inch without holes or tears.

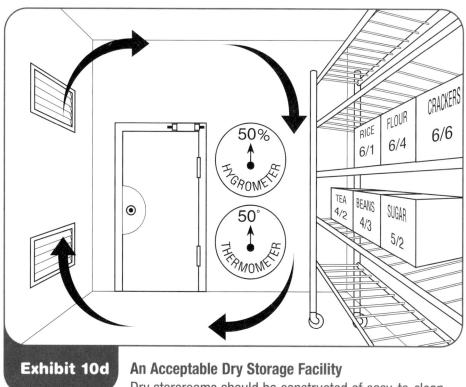

Exhibit 10d **An Acceptable Dry Storage Facility**
Dry storerooms should be constructed of easy-to-clean materials that allow good air circulation.

Restrooms and Handwashing Stations

Local building and health codes usually specify how many sinks, stalls, toilets, and urinals are required in an establishment to meet the needs of both customers and employees. It is best if separate restrooms are provided for employees and customers. If this is not possible, the establishment must be designed so that patrons do not pass through food-preparation areas to reach the restroom, since they could contaminate food or **food-contact surfaces.**

Restrooms should be convenient and sanitary and have a fully equipped handwashing station and self-closing doors. They must be adequately stocked with toilet paper, and trash receptacles must be provided if disposable paper towels are used. Covered waste containers must be provided in women's restrooms for the disposal of sanitary supplies.

Restrooms must be cleaned and inspected on a regular basis, ideally once a day. Cleaning should proceed from top to bottom, starting with the mirrors, followed by walls, sinks, toilets, urinals, and finally the floor.

Handwashing Stations

Handwashing stations must be conveniently located so that employees will be encouraged to wash their hands often. These stations must be operable and must be stocked and maintained. They are required in food preparation areas, service areas, equipment-washing areas, and restrooms. A handwashing station must be equipped with the following items *(see Exhibit 10e):*

○ **Hot and cold running water.** A water temperature of at least 110°F (43°C), supplied through a mixing valve or combination faucet, is required.

○ **Soap.** The soap may be liquid, bar, or powder. Liquid soap is generally preferred, and some local codes require liquid soap.

○ **A means to dry hands.** Most local codes require establishments to supply disposable paper towels in handwashing stations. Continuous-cloth towel systems, if allowed, should be used only if the unit is working properly and the towel rolls are checked and changed regularly. Installing at least one hot-air dryer in a handwashing station may provide an alternate method for drying hands if paper or cloth towels run out. The use of common cloth towels is not permitted because they can transmit contamination from one person's hands to another.

○ **A waste container.** Waste containers are required if disposable paper towels are provided.

Key Point

Handwashing stations must be conveniently located in food preparation, service, equipment washing, and restroom areas.

Sinks

Each sink in an establishment must be used for its intended purpose only. Handwashing sinks are used for handwashing. Food preparation sinks are used for food preparation. Warewashing sinks are used for washing equipment and utensils. **Service sinks** that are used for cleaning mops and disposing of waste water must be kept separate. At least one service sink or one curbed drain area to dispose of soiled water is required in an establishment.

Dressing Rooms and Lockers

Dressing rooms are not required. If they are available, they must be used only as dressing rooms and may not be used for food preparation, storage, or utensil washing. Lockers, if available, should be adequate to contain all employee belongings. They should be located in a separate room or a room where contamination of food, equipment, utensils, linens, and single-service articles will not occur.

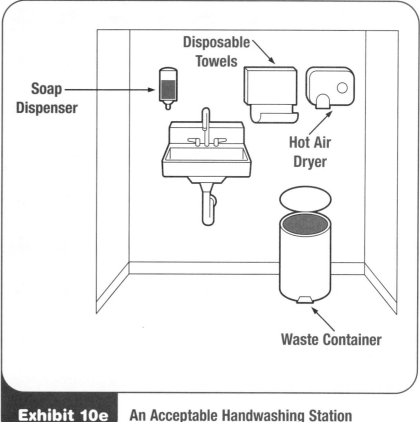

Exhibit 10e **An Acceptable Handwashing Station**
A handwashing station must be equipped with hot and cold running water, soap, a means to dry hands and a waste container (if disposable towels are used).

Premises

Walkways and the parking lot should be kept free of litter and graded so standing pools of water do not form. In addition, they must be surfaced so that dirt and dust blowing inside the building is minimized. It is recommended that concrete and asphalt be used for walkways and parking lots. Gravel, while acceptable, is not recommended.

Patron traffic through the food preparation area is prohibited, although guided tours are allowed. The premises may not be used for living or sleeping quarters.

SANITATION STANDARDS FOR EQUIPMENT

The task of choosing equipment designed for sanitation has been simplified by organizations such as **NSF International,** formerly the National Sanitation Foundation. NSF International develops and publishes standards for sanitary equipment design. **Underwriters Laboratories (UL)** also provides sanitation classification listings for equipment found in compliance with NSF International standards. Foodservice managers should look for the NSF International mark or the UL Sanitation Classification mark on commercial foodservice equipment. Exhibit 10f shows examples of the NSF International and UL Sanitation Classification marks. The regular UL listing mark indicates compliance of equipment to UL safety standards *(see Exhibit 10g).*

NSF International Standards

It is the responsibility of the manufacturer to know safety standards for specific equipment. However, the foodservice manager should know the general features on which NSF International bases its standards.

○ **Equipment must be easy to clean.** All food-contact areas must be easy to access to ensure thorough cleaning.

○ **All food-contact surfaces must be smooth, nontoxic, nonabsorbent, corrosion resistant, and stable and must not cause changes in the color, odor, or taste of the food.** Food-contact surfaces should not react in any way with food products or cleaning compounds. Examples of materials that may not be used for food-contact surfaces include galvanized steel, lead, and copper.

○ **Internal corners and edges exposed to food must be rounded off (coved).** Solder and caulking are not acceptable rounding materials. External corners and angles are to be sealed and finished smooth.

○ **All food-contact surfaces must be smooth and free of pits, crevices, ledges, inside threads and shoulders, bolts, and rivet heads.** Surfaces must be easy to clean and corrosion resistant.

○ **Coating materials must be nontoxic and cleanable and must resist cracking and chipping.** Coating materials must not be used in food-contact areas.

○ **Equipment must be easy to disassemble to encourage frequent, thorough cleaning.** Only commercial foodservice equipment should be used in establishments. Household equipment is not built to withstand the heavy use found in establishments and is not subject to NSF International standards.

Although all equipment used in an establishment must meet standards such as those set by NSF International, certain equipment requires particular attention.

Warewashing Machines

Warewashing machines vary widely by size, style, and method of sanitizing. Hot water machines sanitize with extremely hot water; chemical-sanitizing machines use a chemical solution. The size and type of machine you choose depends upon the nature and volume of the items to be cleaned and upon the required turnaround time for clean tableware and utensils.

Because a warewashing machine is a big investment, the manager needs to carefully match the machine to the needs of the establishment. Chemical-sanitizing machines, for example, usually require more drainboard space to air dry utensils and also require that sanitizing solutions be monitored (volume, concentration). Batch-type dump models drain wash and rinse water after each cycle. Heated multiple-tank models reuse wash and rinse water and sanitizer solution.

Consider the following general guidelines regarding installation and use of warewashing machines:

○ Water pipes to the warewashing machine should be as short as possible to prevent the loss of heat from the water entering the machine.

○ The machine must be raised at least six inches off the floor to permit easy cleaning underneath.

○ Materials used in warewashing machines should be able to withstand wear, including the action of detergents and sanitizers.

○ Information should be posted on or near the machine regarding proper water temperature, conveyer speed, water pressure, and chemical concentration.

○ The machine's thermometer should be located so that it is readable. The thermometer should have a scale in increments no greater than 2°F or 1°C.

○ Chemical-sanitizing machines must be equipped with a device which indicates audibly or visually when more chemical sanitizer needs to be added.

In general, there are two types of warewashing machines, the high-temperature machine and the chemical-sanitizing machine.

Warewashing machines include the following models:

○ **Single-tank, stationary-rack, with doors.** This machine holds a stationary rack of tableware and utensils. Items are washed by detergent and water from below and sometimes from above the rack. The wash cycle is followed by a hot-water final rinse or chemical sanitizer.

○ **Conveyor machine.** With this machine, a moving conveyor moves racks of items through the various cycles of washing, rinsing, and sanitizing. The machine may have a single tank or multiple tanks.

○ **Carousel or circular conveyor machine.** This multiple-tank machine moves tableware and utensils on a peg-type conveyor or in racks. Some models have an automatic stop after items go through the final rinse cycle. In other models, items must be removed after the final rinse, or they will continue to travel through the machine.

○ **Flight-type.** This is a high-capacity, multiple-tank machine with a peg-type conveyor. It may also have a built-in dryer. It is commonly used in institutions and very large establishments.

○ **Batch-type, dump.** This stationary-rack machine combines the wash and rinse cycle in a single tank. Each cycle is timed, and the machine automatically dispenses both the detergent and the sanitizing chemical or hot water. Wash and rinse water are drained after each cycle.

○ **Recirculating, door-type, non-dump.** This stationary-rack machine is not completely drained of water between cycles. The water is diluted with fresh water and reused from cycle to cycle.

○ **Conveyor.** A moving conveyor moves racks of tableware and utensils through the various cycles of washing, rinsing, and sanitizing. Some models include a power pre-rinse.

Some states require the approval of the local regulatory agency to install a chemical warewashing system.

Clean-in-Place Equipment

Some equipment, such as certain automatic ice-making machines and soft-serve ice cream and frozen yogurt dispensers, is designed to be cleaned by having a detergent solution, hot-water rinse, and sanitizing solution passed through it. These machines must be constructed so that the cleaning and chemical-sanitizing solution remains within the fixed system of tubes and pipes for a predetermined amount of time. All food-contact surfaces must be reached by the solutions, but they must not leak into the rest of the machine.

Clean-in-place equipment must be self-draining. Some means of inspection must be provided to make sure that the machine has been completely emptied of cleaning solution and has been thoroughly rinsed with fresh water. Manufacturers' instructions should be carefully followed.

Refrigerators and Freezers

There are several types of foodservice refrigerator and freezer units. These include under-the-counter units, open units, and display cases. The two most common types are walk-in and reach-in refrigerators and freezers. *(See Exhibit 10h and Exhibit 10i.)* These units should be made of stainless steel or a combination of stainless steel and aluminum. Some may have plastic or galvanized interior liners. The doors should be constructed to withstand heavy use and should close with a slight nudge. The door gaskets may be fixed, or removable for easy cleaning. A drain must be provided and maintained for disposal of condensation and defrost water. A properly plumbed, indirect drain may be used in the walk-in refrigerator. Excess condensation can be minimized by maintaining a flush-fitting floor sweep (gasket) under the door.

Walk-in units must include an inside safety release to prevent employees from locking themselves in. Walk-in refrigerators and freezers that have windows in the door may reduce unnecessary opening. Forced-air circulating fans are essential to help provide a quick recovery time so refrigerator and freezer temperatures remain at the appropriate level.

When purchasing a refrigerator or freezer, make sure that it is NSF International certified or UL classified for sanitation (or the equivalent). This will ensure that the units are properly designed to protect food and simplify cleaning. In addition to the NSF International standards mentioned earlier in this chapter, consider these factors when purchasing a refrigeration or freezer unit.

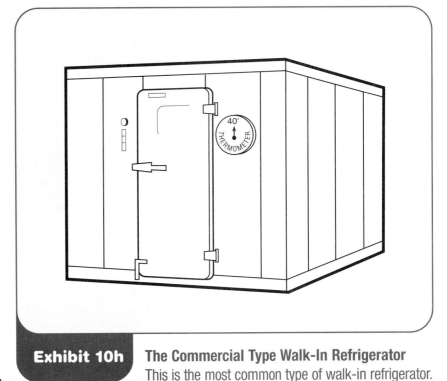

Exhibit 10h **The Commercial Type Walk-In Refrigerator**
This is the most common type of walk-in refrigerator.

Courtesy of Hobart Corporation

Exhibit 10i **The Commercial Type Reach-In Freezer**
Doors of a freezer should be constructed to withstand heavy use.

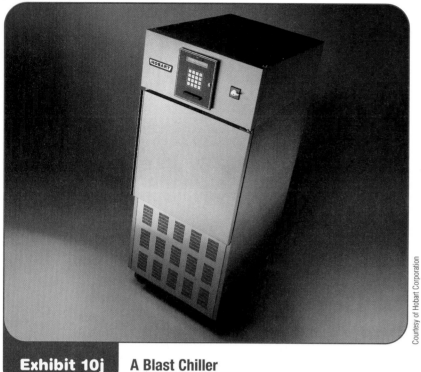

Exhibit 10j **A Blast Chiller**
Many blast chillers can cool foods from 140°F
to 37°F (60°C to 3°C) within ninety minutes.

Exhibit 10k **A Tumble Chiller Unit**
A tumble chiller unit chills prepackaged food by
tumbling it in chilled water.

○ **Sufficient storage space is important.**
Adequate space helps maintain
required holding temperatures. An
uncrowded unit allows for frequent
cleaning, prevents moisture buildup,
and minimizes breakdowns from
overloading.

○ **Reach-in refrigerators and freezer
units should be elevated six inches
off the floor on legs or be mounted
and sealed on a masonry base.**
Casters that make it easy to move the
unit for cleaning are often preferred or
required by local regulatory agencies.

○ **Walk-in units should be sealed to
the floor and wall.** They should
offer no access to moisture or rodents.
Floor materials must be able to
withstand heavy impact.

○ **Refrigerators and freezers must
meet temperature requirements for
specific foods.** Built-in thermometers
should be easy to locate and read,
and be accurate to within 3°F (2°C).

Blast Chillers and Tumble Chillers

Blast chillers are designed to
move food through the temperature
danger zone quickly *(see Exhibit 10j)*.
Many blast chillers are able to cool
food from 140°F to 37°F (60°C to
3°C) within ninety minutes. Most
units allow the operator to set target
chill temperatures and monitor the
temperature of food throughout the
chill cycle. Once chilled to safe
temperatures, the food can then be
stored in conventional refrigerators
or freezers.

Tumble chillers are also designed to cool food quickly *(see Exhibit 10k).* Prepackaged hot food is placed into a drum, which rotates inside a reservoir of chilled water. The tumbling action increases the effectiveness of the chilled water in cooling the food. Some tumble chillers are capable of cooling 160 gallons of food in forty-five minutes.

Cook-Chill Equipment

Some operations prepare food using a cook-chill system. By this method, foods are partially cooked, rapidly chilled, and then held in refrigerated storage. When needed, these foods are simply reheated. A cook-chill unit is an integrated piece of equipment, which is capable of cooking, cooling, and reheating the food.

Cutting Boards

Cutting boards are an important part of the establishment. Many jurisdictions allow the use of either wooden or synthetic boards; however, some experts prefer synthetic boards, which can be cleaned and sanitized in a warewashing machine or by immersion in a compartment sink.

If wooden cutting boards and baker's tables are allowed by local codes, they must be made from a nonabsorbent hardwood, such as maple or oak. They must also be free of seams and cracks, be nontoxic, and must not transfer any odor or taste to the food.

Separate cutting boards should be used for raw and cooked foods in order to prevent cross-contamination. Cutting boards must be washed, rinsed, and sanitized between every use. Due to the high risk of cross-contamination, directions for cleaning and sanitizing cutting boards should be included in your standard operating procedures.

CHOOSING AND INSTALLING KITCHEN EQUIPMENT

Well-designed kitchens make the job of keeping food safe easier. Generally, an efficient kitchen design is a more sanitary kitchen design.

Layout

A well-designed kitchen will address the following factors.

○ **The work flow.** A work flow must be established that will minimize the amount of time that food spends in the temperature danger zone. It must also minimize the number of times that food is handled *(see Exhibit 10l).* For example, storage areas should be located near the receiving area to avoid delays in storing perishable food. Prep tables should be located near refrigerators and freezers for the same reason.

Synthetic cutting boards are generally preferred because they can be cleaned and sanitized in a warewashing machine.

Separate cutting boards should be used for raw and cooked foods.

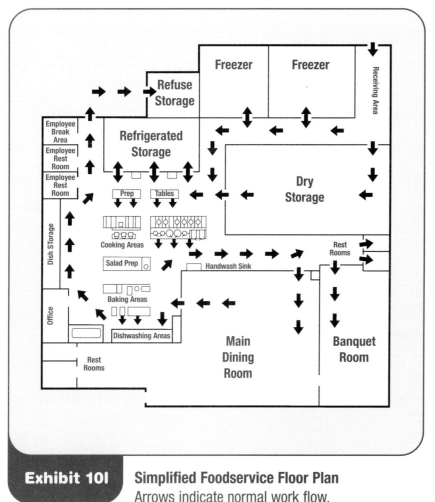

Exhibit 10l **Simplified Foodservice Floor Plan**
Arrows indicate normal work flow.

Key Point

The primary criteria for installing equipment are ease of cleaning and elimination of hiding places for pests.

○ **Contamination.** A good layout will minimize the risk of cross-contamination. Dirty equipment should not be placed where it will touch food or clean equipment. For example, it is not a good practice to place the soiled-utensil table next to the salad preparation sink.

○ **Equipment accessibility.** A well-planned layout will ensure that equipment is accessible for cleaning. Hard-to-reach areas are less likely to be cleaned than those with easy access.

Equipment

Portable equipment is equipment that can be carried or rolled on casters by one person. Equipment that can be moved allows for better cleaning of the equipment and of the surrounding walls and floors.

Equipment that cannot be moved must be installed so that the equipment and the surrounding area can be easily cleaned. It must be either mounted on legs at least six inches off the floor or sealed to a masonry base *(see Exhibit 10m)*. If sealed to the floor, a toe space of one to four inches should be allowed. The recommended distance from the wall, or between pieces of equipment, depends on the size of the equipment and the amount of surface to be cleaned. Manufacturer's directions should be used for exact specifications.

Table-mounted equipment can also pose cleaning problems if the equipment cannot be moved easily by one person. Immobile table-mounted equipment should be mounted on legs that will provide a minimum clearance of four inches between the base of the equipment and countertop, to allow for cleaning. Alternatively, such equipment should be tiltable or should be sealed to the countertop with a non-toxic, food-grade sealant.

When equipment is sealed to the floor, wall, or counter, it is important that the sealant not be used to cover wide gaps caused by faulty construction or repairs. Any opening, crack, or seam greater than ⅓₂″ (1mm) must be filled.

Cracks or seams should be sealed with a non-toxic sealant. Renovation should correct any gaps by closing them entirely.

Cantilever-Mounted Equipment

Some pieces of equipment may be attached to the wall with a bracket—**cantilever mounted**—to allow for easier cleaning behind and underneath it *(see Exhibit 10n)*.

UTILITIES

Establishments could not operate without water and plumbing, electricity and gas, lighting, ventilation, sewage, and waste handling. Sanitary design of these utilities and services is important. Two basic goals must be met in their design. There must be enough utilities to meet the cleaning needs of the establishment. The utilities themselves must not contribute to contamination.

Water Supply

Safe water is vital in an establishment. Safe water is used as a beverage, in ice, as an ingredient in food, and for handwashing, restrooms, cleaning, laundry, and showers. Unsafe water can carry bacteria, viruses, and parasites.

Water that is safe to drink is called **potable water.** Potable sources of water include approved public water mains, private water sources that are regularly maintained and tested, and bottled drinking water. Other sources include closed portable water containers filled with potable water, on-premise water storage tanks, and water transport vehicles that are properly maintained.

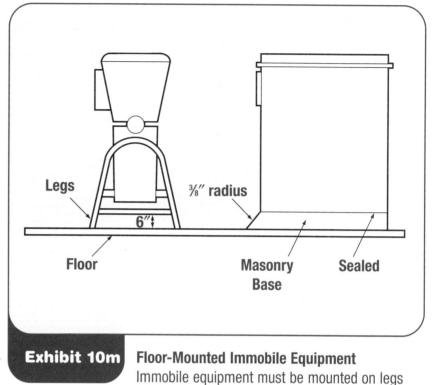

Exhibit 10m **Floor-Mounted Immobile Equipment**
Immobile equipment must be mounted on legs at least six inches off the floor or sealed to a masonry base.

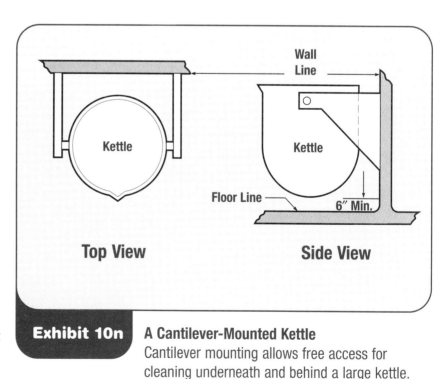

Exhibit 10n **A Cantilever-Mounted Kettle**
Cantilever mounting allows free access for cleaning underneath and behind a large kettle.

If an establishment uses a private water supply, such as a well, rather than an approved public source, the establishment should check with its regulatory agency for information on inspections and testing and other requirements. Generally, nonpublic water systems should be tested at least annually and the report kept on file in the establishment.

The use of nonpotable water is extremely limited. If nonpotable water is allowed by local codes, the uses are generally limited to air conditioning, cooling equipment not used for cooling food, fire protection, and irrigation (for outdoor grass and plants).

Water Emergencies

Occasionally an emergency occurs which causes the water supply to become unusable. Some examples would include natural disasters (hurricanes, floods, tornadoes), problems with the city water supply, and local problems with broken pipes. If there is a problem with the potable water supply, the establishment may have to be closed or an alternate source of water may need to be found.

Establishments may not want to close because they do not want to lose business. Some establishments that serve institutions such as hospitals or schools may want to stay open in an emergency. In these cases, the regulatory agency may allow the establishment to continue to function if certain precautions are followed. The following discussion provides some examples of things that an establishment must consider, in order to continue to serve food safely during a water emergency.

Water Used as a Beverage or Ingredient

If potable water must be obtained from an alternate source for use as a beverage or ingredient, there are several options. These include buying bottled water, or boiling water (the local regulatory authority should be contacted to determine the proper length of time for boiling). You can also store a supply of potable water in advance of an upcoming problem, such as before water is scheduled to be turned off for repairs, or when a flood or hurricane is predicted. When storing water, use closed food-grade water containers or on-site or enclosed vehicular water tanks.

Ice

Most ice machines make new ice and drop it onto previously made ice. Ice made from contaminated water will therefore contaminate any previously made ice already in the bin. If it is known in advance that the water supply may become unsafe, previously made ice can be stored before the emergency occurs. This might be possible for some predicted natural disasters such as floods or hurricanes. In an unexpected water emergency, ice can be bought or water can be boiled and used to make ice. If purchased, ice must be contained in single-use food-grade plastic bags or wet-strength paper bags filled and sealed at the point of manufacture.

Cleaning

Water should be boiled for use when doing essential cleaning tasks such as cleaning pots and pans. Establishments may consider using disposable or **single-use items** to eliminate the need for using a warewasher, which requires large amounts of water. If a water emergency occurs, non-essential cleaning in the establishment should be minimized.

Handwashing

Warm, potable water is required for handwashing. Boiled water can be placed in large plastic containers which dispense water through a spout. Replenish the supply often to keep it warm.

Restrooms, Showers, and Laundry

Restrooms, showers, and laundries require large amounts of water. Managing them would be difficult without a potable water system. Portable toilets may be used, but it may not be possible to supply water to showers and laundry facilities. The local regulatory agency should be contacted to determine possible alternatives. Some codes may allow the use of nonpotable water for flushing toilets and operating washing machines.

If there is any question of safety regarding a particular practice, the local regulatory agency should be contacted. Any time there has been an emergency situation, such as a flood, or after the repair of broken plumbing, water systems and equipment that use water (e.g. ice machines) should be flushed and disinfected before resuming use.

Hot Water

Providing a continuous supply of hot water can be a problem for many establishments serving the public. Water heaters should be evaluated regularly to make sure that they can meet demand. Consider how quickly the heater produces hot water; the size of the holding tanks; and the location of the heater in relation to the sinks or warewashing machine.

Since most general-purpose water heaters will not heat water to temperatures required for hot-water sanitizing, a **booster heater** may be needed to maintain a water temperature of 180°F (82°C) for heat-sanitizing tableware and utensils. Many warewashing machines now come with booster heaters inside the machine itself.

Plumbing

In almost every community in the United States, plumbing design is regulated by law, and with good reason. Improper plumbing design can cause serious health concerns. Improperly installed or poorly maintained plumbing, which allows the mixing of potable and nonpotable water, has been implicated in outbreaks of typhoid fever, dysentery, Hepatitis A, Norwalk virus, and other gastrointestinal illnesses. Improperly installed water pipes can also lead to contamination from metals, such as copper poisoning from beverage dispensers or contamination from chemicals used in the system, such as detergents, sanitizers, or drain cleaners.

Local plumbing codes vary widely regarding the types of connections permitted and what kinds of protection are required. If in doubt about the plumbing regulations for an establishment, contact the local regulatory agency. Only licensed plumbers should install and maintain plumbing systems in an establishment.

Cross-Connections

The greatest challenge to water safety comes from **cross-connections.** A cross-connection is a physical link through which contaminants from drains, sewers, or waste water can enter a potable water supply. A faucet located below the **flood rim** of a sink, or a hose in a mop bucket, are examples of a cross-connection.

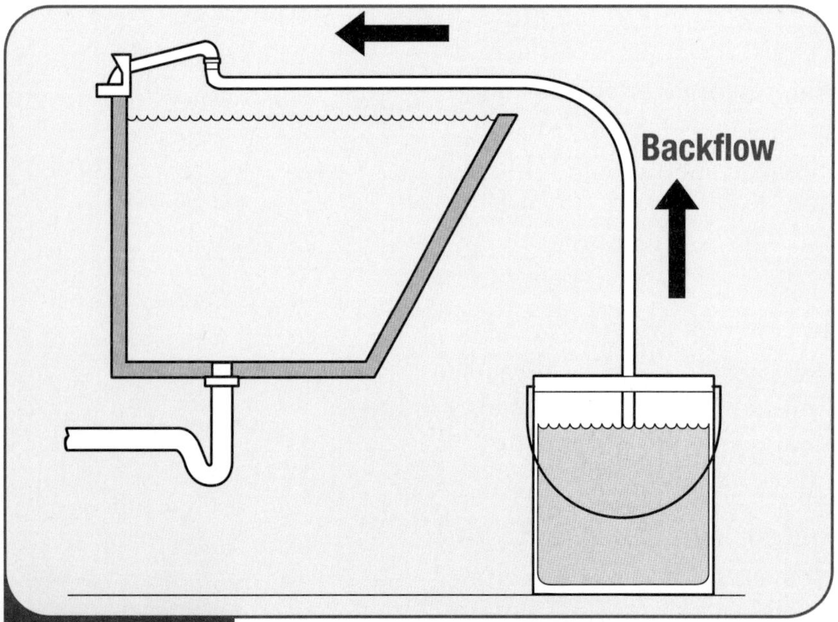

Cross-connection is dangerous because it allows the possibility of **backflow.** Backflow is the unwanted reverse flow of contaminants through a cross-connection into a potable water system. It can occur whenever the pressure in the potable water supply drops below the pressure of the contaminated supply.

In *Exhibit 10o,* an employee has attached a hose to the faucet of a utility sink to add hot water to a partially filled mop bucket. He leaves the nozzle of the hose submerged in the bucket of dirty water, accidentally creating a cross-connection. Because of heavy water usage somewhere else in the facility, the water pressure could drop low enough that contaminated water from the mop

Backflow

Exhibit 10o A Common Cross-Connection
A hose connected to a faucet and left submerged in a mop bucket creates a dangerous cross-connection.

bucket would be drawn back through the hose and into the potable water supply.

To prevent cross-connections like this, do not attach a hose to a faucet unless a backflow prevention device, such as a **vacuum breaker,** is attached. Threaded faucets and connections between two piping systems must have a vacuum breaker or other approved backflow prevention device *(see Exhibit 10p).*

The only completely reliable device to prevent backflow is the **air gap.** An air gap is an air space that is used to separate a water supply outlet from any potentially contaminated source. A sink may make use of air gaps to prevent backflow. The air space between the faucet and the flood rim of the sink is one air gap. Another may be located between the drain pipe of the sink and the floor drain of the establishment *(see Exhibit 10q).* Typically, the size of the air gap should be twice the diameter of the water supply outlet. For example, if a faucet has an opening with a diameter of one inch, the air gap between the faucet and the flood rim of the sink must be at least two inches.

Grease Condensation and Leaking Pipes

Grease condensation in pipes is another common problem in plumbing systems. Grease traps are often installed to prevent a buildup of grease from creating a drain blockage. The trap must be cleaned and the grease removed periodically. If this is not done, or is not done properly, an overflow could lead to problems with odor and contamination.

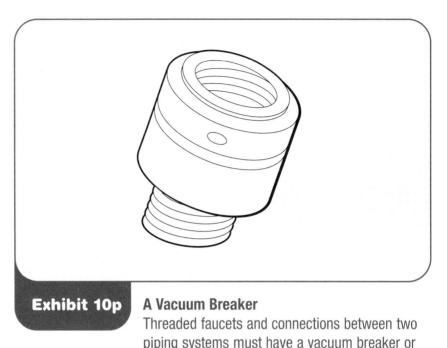

Exhibit 10p   **A Vacuum Breaker**
Threaded faucets and connections between two piping systems must have a vacuum breaker or other approved backflow prevention device.

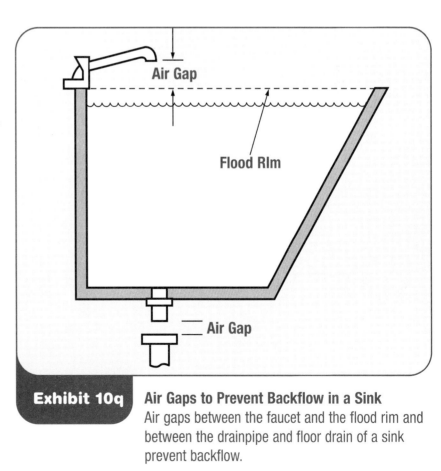

Exhibit 10q   **Air Gaps to Prevent Backflow in a Sink**
Air gaps between the faucet and the flood rim and between the drainpipe and floor drain of a sink prevent backflow.

Overhead waste-water pipes or fire safety sprinkler systems can leak and become a source of contamination. Even overhead lines carrying potable water can be a problem, since water can condense on the pipes and drip onto food. All piping should be serviced immediately when leaks occur.

Sewage

Sufficient drainage must be provided to handle waste water. Any area subjected to heavy water exposure should have its own floor drain. Waste water from equipment and from the potable supply should be channeled into an open, accessible waste sink or floor drain. The drainage system should be designed to keep floors from being flooded, which is both unsanitary and a safety hazard.

Sewage and waste water are dangerous reservoirs of pathogens, soil, and chemicals. It is absolutely essential to prevent any possible contamination of food or food-contact surfaces from waste water. The pipes of any nonpotable water system, such as those from toilets or sinks, should be clearly identified by a licensed plumber so that they can be distinguished from pipes carrying potable water.

If there is a backup of waste water, prompt action needs to be taken. The action taken depends on the type of backup. A backup of raw sewage on the floor is cause for immediate closure of the establishment, correction of the problem, and thorough cleaning. Other backups are not as serious but will require an immediate correction of the problem and a clean-up.

Lighting

Building and health codes usually set minimum acceptable levels of lighting, typically based on **foot-candles.** A foot-candle is a unit of illumination one foot from a uniform source of light. Other units of measurement for light include lumens, luxes, and luminaires. Good lighting generally results in improved employee work habits, easier and more effective cleaning, and a safer work environment. Lighting requirements are different for various areas of the establishment. Here are some recommendations.

○ **Provide a minimum of fifty foot-candles of light (540 lux)** in food prep areas and where employees are working with utensils or equipment.

○ **Provide a minimum of twenty foot-candles of light (220 lux)** in handwashing or warewashing areas; at buffets and salad bars and where produce is displayed for sale; inside certain equipment (e.g. a reach-in refrigerator); in utensil storage areas and wait stations; and in restrooms.

○ **Provide a minimum of ten foot-candles of light (110 lux)** inside walk-in refrigerator and freezer units, in dry storage areas, and in the dining room during cleaning.

Overhead or ceiling lights over work stations should be positioned so that employees do not cast shadows on the work surface. Using fluorescent lights helps to minimize such shadows. Locate lighting fixtures to minimize the chance of contamination that may result from shattered glass bulbs or fluorescent tubes. Use shatter-resistant light bulbs and protective covers made of metal mesh or plastic to prevent broken glass. Shields should also be provided for heat lamps.

Ventilation

Proper ventilation helps maintain an establishment's indoor air quality by removing steam, smoke, grease and heat from the establishment. Adequate ventilation is particularly important in the food-preparation area, because it reduces the level of odors, gases, airborne dirt, mold, humidity, grease, and fumes that are present in the air, all of which can contribute to contamination. Excess humidity can cause condensation on walls and ceilings, which may drip onto food. An accumulation of grease can cause fires. If ventilation is adequate there will be little or no buildup of grease and condensation on walls and ceilings.

Mechanical ventilation must be used in areas for cooking, frying, and grilling. Exhaust hoods are used over cooking equipment, steam tables, and warewashing machines.

Ventilation must be designed so that hoods, fans, guards, and ductwork do not drip onto food or equipment. Hood filters or grease extractors must be tight fitting, easily removable, and should be cleaned on a regular basis. Thorough cleaning of the hood and ductwork should also be done periodically by a professional company.

Since so much air is moved through exhaust hoods, clean air must be taken in to replace it. This replacement air is called make-up air. Air must be replaced without creating drafts. All outside air intakes must be screened to keep pests out.

In many areas of the United States, clean-air ordinances restrict the use of exhaust fans. Exhaust air which contains food odors, smoke, and grease may have to be purified by filters or other devices. It is the establishment's responsibility to see that the ventilation system meets local regulations.

Solid Waste Management

Waste management is an important issue today in establishments. There are many things that foodservice managers may do to improve the waste management problem.

The **Environmental Protection Agency (EPA)** has recommended three approaches to manage **solid waste.**

○ **Reduce the amount of waste produced.** Eliminate unnecessary packaging.

○ **Reuse when possible.** Reused containers must be cleaned and sanitized. Never reuse chemical containers for food items.

○ **Recycle materials.** Store recyclables so that they can't contaminate food or equipment or attract pests.

When these systems are used, the amount of waste can be greatly reduced.

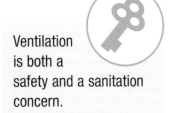

Ventilation is both a safety and a sanitation concern.

Garbage Disposal

Garbage can be a hazard to an establishment. **Garbage** is wet waste matter, usually containing food, that cannot be recycled. It attracts pests and has the potential to contaminate food items, equipment, and utensils.

Garbage containers must be leak-proof, waterproof, pest-proof, easily cleanable, and durable. They may be made of galvanized metal or an approved plastic. They must have tight-fitting lids and must be kept covered when not in use. Plastic bags and wet-strength paper bags may be used to line these containers.

Garbage should be removed from food-preparation areas as soon as possible. Garbage storage areas, inside or outside, should be large enough to contain all garbage and must be located away from food preparation areas. Frequent disposal prevents odor and pest problems. When removing garbage, employees should not carry it above or across a food preparation area. For safety purposes, do not block exit doors with trash.

All garbage containers should be cleaned frequently and thoroughly. Both the inside and the outside of the containers must be cleaned. A cleaning area equipped with hot and cold water and a floor drain is recommended for indoor cleaning. This area must be located so that food in preparation or storage will not be contaminated when garbage containers are being cleaned.

Exhibit 10r   **Outdoor Trash Receptacles**
Receptacles and compactor systems should be located on or above a smooth surface of nonabsorbent material such as concrete or machine-laid asphalt.

Food waste may also be disposed of through the use of in-drain garbage disposals, which reduce the amount of waste that goes into garbage containers. However, local regulations often limit their use in establishments. These disposals create huge amounts of food and water waste, which may overload local waste-water systems.

Pulpers or grinders have been suggested as a better alternative. Pulpers grind food and some other types of waste (such as paper) into small parts that are flushed with water. The water is then removed so that the processed solid wastes weigh less and are more compact for easier disposal.

Outdoor trash receptacles should be kept covered (with their drain plugs in place) at all times, except during cleaning. Receptacles and compactor systems should be located on or above a smooth surface of nonabsorbent material such as concrete or machine-laid asphalt. The area must be kept clean *(see Exhibit 10r).*

SUMMARY

An establishment that is difficult to clean will not be cleaned well. Sanitation efforts will be more effective if the establishment is designed and equipped with easy cleaning in mind.

The manager must give sanitation a high priority when planning to build a new establishment or remodeling an existing one. In most communities, plans for new construction or extensive remodeling are subject to review and approval by local regulatory agencies.

The manager can create built-in sanitation in three aspects of the facility: construction of floors, walls, and ceilings; choice and placement of equipment; and planning of utilities.

Coverings for floors, walls, and ceilings should be selected for ease of cleaning and durability, as well as appearance.

Equipment must meet the sanitation standards set by NSF International or the equivalent. Electrical equipment should also be listed by Underwriters Laboratories or other nationally recognized certifiers for safety considerations.

Equipment should be placed so that all areas around and under it can be cleaned effectively. Crevices or surfaces that catch dirt are difficult to clean and can harbor microorganisms and pests.

An establishment with built-in sanitation will also have utilities that are designed to be sanitary and promote easy cleaning. Utilities include the water supply, plumbing and sewage installations, ventilation, lighting systems, toilet and lavatory facilities, and solid waste management.

A CASE IN POINT

Case Study

Several people became ill shortly after drinking beverages at a local restaurant with a bar lounge. They all complained that their iced drinks had an odd taste. At this time in the lounge, the glassware washing machine had been out of service.

When interviewed, the manager explained that the machine's wash cycle was functional but that the unit could not be used because the large volume of water discharged after each wash load was worsening a recent drain blockage problem. A maintenance worker mentioned that there had been intermittent backups in the plumbing during the previous week and a large pool of water was found under the glassware washing machine. Drain cleaners had been used repeatedly with no change in the blockage.

The beverage ice-making machine shared piping with the glassware washing machine in the bar and the grease trap on the sink in the restaurant. The manager revealed that he had installed the plumbing fixtures himself.

When ice cubes were removed from the beverage ice bin, congealed food grease and food debris were found on them. Grease and debris also covered the bottom of the ice bin.

Why do you think the people became ill? What should the manager do to correct the problem?

TRAINING TIPS

Training Tips for the Classroom

1. Expect the Unexpected: Crisis Management Group Activity

Objective: *After completing this activity, class participants will be able to identify food-safety risks associated with various operational emergencies and develop action plans to address them.*

Directions: List a number of possible crisis situations that could occur in an establishment related to the following topics in Chapter 10. These may include the following:

○ **Plumbing:** broken or stopped-up pipes or drains; cross-connections; backup of waste water or raw sewage

○ **Water:** water turned off for eight hours or longer; contamination of city or private water supply

○ **Electric:** a power outage for eight hours or longer

○ **Equipment failure:** a broken booster heater in warewashing machine; freezer or cooler malfunction; hot water heater breakdown

○ **Natural disaster:** major flooding, hurricane, earthquake

○ **Waste management:** a strike by garbage workers

Break the class into groups. Assign each group a specific crisis and give them this set of directions.

○ Determine the specific food-safety risk or risks, if any, involved with the crisis.

○ Create an action plan to address the crisis.

○ Develop a plan to deal with, or prevent a future occurrence of, that specific crisis.

Allow twenty minutes for this activity, then have a spokesperson for each group present their work. Solicit feedback and additional suggestions from the other groups after each spokesperson has finished speaking.

Allow time for students to tell about any personal experiences related to crisis situations such as those discussed. Conclude the activity by asking students if their establishment has adequate crisis management plans, and if not, what steps they will take to implement such plans.

2. Let's Design and Build a Restaurant

Objective: *After completing this activity, class participants will be able to identify food-safety issues that should be considered when designing, constructing and equipping an establishment.*

Directions: Present an establishment concept (example: full-service Italian) and menu to the class. Provide the class with the proposed establishment's hours of operation and the projected volume of business (number of meals served). Include the possibility of additional business from carryout and catering sales.

Break the class into five groups, and assign each group one of the following topics that were discussed in Chapter 10.

○ Design and layout

○ Equipment selection

○ Plumbing, water, and electric

○ General construction (floors, walls, ceiling)

○ Heating, venting, air conditioning

Direct the groups to develop an action plan that includes a list of food-safety concerns and issues related to their topic. Allow ten minutes for this activity. Then allow time for members of each group to share ideas with other groups, to ask questions about each other's plans, and so on, for another ten minutes.

Ask each group to make a brief presentation on its action plan. Solicit class discussion. Conclude this activity by reminding the class of all the food-safety issues related to these areas, and reiterate how a well-designed and efficiently operated establishment can contribute to a successful food-safety program.

3. Facility Food-Safety Pop Quiz

Objective: *After completing this activity, class participants will be able to identify possible ways in which facilities and equipment may contaminate food if not properly designed, constructed or installed.*

Directions: Design a quiz with the following categories at the top of a sheet of paper:

Area or Item   **Cause of Contamination**   **Prevention**

Break the class into teams of two. Instruct the class that they are to list ways that food may become contaminated by facilities and equipment that are not properly designed, constructed or installed. Use the following example to show students how to organize their lists.

Area or Item

○ Plumbing

Cause of Contamination

○ Cross-connection

○ Overhead sewage pipe dripping on foods

○ Grease trap backup

Prevention

○ Install anti-backflow device

○ Create an air gap

○ Repair pipes

○ Clean grease trap weekly

Limit the activity to five to ten minutes. When time expires, have each team give itself one point for each Area or Item, Cause and Prevention. Ask the teams to add up the number of points and have the team with the highest number read its list. Discuss any missed items with the class.

Training Tips on The Job

1. Develop a "Waste Management Task Force"

Purpose: *To involve staff members in an effort to improve the waste-handling program at your establishment. Note: Involving your staff and allowing them to initiate approved changes in the existing program will create buy-in.*

Directions: Solicit a team of employees to perform the following research and development assignment.

○ Assess the waste-handling methods currently being used in the establishment. Consider all aspects: number, type, and conditions of garbage cans; location and condition of outside garbage bins; other waste-handling equipment; the recycling program, and so on.

○ Research and compare waste-handling methods in other operations similar to yours.

○ Read up on local health department codes and regulations regarding waste handling.

○ Develop recommendations for improved waste-handling techniques and systems, along with a list of expected benefits from the proposed improvements.

Give the team some clear direction and a deadline, and keep in touch with its progress along the way. When the report and recommendations are complete, schedule a formal meeting with the team in order to receive its proposals.

Authorize the team to initiate any approved changes. Provide funding if necessary. Monitor the results, offer suggestions, and solicit feedback from the team. Acknowledge the team's effort with an appropriate reward.

2. "Find the Seal" Search

Purpose: *To identify equipment in your operation that has met sanitation standards developed by NSF International and/or Underwriters Laboratory.*

Directions: Invite your kitchen staff to participate in a competition.

Over the course of a week, have each participant search for NSF labels and/or UL Sanitation Classification labels throughout the kitchen and record on paper where the label was located. You may limit the search to the receiving, storing, kitchen, and warewashing areas, or you may extend it to other areas of your operation.

At the end of the week, collect the lists, record the number of items found by each participant, and announce a winner at the next kitchen staff meeting. Offer the winner an appropriate reward. At the meeting, discuss with the staff the significance of NSF and UL, and explain why it is important to use equipment and utensils that are designated for commercial food service.

DISCUSSION QUESTIONS

1. What is one of the most important sanitation requirements to consider in choosing a floor-covering material for food preparation areas?

2. What action must be taken in the event of a backup of raw sewage in an establishment?

3. What can be done to prevent backflow in an establishment?

4. What can an establishment do to continue to serve safe food during an interruption in the water supply?

5. What are some potable water sources for an establishment? When might an establishment use non-potable water?

6. What are the requirements of a handwashing station? In what areas of an establishment are handwashing stations required?

7. How should restrooms be cleaned and maintained?

8. What are the requirements for installing immobile equipment?

MULTIPLE-CHOICE STUDY QUESTIONS ???

1. A commercial warewashing machine has a NSF International seal on it. What does the seal tell you about the cleaning and maintenance of the machine?

   A. The machine is designed to be easy to take apart for routine cleaning purposes.
   B. The machine has a self-sanitizing cycle, making routine cleaning unnecessary.
   C. All joints are rounded off with a silicone caulk to prevent water contamination.
   D. All food-contact surfaces are made of pure copper sheeting for quick sanitizing.

2. You are opening a restaurant in an area not served by a city water supply. Which of the following is not a safe source of drinking water for your restaurant?

   A. A private underground well
   B. A creek with clear water that runs through the property
   C. A fixed above-ground water storage tank
   D. An approved mobile water storage tank

3. Which of the following is usually an approved use of non-potable water?

   A. Sanitizing the restroom walls and floors
   B. Laundering the kitchen linen supply
   C. Operating the fire safety sprinkler system
   D. Cleaning the warewashing machine

4. Due to a broken water main, your coffee shop has no running water. It will be about two hours before service will be restored. What should you do in response to the situation?

   A. Drain and disconnect the water supply to the coffee makers.
   B. Cross-connect the hot and cold water lines to prevent backflow.
   C. Close the coffee shop for the day.
   D. Bring in bottled water for coffee making.

5. Why does a grease trap need to be cleaned regularly?

   A. A blocked grease trap can cause an overflow.
   B. Grease in a water supply line can give potable water a bad taste.
   C. Grease in a fire sprinkler system can corrode pipes and cause leaks.
   D. Grease in floor washing water can leave a film on the floor and cause an accident.

6. What is the primary reason for protecting the water supply from contamination?

   A. Microorganisms in contaminated water can make people ill.
   B. Sediment in contaminated water can cause service pipes to leak.
   C. Rust in contaminated water can permanently stain foodservice towels.
   D. The bad odor of contaminated water can make people ill.

7. Which of the following will not prevent backflow?

   A. An air gap between the sink drain and the floor drain
   B. The space between the faucet and the flood rim of a sink
   C. A vacuum breaker
   D. A cross-connection

8. You manage the food service for a college dormitory. One morning you notice standing water around a floor drain. You think it is from the defrost system of the refrigerator. What should you do?

   A. Call in a plumber to inspect the refrigerator drain pipe and the floor drain.
   B. Turn off the refrigerator immediately.
   C. Pour drain cleaner down the drain, then mop up the water and dispose of it in the service sink.
   D. Arrange for the college students to eat in another facility.

9. Which of the following is not a problem associated with improper drainage of waste water?

   A. Waste water can back flow into water supply lines and contaminate the potable water supply.
   B. Damp floors can cause slip-and-fall accidents.
   C. Foods can be exposed to dangerous microorganisms in waste water.
   D. Exposed sewer water can cause the establishment to have to close early.

10. You are installing an exhaust fan hood over your stove. What do you need to do to ensure its safe operation?

    A. The condensation collected must go directly into a floor drain.

    B. The hot air must be purified before being exhausted to the outside.

    C. The filters and ductwork must be cleaned regularly.

    D. The foods cooked under the hood must be covered at all times.

11. Which is a proper container for garbage?

    A. A covered plastic bin

    B. An enclosed wire-mesh bin

    C. A stainless-steel storage container

    D. A ceramic tile-lined container

12. Where should kitchen trash containers be cleaned?

    A. In the "scrape" sink of a commercial three-compartment sink

    B. In a designated area away from food preparation areas

    C. In a special area equipped with a bleach bath

    D. Any place in the kitchen where there is a floor drain

13. Why must handwashing stations be located close to food preparation areas?

    A. To allow the handwashing sinks to also serve for washing vegetables if necessary

    B. To make it easy for workers to wash their hands frequently while preparing food

    C. To give workers easy access to a supply of paper towels for food preparation

    D. To allow the floor drain in the food preparation area to also serve the handwashing station

14. Which statement below describes the proper method for cleaning the employee restroom?

    A. The mirror needs to be cleaned with glass polish.

    B. The disinfectant spray should remain in the sink for one minute.

    C. The toilet should be cleaned before the sink.

    D. The floor should be cleaned after the sink and toilet.

15. Which hand-drying method is not recommended for use in a handwashing station?
    A. Forced-air dryer
    B. Single-use paper towels
    C. A common hand towel
    D. A continuous-cloth towel system

ADDITIONAL RESOURCES

Books and Periodicals

Bjerklie, S. (1997). The super soaker. *ID, 50*(11), 35.

Cichy, R. F. (1993). *Sanitation management.* East Lansing, MI: Educational Institute of the American Hotel & Motel Association.

Durocher, J. (1996). A clean sweep. *Restaurant Business, 95*(4), 202.

Durocher, J. (1995). Working with sanity. *Restaurant Business, 94*(9), 154.

Equipped for food safety. (1998). *Best Practices, 2*(4), 6-10.

Frable, F., Jr. (1998). Cleanability of equipment should be a factor in your purchasing decisions. *Nation's Restaurant News, 32*(33), 19.

Frable, F., Jr. (1997). 10 common errors in kitchen planning and ways to avoid them. *Nation's Restaurant News, 31*(6), 22.

Kratt, D. T. (1997). Running a clean operation. *Nightclub & Bar, 13*(11), 79.

Longree, K & Armbruster, G. (1996). *Quantity food sanitation.* New York: Wiley.

Marriott, N. G. (1994). *Principles of food sanitation.* New York: Chapman & Hall.

Riell, H. (1997). Beverage Dispenser Sanitation. *FoodService Director, 10*(12), 166.

The right way to…calibrate equipment. (1997). *Best Practices, 1*(3), 12.

The right way to…organize workspaces. (1998). *Best Practices, 2*(3), 15-16.

Sanson, M. (1996). A blueprint for safety: Restaurant kitchen design. *Restaurant Hospitality, 80*(8), 65.

Secrets of self-inspection. (1998). *Best Practices, 2*(2), 6-9.

Troller, J. A. (1993). *Sanitation in food processing.* San Diego, CA: Academic Press.

Web Sites

1999 FDA Model Food Code

http://vm.cfsan.fda.gov/~dms/fc99-toc.html

Complete outline of the FDA's latest code for regulating operations that provide food directly to consumers. Also includes a quick synopsis of changes from the 1997 Food Code.

FDA-CFSAN Chemistry Information via the Internet

http://vm.cfsan.fda.gov/~dms/chemist.html

FDA food-safety information with a chemistry focus. Includes research information, as well as publications and databases.

FDA-CFSAN Pesticides, Metals, and Chemical Contaminants

http://vm.cfsan.fda.gov/~lrd/pestadd.html

Technical information on pesticides, toxic metals, and chemicals that can adversely affect food.

FDA Plan Review Guide

http://vm.cfsan.fda.gov/~dms/prev-toc.html

The document on this site was developed to serve as a guide to facilitate greater uniformity and ease in conducting plan review whether your position is a regulator or an owner/operator wishing to build or expand.

Food Online News

http://www.foodonline.com

This site provides a wealth of information for professionals and vendors in the foodservice equipment industry, including latest news and a searchable database of suppliers and products.

Hobart

http://www.hobartcorp.com

Hobart designs food equipment such as mixers and slicers, and warewashing, cooking, bakery, refrigeration, and weighing and wrapping equipment. This site includes information on all the equipment Hobart manufactures.

National Institute of Standards and Technology

http://www.nist.gov/public_affairs/welcome.htm

The National Institute of Standards and Technology is the government agency that works with industry to develop and apply technology, measurements, and standards. The site includes links to nontechnical information as well.

National Pest Control Association

http://www.pestworld.org

The association representing pest management firms offers resources and information for homeowners, professionals, and the media.

National Restaurant Association

http://www.restaurant.org

The National Restaurant Association site provides information on government agencies affecting the restaurant industry, the latest training and certification updates, and links to state restaurant associations and hospitality schools and universities.

North American Association of Food Equipment Manufacturers

http://www.nafem.org

The North American Association of Food Equipment Manufacturers (NAFEM) represents approximately seven hundred companies throughout the United States, Canada, and Mexico that manufacture commercial foodservice equipment and supplies. This site allows you to access Web sites for many foodservice equipment manufacturers.

NSF International

http://www.nsf.org

This site offers product information, resources, and publications regarding public health safety.

Occupational Safety and Health Administration (OSHA)

http://www.osha.gov

OSHA's Web site provides news releases, regulations, and a library to search related topics and past publications.

Sanitation Standard Operating Procedures Reference Guide

http://bbq.tamu.edu/USDA/ssop/ssop.html

This USDA site includes a reference guide, additional resources, and flow chart for sanitation compliance.

Chapter 11
Cleaning and Sanitizing

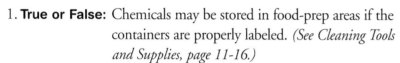

TEST YOUR FOOD-SAFETY KNOWLEDGE

1. **True or False:** Chemicals may be stored in food-prep areas if the containers are properly labeled. *(See Cleaning Tools and Supplies, page 11-16.)*

2. **True or False:** Rinse water for a high-temperature warewashing machine should be as hot as possible. *(See High-Temperature Machines, page 11-8.)*

3. **True or False:** A surface that has been cleaned is safe to use for food preparation. *(See Cleaning and Sanitizing, page 11-2.)*

4. **True or False:** A properly equipped manual warewashing station should have an adequate supply of clean linens for drying. *(See Manual Warewashing, page 11-10.)*

5. **True or False:** A machine warewashing detergent can also be used for manual warewashing. *(See Cleaning Agents, page 11-3.)*

Table of Contents

Learning Objectives

After completing this chapter, you should be able to:

○ Explain the difference between cleaning and sanitizing.

○ Identify factors that affect the efficiency of sanitizers.

○ Choose appropriate cleaners and sanitizers and safely store and handle them.

○ Manually clean and sanitize tableware and equipment.

○ Use proper machine warewashing techniques.

Key Terms

Cleaning
Sanitizing
Clean
Cleaning agent
Detergent
Solvent cleaners
Abrasive cleaners
Acid cleaners
Surfactants
Heat sanitizing

Chemical sanitizing
Sanitizer
Chlorine
Iodine
Quaternary ammonium compounds (Quats)
Hazard Communication Standard (HCS)
Master cleaning schedule

In Chapter 9, you learned that a HACCP plan depends on prerequisite programs. One of the most important of these is cleaning and sanitizing.

If you don't maintain a high standard of cleanliness and sanitation, food can easily become contaminated. No matter how carefully you prepare and cook food, without a clean and sanitary environment, bacteria and viruses, such as those that cause Salmonellosis and Hepatitis A, can quickly spread to both cooked and uncooked food. Cleaning and sanitizing must be done carefully and correctly, however. If not used properly, cleaning and sanitizing chemicals can be just as harmful to customers and employees as the diseases they help prevent.

CLEANING AND SANITIZING

It is important to understand the difference between **cleaning** and **sanitizing**. Cleaning is the process of removing food and other types of soil from a surface such as a countertop or plate. Sanitizing is the process of reducing the number of microorganisms on that surface to safe levels. To be effective, cleaning and sanitizing must be a two-step process. Surfaces must first be cleaned and rinsed before being sanitized.

Everything in your operation must be kept **clean**; however, any surface that comes in contact with food must be cleaned *and* sanitized. All food-contact surfaces must be washed, rinsed, and sanitized:

○ After each use.

○ When you begin working with another type of food.

○ Any time you are interrupted during a task and the tools or items you have been working with may have been contaminated.

○ At four-hour intervals if the items are in constant use.

Cleaning

Several factors affect the cleaning process. The table in *Exhibit 11a* lists these factors and provides a brief explanation of each.

Cleaning Agents

Cleaning agents are chemical compounds which remove food, soil, rust, stains, minerals, or other deposits. Choose cleaning agents for specific cleaning properties. Check with suppliers to find out which compounds are suitable for your needs. A product that works for one task may not be effective for another. Cleaning agents must be stable, noncorrosive, and safe for employees to use.

Cleaning agents work best when used as directed. They can be ineffective and even dangerous if misused. Follow manufacturers' instructions carefully.

Key Point

Surfaces must first be cleaned and rinsed before being sanitized.

| **Exhibit 11a** | Factors That Affect the Cleaning Process |

Factor	Effect on Cleaning Process
Type of Soil	Certain types of soil require special cleaning methods.
Condition of Soil	The condition of the soil or stain affects how easily it can be removed. Dried or baked-on stains will be more difficult to remove than soft, fresh stains.
Water Hardness	Cleaning is more difficult in hard water because minerals react with the detergent, decreasing its effectiveness. Hard water can cause scale or lime deposits to build up on equipment, requiring the use of lime-removal cleaners.
Water Temperature	In general, the higher the water temperature, the better a detergent will dissolve, and the more effective it will be in loosening dirt.
Cleaning Agent and Surface Being Cleaned	Different surfaces require different cleaning agents. Some cleaners work well in one situation, but might not work well or may even damage equipment when used in another.
Agitation or Pressure	Scouring or scrubbing a surface helps remove the outer layer of soil, allowing a cleaning agent to penetrate deeper.
Length of Treatment	The longer soil on a surface is exposed to a cleaning agent, the easier it is to remove.

Employees should never combine compounds or attempt to make up their own cleaning agents. Combining ammonia and chlorine bleach, for example, produces chlorine gas, which can be fatal. Also, do not substitute one type of detergent for another unless the intended use is clearly stated on the label. Detergents used for warewashing machines, for example, can cause severe burns to the skin if used for manual warewashing.

Cleaning agents are divided into four categories: **detergents, solvent cleaners, abrasive cleansers,** and **acid cleaners.** Some categories may overlap. For example, most abrasive cleansers and some acid cleaners contain detergents. Some detergents also may contain solvents.

Detergents

There are different types of detergents for all types of cleaning jobs. All detergents contain **surfactants** (surface acting agents) that reduce surface tension between the soil and the surface, so the detergent can quickly penetrate and soften the soil. General-purpose detergents are mildly alkaline and are used to clean fresh soil from floors, walls, ceilings, prep surfaces, and most equipment and utensils. Heavy-duty detergents are highly alkaline and are used to remove wax, aged or dried soils, and baked-on or burned-on grease. Warewashing detergents also are highly alkaline.

Solvent Cleaners

Solvent cleaners, often called degreasers, are alkaline detergents that contain a grease-dissolving agent. These cleaners work well in areas where grease has been burned on, such as grill backsplashes, oven doors, and range hoods. Solvents are usually effective only at full strength, so they are costly to use on large areas.

Acid Cleaners

Acid cleaners are used on mineral deposits and other soils that alkaline cleaners can't remove. These cleaners are often used to remove scale in warewashing machines and steam tables, and rust stains and tarnish on copper and brass. The type and strength of the acid varies with the cleaner's purpose. Follow the instructions carefully and use acid cleaners with caution.

Abrasive Cleansers

Abrasive cleaners contain a scouring agent such as silica that helps scrub off hard-to-remove soils. These cleaners are often used on floors or to remove baked-on or burned-on foods in pots and pans. Use abrasives with caution since they can scratch many smooth surfaces such as plastic and even stainless steel.

Sanitizing

There are two methods that can be used to sanitize surfaces, **heat sanitizing** and **chemical sanitizing.** Which you use depends on the application.

Heat Sanitizing

The higher the heat, the shorter the time required to kill microorganisms. The most common way to heat sanitize tableware, utensils, or equipment is to immerse or spray the items with hot water. Use a thermometer to check water temperature when heat sanitizing by immersion. Another way to check the water temperature of a machine is to attach temperature-sensitive labels and tapes, or a high-temperature probe, to the items being cleaned and sanitized.

Chemical Sanitizing

Chemical-sanitizing solutions are widely used in establishments because they are effective, reasonably priced, and easy to use. Chemical sanitizers are regulated by federal and state Environmental Protection Agencies (EPAs). Do not use any sanitizer on a food-contact surface unless it is EPA approved.

Chemical sanitizing is done in two ways: either by immersing a clean object in a specific concentration of sanitizing solution for a required period of time; or by rinsing, swabbing, or spraying the object with a specific concentration of sanitizing solution.

The three most common types of **sanitizer** used in the restaurant and foodservice industry are **chlorine, iodine,** and **quaternary ammonium compounds (quats)**. There are advantages and disadvantages to using each type. *See Exhibit 11b* for a list of these advantages and disadvantages.

Key Point

Do not use any sanitizer on a food-contact surface unless it is EPA approved.

Exhibit 11b — Advantages and Disadvantages of Different Sanitizers

Types	Advantages	Disadvantages
Chlorine	Most commonly used sanitizer	Less effective in pH ranges outside 6 to 7.5
	Kills a wide range of microorganisms	Dirt quickly inactivates these solutions
	Leaves no film on surfaces	Corrosive to some metals such as stainless steel and aluminum when used improperly
	Least expensive	
	Effective in hard water	Adversely affected by temperatures above 115°F (46°C)
Iodine	Effective at low concentrations	Less effective than chlorine
	Not as quickly inactivated by dirt as chlorine	Less effective at pH levels above 5.0
	Color indicates presence	Becomes corrosive to some metals at temperatures above 120°F (49°C)
		More expensive than chlorine
		May stain surfaces
Quaternary ammonium compounds (Quats)	Not as quickly inactivated by dirt as chlorine	Leaves a film on surfaces
	Remains active for a short period of time after it has dried	Does not kill certain types of microorganisms
	Noncorrosive	Hard water reduces effectiveness
	Nonirritating to skin	
	Works in most temperature and pH ranges	

In some instances, detergent-sanitizer blends may be used to sanitize surfaces, but items still must be cleaned and rinsed first. These compounds are usually more expensive and may not be meant for food-contact surfaces, since they may leave a residue. Scented or oxygen bleaches are not acceptable as sanitizers for food-contact surfaces. Household bleaches are acceptable only if the labels indicate they are EPA registered.

Factors Influencing the Effectiveness of Sanitizers

Different factors influence the effectiveness of chemical sanitizers. The most critical include contact time, selectivity, temperature, and concentration *(see Exhibit 11c)*.

Contact Time

In order for a sanitizing solution to kill microorganisms, the sanitizing solution must make contact with the object for a specific amount of time. Minimum times may differ for each sanitizer. Check your local regulatory agency for requirements in your jurisdiction.

Selectivity

Some sanitizers may be more effective in their ability to kill certain microorganisms.

Temperature

Generally, all sanitizers work best at temperatures between 75°F to 120°F (24°C to 49°C). Solutions at the low end of this temperature range will last longer; however, at temperatures below this range they may not sanitize effectively. At temperatures above this range, chlorine may corrode metals or evaporate.

Concentration

Concentrations below those required in your jurisdiction may fail to sanitize objects. Concentrations higher than recommended can be unsafe, may leave an odor or bad taste on objects, and may corrode metals.

The concentration of a sanitizing solution must be checked frequently. This is important since a sanitizer is depleted in the process of killing microorganisms, or may become bound up by food particles or detergent that is not adequately rinsed off of a surface.

A test kit designed for the specific sanitizer should be used to test its concentration in the sanitizing solution. Test kits are usually available from the manufacturer or a restaurant supplier *(see Exhibit 11d)*. A sanitizing solution must be changed when it is visibly dirty, or when the concentration of the sanitizer has dropped below the amount required.

Key Point

The effectiveness of a chemical-sanitizing solution depends upon its temperature, concentration, and contact time.

Exhibit 11c General Guidelines for Chemical Sanitizers

	Chlorine	Iodine	Quaternary Ammonium
Minimum Concentration			
○ For Immersion	50 parts per million (PPM)	12.5-25.0 PPM	220 PPM
○ For Spray Cleaning	50 PPM	12.5-25.0 PPM	220 PPM
Temperature of Solution	Above 75°F (24°C) Below 115°F (46°C)	75°F (29°C) Iodine will leave solution at 120°F (49°C)	Above 75°F (24°C)
Contact Time			
○ For Immersion	7 seconds	30 seconds	30 seconds—some products require longer contact time—read label
○ For Spray Cleaning	Follow manufacturer's directions	Follow manufacturer's directions	
pH (detergent residue raises pH, so rinse completely)	Must be below 8.0	Must be below 5.0	Most effective at 7.0, but varies with compound
Corrosiveness	Corrosive to some substances	Noncorrosive	Noncorrosive
Reaction to Organic Contaminants in Water	Quickly inactivated	Made less effective	Not easily affected
Reaction to Hard Water	Not affected	Not affected	Some compounds inactivated—read label; hardness over 500 PPM is undesirable
Indication of Proper Strength	Test kit required	Amber color indicates presence. Use test kit to determine concentration.	Test kit required; follow directions closely

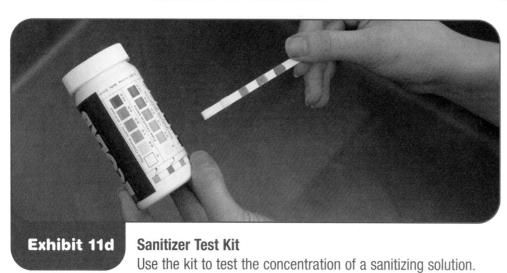

Exhibit 11d Sanitizer Test Kit
Use the kit to test the concentration of a sanitizing solution.

MACHINE WAREWASHING

Most tableware, utensils, and even pots and pans can be cleaned and sanitized in warewashing machines. Warewashing machines are designed for all types of operations. They range in size from single-tank, stationary-rack machines to multi-tank flight-type machines capable of washing thousands of pieces per hour. Most warewashing machines sanitize by using either hot water or a chemical-sanitizing solution.

High-Temperature Machines

These machines rely on hot water to clean and sanitize. Water temperature is critical and may vary with the model. The temperature of the final sanitizing rinse must be at least 180°F (82°C). For stationary-rack single-temperature machines, the temperature of the final sanitizing rinse must be at least 165°F (74°C). At higher temperatures, the water will vaporize before sanitizing the items. Water temperatures that are too high will also bake food onto tableware and utensils, making it even harder to get them clean. You may need a booster heater to provide enough hot water to sanitize a high volume of tableware. Make sure your warewasher has a built-in thermometer to measure the temperature of water at the manifold, where it sprays into the tank.

Chemical-Sanitizing Machines

Warewashing machines that use chemical sanitizing often wash at much lower temperatures, but not less than 120°F (49°C). Rinse-water temperature in these machines should be between 75°F and 120°F (24°C to 49°C) for the sanitizer to be effective. Items washed and rinsed at these lower temperatures may take longer to air dry, so you may need more room at the clean end of the machine and more tableware during peak times.

The effectiveness of your warewashing program will depend on a number of factors.

○ A sufficient water supply, especially hot water.

○ A well-planned layout *(see Exhibit 11e)* in the warewashing area with a scraping and soaking area and adequate space for both soiled and clean items.

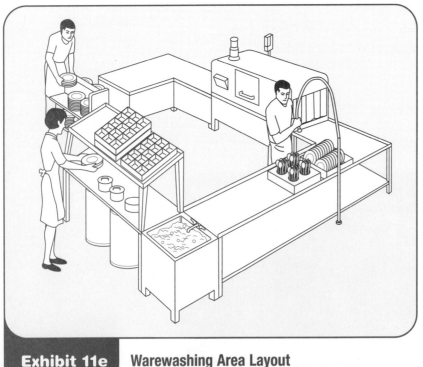

Exhibit 11e   **Warewashing Area Layout**
A well-planned layout will include a scraping and soaking area and adequate space for both soiled and clean items.

○ A separate area for cleaning pots and pans.

○ Devices that indicate water pressure and water temperature of both wash and rinse cycles.

○ Devices that automatically dispense detergent and sanitizer into the machine and indicate when reserves are low.

○ Protected storage areas for clean tableware and utensils.

○ Employees who are trained to operate and maintain the equipment and use the proper chemicals.

All warewashing machines should be operated according to manufacturers' instructions. These instructions will typically be located on the machine. No matter what type of machine you use, there are some general procedures to follow to clean and sanitize tableware, utensils, and related items.

○ **Check the machine for cleanliness and clean it as often as needed, at least once a day.** Fill tanks with clean water. Clear detergent trays and spray nozzles of food and foreign objects. Use an acid cleaner on the machine at least once a week to remove mineral deposits caused by hard water.

○ **Make sure detergent and sanitizer dispensers are properly loaded.**

○ **Scrape, rinse, or soak items before washing.** Pre-soak items with dried-on food.

○ **Load warewasher racks correctly and use racks designed for the items being washed.** Make sure that all surfaces are exposed to the spray action. Never overload racks.

○ **Check temperatures and pressure.** Follow manufacturers' recommendations.

○ **Check each rack as it comes out of the machine for soiled items.** Run dirty items through again until they are clean. Most items will need only one pass if you use proper equipment and procedures.

○ **Air dry all items.** Towels may contaminate items.

○ **Keep your warewashing machine in good repair.** Some common problems can be solved easily by changing procedures or chemicals.

Key Point

All warewashing machines should be operated according to manufacturers' instructions.

MANUAL WAREWASHING

Establishments that do not have a warewashing machine may use a three-compartment sink to wash items (some local regulatory agencies allow the use of two-compartment sinks; others require four-compartment sinks). These sinks may also be used to wash larger items as well. A properly set up warewashing station will include:

○ An area for scraping or rinsing food into garbage containers.

○ Drain boards to hold both soiled and clean items.

○ A thermometer in each sink to measure water temperature.

○ A clock with a second hand. This will allow employees to time how long items have been immersed in the sanitizing sink.

Before washing items, clean and sanitize each sink and all work surfaces. Follow these steps when washing and sanitizing all tableware, utensils, and equipment (see Exhibit 11f).

Step 1: Rinse, scrape, or soak all items before washing.

Step 2: Wash items in the first sink in detergent solution that is at least 110°F (43°C). Use a brush, cloth, or nylon scrub pad to loosen the remaining soil. Replace the detergent solution when the suds are gone or the water is dirty.

Step 3: Immerse or spray-rinse items in the second sink, using water that is at least 110°F (43°C). Remove all traces of food and detergent. If you are using the immersion method replace the rinse water when it becomes cloudy or dirty.

Step 4: Immerse items in the third sink in hot water or a chemical-sanitizing solution. If hot-water immersion is used, the water must be at least 171°F (77°C) (some health codes require a temperature of 180°F) and the items must be immersed for thirty seconds. If chemical sanitizing is used, the sanitizer must be mixed at the proper concentration and the water temperature must be correct. Check the concentration of the sanitizing solution at regular intervals with a test kit.

Step 5: Air dry all items on a drain board.

Wood surfaces, such as cutting boards, handles, and bakers' tables, need

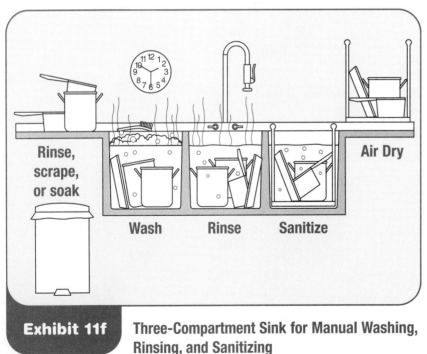

Exhibit 11f   **Three-Compartment Sink for Manual Washing, Rinsing, and Sanitizing**

special care. Scour them in detergent solution with a stiff-bristle nylon brush, rinse in clean water, and sanitize after each use. Do not soak wood surfaces in detergent or sanitizing solution.

CLEANING AND SANITIZING EQUIPMENT

Equipment must be kept clean. All food-contact surfaces must be sanitized as well. Teach all employees how to properly clean and sanitize each type of equipment.

Clean-in-Place Equipment

Some pieces of equipment, such as soft-serve yogurt machines or milk and beverage dispensers, are designed to have cleaning and sanitizing solutions pumped through them. Since many of them hold and dispense potentially hazardous foods, they must be cleaned and sanitized every day unless otherwise indicated by the manufacturer.

Stationary Equipment

Equipment manufacturers will usually provide cleaning instructions. Food-contact surfaces may require a different cleaning or sanitizing solution than non-food-contact surfaces. In general, follow these steps.

○ Turn off and unplug equipment before cleaning. Refrigerators and freezers should be turned off but can be left plugged in.

○ Remove food and soil from under and around the equipment.

○ Remove detachable parts and manually wash, rinse, and sanitize them, or run them through a warewasher if permitted. Allow them to air dry. When washing sharp parts such as slicer blades, turn the blades away from yourself and wipe away from sharp edges.

○ Wash and rinse fixed food-contact surfaces, then wipe with chemical-sanitizing solution.

○ Keep cloths used for food-contact and non-food-contact surfaces in separate, properly marked containers of sanitizing solution.

○ Air dry all parts, then reassemble according to directions. Tighten all parts and guards. Test equipment at recommended settings, then turn off.

○ Resanitize food-contact surfaces that were handled when putting the units back together, by wiping with a cloth that has been submerged in sanitizing solution.

In some cases, you may be able to spray-clean fixed equipment. Check with the manufacturer first. If allowed, spray each part with solution in the right concentration for two or three minutes.

Key Point

Clean-in-place equipment used to hold and dispense potentially hazardous food must be cleaned and sanitized every day unless otherwise indicated by the manufacturer.

Key Point

Cloths used for food-contact and non-food-contact surfaces should be stored between use in separate, properly marked containers of sanitizing solution.

Refrigerated Units

As mentioned in Chapter 6, clean up spills in refrigerators and freezers immediately. These units should be thoroughly cleaned and sanitized regularly to remove soil, mold, and odors. When cleaning and sanitizing refrigerators, follow these suggestions.

○ Clean before taking deliveries so less food has to be moved.

○ Move food to another unit before you start cleaning.

○ Wash, rinse, and sanitize shelves regularly. Thoroughly clean walls, floors, door edges, and gaskets.

CLEANING THE KITCHEN

Kitchen floors, walls, shelves, ceilings, light fixtures, and drains are non-food-contact surfaces, but they still require regular cleaning. Sanitizing these surfaces is not required, but may be a good idea. How often these surfaces are cleaned will depend on factors such as the type of food served, surfaces to be cleaned, and rate of ventilation in the kitchen. Spills should be cleaned up immediately. Floors and walls around food-prep and cooking areas should be cleaned at least daily—before or after each shift is preferable. Other non-food surfaces may need cleaning less often.

Floors

Floors are a safety hazard if they aren't cleaned regularly or rinsed properly. They can also be a source of cross-contamination. Soils and spills on the kitchen floor can be tracked through the entire operation. To clean floors, follow these steps.

○ Mark the area being cleaned with signs or safety cones to prevent slips and falls.

○ Sweep the floor first. Use a dust-free compound or a vacuum cleaner.

○ Use a deck or scrub brush and full-strength detergent on extra-soiled areas to remove grease and dirt.

○ Mop or pressure-spray the area, working from the walls toward the floor drain. Soak the mop in a bucket of detergent solution and wring it out. Clean a ten-foot by ten-foot area with both sides of the mop using a figure-eight motion. Soak and wring out the mop, then clean the same area again.

○ Remove excess water with a damp mop or squeegee, working away from the walls and toward the floor drain.

○ Rinse thoroughly with clean water using the same mopping procedure.

Walls and Shelves

Clean tile and stainless-steel surfaces by spraying or sponging with a detergent solution. Use a nylon scrub brush to clean dried-on soil or grease. Rinse with clean water, then wipe down surfaces with a sanitizing solution. When spray-cleaning, take care not to damage walls, and protect food, equipment, or nearby supplies. Use a sponge or wet cloth to clean other wall surfaces, such as painted drywall.

Ceilings and Light Fixtures

Ceilings do not need to be cleaned as often as floors or walls. Check ceilings and light fixtures daily to ensure that cobwebs, dust, dirt, or condensation don't fall and contaminate food or food-contact surfaces below. Wipe and rinse ceilings and light fixtures with a sponge or cloth.

CLEANING THE PREMISES

Although the kitchen requires the most attention, all areas of your operation need to be kept clean. Areas such as restrooms, busing or serving stations, and tables and booths need to be kept clean and sanitary as well. Pay attention to the exterior of the facility, too. Dirty receiving docks and garbage areas can attract pests. A dirty parking lot or exterior is unsightly and can hurt business.

Tables

Strictly speaking, tables, booths, and counters are non-food-contact surfaces, since food is usually served on dishes. However, tables should be kept clean. If you use table linens or butcher's paper, change them before seating new customers. If flatware is placed directly on the table surface, tables should be kept sanitary, too.

- When seating customers, remove extra tableware. If seating three people at a four-top, for example, take away the extra place setting so it won't be contaminated.

- Use a dry wiping cloth to clean crumbs and dry food spills from tables. Replace the cloth whenever it becomes soiled or sticky, so it doesn't transfer soil and microorganisms to other surfaces.

- Use a moist cloth to clean up other types of food spills. Keep moist cloths in a bucket of chemical-sanitizing solution. Replace both the sanitizing solution and wiping cloths as necessary. Check the concentration of the sanitizing solution often.

Serving Stations

Serving stations may be little more than a small space to store tableware and linens. In many operations, however, they are also used to dispense water, coffee and other beverages; to prepare and serve bread and condiment baskets; and to serve desserts. Serving stations often contain a sink, a coffee brewer, beverage dispensers, ice makers or bins, and even small coolers to hold butter, cream, condiments, and more. To keep serving stations clean, follow these procedures.

○ Clean up spills immediately. Wipe or sweep up dry foods with a dry cloth or broom. Clean wet and sticky spills with a damp cloth kept in sanitizing solution.

○ Wash, rinse, and sanitize sinks and countertops at least daily or after each shift. If work areas are used to prepare potentially hazardous foods, such as cheesecake, clean at least every four hours.

○ Clean equipment daily or as often as recommended by the manufacturer. Iced tea dispensers and utensils used for self-service must be cleaned daily. Items such as ice bins, beverage-dispensing nozzles and lines, and coffee grinders should be cleaned as often as needed to prevent dirt or mold from accumulating.

○ Wash, rinse, and sanitize bus tubs manually or in the warewashing machine at least daily or after every shift.

Public Restrooms

Unsanitary restrooms pose a danger to your customers and your business. Never underestimate cleaning needs in this area. Restrooms can quickly become visibly dirty and may harbor unpleasant odors and disease-causing microorganisms. Clean restrooms are important to customers. Many directly associate cleanliness of the restrooms with the establishment's commitment to food safety. When maintaining restrooms, consider the following suggestions.

○ Check public and employee restrooms regularly.

○ Restock soap, toilet paper, and towel supplies before they run out.

○ Clean sinks, mirrors, walls, floors, counters, dispensers, toilets, urinals, and waste receptacles at least daily. Sanitize toilets and urinals at least once daily. Clean up accidents and spills as often as necessary.

○ Remove trash at least once daily, or as often as necessary.

Key Point

Many customers associate the cleanliness of the restroom with the establishment's commitment to food safety.

Exterior Premises

The exterior of the premises should be kept in good, clean condition. Windows, walls, landscaping, and fixtures should be cleaned on a regular rotating basis, and should be included on the master cleaning schedule. Check the grounds at least daily and pick up trash and sweep walkways. Clean garbage areas as often as necessary to prevent odors or trash from attracting pets.

TOOLS FOR CLEANING

Cleaning is easier when you have the right tools for the job at hand. Worn-out or wrong-size tools won't give you the pressure or friction needed for cleaning. Tools also should be used only for the job for which they were designed. Keep tools used for food-contact surfaces separate from tools used to clean non-food-contact surfaces. Keep tools used to sanitize away from tools used to clean. Use a separate set of tools for the restrooms. Color-coding is one way to keep tools separate for use in specific areas.

Key Point

Use a separate set of cleaning tools for the restroom.

Brushes

Brushes apply more effective pressure than wiping cloths, and the bristles loosen soil more easily. Worn brushes will not clean effectively and may be a source of contamination.

Lacquered wood or plastic brushes with synthetic bristles are preferred. They don't absorb moisture, are nonabrasive, and last longer.

Use the right brush for the job. Brushes come in different shapes and sizes for each task. Crimped-bristle brushes, for example, hold more water for wet cleaning jobs. Brushes designed to clean grills or pizza ovens have stiff bristles made of brass or stainless steel.

Scouring Pads

Steel wool and other abrasives are sometimes used to clean heavily soiled pots and pans, equipment, or floors. However, metal scouring pads can break apart and leave residue on surfaces, which may later contaminate food. Nylon scouring pads can provide an alternative.

Mops and Brooms

Keep both light-duty and heavy-duty mops and brooms on hand. Mop heads can be all-cotton or synthetic blends. It makes sense to have a bucket and wringer for both the front and back of the facility. You'll need several smaller buckets for cleaning and sanitizing. Both vertical and push-type brooms will come in handy.

Key Point

Store flatware and utensils with handles up so employees can pick them up without touching the food-contact surface.

NON-FOOD STORAGE

Once tableware, utensils, and equipment are clean and sanitary, store them so they stay that way. Storage areas for cleaning supplies should be out of the way of kitchen traffic and potential cross-contamination. Keep storage areas clean and sanitary, too.

Tableware and Equipment

- Store tableware and utensils at least six inches off the floor. Keep them covered or otherwise protected from dirt and condensation.
- Clean and sanitize drawers and shelves before clean items are stored.
- Clean and sanitize trays and carts used to carry clean tableware and utensils to and from the storage area. Do this daily or more often if they become soiled.
- Store glasses and cups upside down. Store flatware and utensils with handles up or out so employees can pick them up by the handles.
- Keep food-contact surfaces of clean-in-place equipment covered until ready to use.

Cleaning Tools and Supplies

Cleaning tools and supplies should be cleaned and sanitized before being put away. Tools and chemicals should be stored in a locked area away from food and prep areas. The area should be well lighted so employees can identify chemicals easily. The area also should be equipped with hooks for hanging mops, brooms, and other cleaning tools; a utility sink for filling and cleaning buckets and tools; and a floor drain *(see Exhibit 11g)*. Never use hand sinks, food-prep sinks, or warewashing sinks to clean mops, brushes, or tools. When storing tools and supplies, consider the following suggestions.

- Air dry wiping cloths overnight.
- Hang mops, brooms, and brushes on hooks to air dry. Do not leave brooms or brushes standing on their bristles.
- Clean, rinse, and sanitize buckets. Let them air dry, and store with other tools.

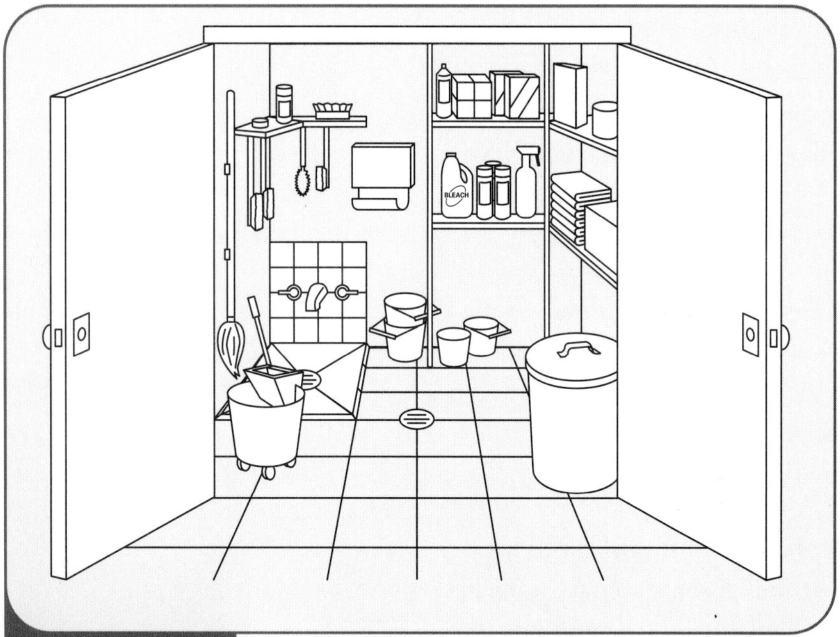

Exhibit 11g Storage Area for Cleaning Tools and Supplies

USING HAZARDOUS MATERIALS

Chemicals are both useful and necessary to keep an establishment clean, sanitary, and pest free. Used properly, they may pose little threat to an employee's safety. Used improperly, they may become a health hazard that can cause injury.

Because of the potential danger of chemicals used in the workplace, the Occupational Safety and Health Administration (OSHA) requires employers to comply with their **Hazard Communication Standard (HCS).** This standard, also known as Right-to-Know or HAZCOM, requires employers to tell their employees about chemical hazards to which they may be exposed to at the establishment. It also requires employers to train employees on how to safely use the chemicals they work with. Employers must comply with OSHA's HCS standard by developing a hazard communication program for their establishment.

A hazard communication program must include the following components.

- An inventory of the hazardous chemicals used at the establishment
- Chemical labeling procedures
- Material Safety Data Sheets (MSDS)
- Employee training
- A written plan addressing hazard communication standards

Inventory of Hazardous Chemicals

A hazardous chemical is any chemical which is a physical or health hazard to humans. Chemicals which are known to have acute or chronic health effects, or are explosive, flammable, or unstable, are considered hazardous. Virtually any chemical with toxic properties should be included in an establishment's HAZCOM program.

Take an inventory of the hazardous chemicals stored in your establishment. List the name of the chemical and where it is stored. Update the list when chemicals are added or no longer used.

Labeling Procedures

OSHA requires chemical manufacturers to clearly label the outside of containers with the chemical name, manufacturer's name and address, and possible hazards. Accept only containers with proper labels, and make sure they remain readable and stay attached to the container.

Key Point

OSHA requires employers to tell employees about the chemical hazards they may be exposed to and to train employees to safely use the chemicals they work with.

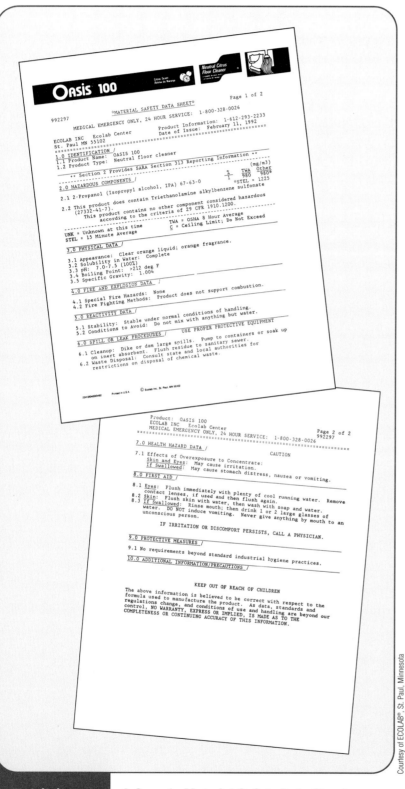

Exhibit 11h   **A Sample Material Safety Data Sheet**

If a chemical is transferred from a manufacturer's container to another container, the new container's label must contain the following information.

○ Chemical name

○ Manufacturer's name and address

○ Potential hazards of the chemical

Material Safety Data Sheets (MSDS)

OSHA requires that chemical manufacturers and suppliers provide a Material Safety Data Sheet (MSDS) *(see Exhibit 11h)* for each hazardous chemical at your establishment. These sheets are a part of employees' right to know about the hazardous chemicals that they work with. MSDS sheets contain the following information about the chemical.

○ Information about safe use and handling

○ Physical, health, fire, and reactivity hazards

○ Precautions

○ Appropriate personal protective equipment (PPE) to wear when using the chemical

○ First-aid information and steps to take in an emergency

○ Manufacturer's name, address, and phone number

○ Date the MSDS was prepared

○ Hazardous ingredients and identity information

MSDS sheets should be collected in a binder and stored in a location accessible to all employees while they are on the job.

Training

OSHA requires that every employee who might be exposed to hazardous chemicals during normal or regular working conditions be told of hazards and trained to properly use chemicals. Employees should receive this training annually, and new employees must receive it when first assigned to a new department or area. The following topics should be covered during training.

○ Existence and requirements of the Hazard Communication Standard

○ How the hazard communication program is implemented in the workplace

○ Operations and processes where hazardous chemicals are used

○ The inventory of chemicals in your establishment

○ The location of MSDS

○ How to read MSDS and product labels

○ Physical and health hazards of all chemicals used

○ Specific procedures adopted to provide protection, such as work practices and the use of engineering controls

○ Using PPE, and steps to prevent or reduce exposure to chemicals

○ Safety and emergency procedures

○ Information on the normal use of chemicals

A Written Plan

OSHA requires employers to develop a written plan which describes how they will meet the requirements of the Hazard Communication Standard in their establishment. The following items should be included in your written plan.

○ List of hazardous chemicals stored on the premises and their amounts

○ Purchasing specifications for chemicals

○ Procedures for receiving and storing chemicals

○ Labeling requirements in your establishment

○ Procedures for accessing MSDS

○ List of personal protection equipment (PPE)

○ Employee training procedures

○ Reporting and record-keeping procedures

○ How the employer will inform employees of the hazards of nonroutine tasks

Key Point

MSDS sheets should be collected in a binder and stored in a location accessible to all employees while they are on the job.

Key Point

OSHA requires employers to develop a written plan that describes how they will meet the requirements of the Hazard Communication Standard in their establishment.

IMPLEMENTING A CLEANING PROGRAM

A clean and sanitary environment is a prerequisite to an effective HACCP-based food-safety program. A cleaning program will give you a system to organize all your cleaning and sanitizing jobs. As with HACCP, an effective cleaning program takes commitment from management and involvement from employees. Here are some basic steps to design and implement a cleaning program.

Identify Cleaning Needs

Walk through every area of the facility. Write down all the surfaces, tools and equipment that need to be cleaned. Often, sanitation problems are overlooked because both the operation and job routines become too familiar.

Look at the way cleaning is currently done. Get input from employees. Ask them how and why they clean a certain way. Find out which procedures can be improved.

Estimate the amount of time and what skills are needed for each task. Some jobs may be done more efficiently with two or more people. Other jobs may require the services of an outside contractor. Determine how often things need to be cleaned. Steam tables, for example, need to be delimed on occasion, in addition to daily cleaning and sanitizing. Kitchen floors may need to be swept and mopped after every meal, not just once a day. Some jobs, such as cleaning out gutters or pressure-washing garbage areas, might be seasonal.

Create a Master Cleaning Schedule

Take all the input from employees during your walk-through and develop a **master cleaning schedule** (see Exhibit 11i). Organize the schedule by area. List all items that need to be cleaned and how often. Assign responsibility by job title. Provide a brief description of how to do the job. The schedule should include the following.

What should be cleaned. Arrange the schedule in a logical way so nothing is left out. List all cleaning jobs in one area, or list jobs in the order they should be performed. Keep the schedule flexible enough so that you can make changes if needed. For example, use dry-erase chartboards to post daily cleaning schedules or names of individuals responsible for that task.

Who should clean it. Assign each task to a specific individual. Since turnover can be high, use job titles. In general, employees should clean their own areas. Other cleaning tasks, however, should be rotated to distribute them more fairly and to prevent people from getting bored.

Key Point
As with HACCP, an effective cleaning program takes commitment from management and involvement from employees.

Key Point
When developing a master cleaning schedule for equipment and the facility, consider what should be cleaned, when, how, and by whom.

Exhibit 11i **Sample Cleaning Schedule (partial) for Food Preparation Area**

Item	What	When	Use	Who
Floors	Wipe up spills	Immediately	Cloth mop and bucket, broom and dustpan	Busers
	Damp mop	Once per shift, between rushes	Mop, bucket, safety signs	
	Scrub	Daily: at closing	Brushes, squeegee bucket, detergent (brand), safety signs	
	Strip, reseal	Every 6 months	See procedure	
Walls and ceilings	Wipe up splashes	As soon as possible	Clean cloth, detergent (brand)	Dishwashing Staff
	Wash walls	Food prep and cooking areas: daily		
		All other areas: first of the month		
Worktables	Clean and sanitize tops	Between uses and at end of day	See cleaning procedure for each table	Prep Cooks
	Empty, clean and sanitize drawers	Weekly	See cleaning procedure for table	

When it should be cleaned. Employees responsible for their own areas should clean as they go and clean and sanitize at the end of their shifts. Schedule major cleaning when food won't be contaminated and service won't be interrupted—that usually means after closing. Schedule work shifts to allow enough time for cleaning. Employees who rush to clean before the end of their shift may cut corners. Schedule additional crew members to come in near closing time to help clean, or try not to schedule too much cleaning after late-night closings.

How it should be cleaned. Provide clearly written procedures for cleaning. Lead employees through the process step by step. Always follow manufacturer's instructions when cleaning equipment. Specify cleaning tools and chemicals by name. Use color-coding to distinguish food-contact from non-food-contact cleaning supplies. Post cleaning instructions and safety warnings near the item to be cleaned *(see Exhibit 11j on the next page)*.

Exhibit 11j	Sample (partial) Cleaning Procedure: How to Clean a Food Slicer

Step	Process	Notes
1	Turn off machine and unplug cord	
2	Set blade control to zero	
3	Remove meat carriage	
4	Remove the back blade guard	
5	Remove the top blade guard	
6	Scrub meat carriage and guards in pot-and-pan sink	Use detergent solution and a scrub brush
7	Rinse parts	Immerse in clean water at 171°F (77°C) for 30 seconds. Use an S-hook to remove parts from water
8	Allow parts to air dry on a clean and sanitary surface	

Choose Cleaning Materials

When selecting cleaning tools for your establishment, consider the following suggestions.

○ **Pick cleaning agents and tools according to the needs identified on the master cleaning schedule.** Talk to suppliers for suggestions on which tools and supplies are appropriate for your operation. Make sure your supplies match the needs listed on the master schedule. If two people are scheduled to mop floors, provide two mops and two buckets. If the dining room is to be vacuumed, provide a supply of replacement vacuum cleaner bags.

○ **Replace tools that are worn out.** Equipment that is too worn or soiled may not clean or sanitize surfaces properly.

○ **Provide employees with the right protective gear.** Make sure there's an adequate supply of rubber gloves, boots, aprons, goggles, and other supplies.

Training Employees

○ **Schedule a kick-off meeting to introduce the program to employees.** Explain the reason behind it. Stress how important cleanliness is to food safety. If people understand why they are supposed to do something, they're more likely to do it properly.

○ **Schedule enough time for proper training.** Work with small groups or conduct training by area. Show employees how to clean equipment and surfaces in each area.

○ **Provide plenty of motivation.** Reward employees for jobs that are well done. Create small incentives for individuals or teams, such as "clean team of the month" awards. Tie performance to specific measurements or goals, such as achieving high marks during health code inspections.

Monitor the Program

○ **Supervise daily cleaning routines.** This should be done on a rotating basis to see if employees are cleaning and sanitizing properly.

○ **Monitor completion of all cleaning tasks daily against the master cleaning schedule.**

○ **Review the master schedule every time there is a change in menu, procedures, or equipment.** Make sure the cleaning program addresses any changes.

○ **Request employee input on the program during staff meetings.** Ask if they need additional equipment, supplies, manpower, time, or training to get cleaning jobs done. Find out if they have suggestions for improving the program.

○ **Conduct spot inspections.** Have employees play the role of health inspectors and check the premises against an inspection form. Employees will feel more involved and will learn more about cleaning other employees' areas.

SUMMARY

All the work that goes into a HACCP-based food-safety system is undermined if you don't keep utensils, equipment, and your facility clean and sanitary. Cleaning is the process of removing visible soil with a cleaning agent and agitation. Sanitizing is the process of reducing the number of harmful microorganisms to a safe level. Surfaces can be sanitized with hot water of at least 171°F (77°C) or with a chemical-sanitizing solution. You must clean and rinse a surface before it can be sanitized effectively.

All surfaces should be cleaned on a regular basis. Food-contact surfaces must be cleaned and sanitized after every use, every four hours if in continuous use, or at least once every day.

Warewashing machines can clean, rinse, and sanitize most tableware and utensils, as well as many pots and pans. Follow manufacturers' instructions, and make sure that your machine is clean and in good working condition. Check temperatures and pressure of wash and rinse cycles daily.

Items that are too large to be placed in a warewashing machine can be cleaned and sanitized manually. This may be done in a compartment sink, or if the items are immovable, by cleaning and then spraying them with a sanitizing solution. Items that are cleaned in a three-compartment sink should be pre-soaked or scraped clean, washed in a detergent solution, rinsed in clear water, and sanitized in either hot water (at 171°F or 77°C) or immersed in a sanitizing solution for a predetermined amount of time. All items should then be air dried.

Key Point

Once a cleaning program has been established, it must be monitored. Supervise cleaning routines, conduct spot inspections, and review the master schedule often.

Clean all non-food-contact surfaces regularly. Areas such as public restrooms, floors, shelves, and floor drains should be cleaned daily (or more often as needed) and sanitized where appropriate. Ceilings, walls and fixtures, as well as exterior areas such as docks, garbage containers, driveways, and parking lots, may be cleaned less frequently.

Keep all cleaning tools clean and sanitary as well. Cleaning tools and supplies should be stored in a well-lighted locked room separate from food storage and preparation areas. Make sure that chemical supplies are clearly labeled. Keep MSDS on hand, inform employees about potential hazards, and train them how to handle and use chemicals correctly.

Develop and implement a cleaning program. Identify cleaning needs by walking through the operation and talking to employees. Create a master cleaning schedule that lists all cleaning tasks and when and how they are to be cleaned. Assign responsibility for each task by job title. Enlist employee support by including their input in the program's design and rewarding good performance. Explain to employees the important relationship between cleaning and sanitizing and food safety.

Monitor the cleaning program to keep it effective. Supervise cleaning procedures. Check completion of each job against the master schedule. Adjust cleaning and sanitizing procedures if necessary when there is a change in menu, equipment, or procedures.

A CASE IN POINT I

Case Study

It was only 9:05 a.m., but already, Tim, the day shift manager, could tell it was going to be a bad day. While the executives from American Widget munched unenthusiastically on complimentary doughnuts, they threw annoyed looks at Tim and the buser, who were cleaning the private room that American Widget had hired for 9:00 a.m. Tim had opened the banquet room at 8:50 a.m. and found, to his dismay, that the remains of the annual banquet of the Pine Valley Martial Arts Club were still strewn about the room.

Later, Tim sat down with his cleaning schedule and tried to determine what had gone wrong. The banquet had been scheduled to end at 11:30 p.m. the previous night. The assistant dishwasher was scheduled to clean the room at midnight when he got out of the dish room and before he went home at 12:30 a.m. But a note from Norman, the night shift manager, told Tim that the banquet had been a wild one and that the last guest had staggered out long after the 1:00 a.m. closing time. The assistant dishwasher had punched out at 12:30 a.m. as he always did. Tim sighed.

What do you think went wrong? How can Tim prevent this from happening again? How should any cleaning policy changes be introduced?

A CASE IN POINT II

Case Study

The student kitchen employees in a university cafeteria had scraped every utensil and piece of tableware and placed all of them in the racks that automatically fed into the high-temperature warewashing machine. When the wash and rinse cycles were complete, one of the employees noticed that the items were spotted. The thermometer that registered the final rinse temperature for the sanitizing cycle had a reading of 140°F (60°C), rather than the required 180°F (82°C) indicated on the manufacturer's label.

The employee went to inform the manager. The manager called the manufacturer's representative to come and examine the machine.

What alternative methods can be used to clean and sanitize the tableware since the machine is not functioning properly? How often should the temperature in the warewashing machine be checked?

TRAINING TIPS

Training Tips for the Classroom

1. The Clean Team Competition

Objective: *After completing this activity, class participants should be able to write step-by-step cleaning and sanitizing procedures for kitchen areas and equipment.*

Directions: Break your class into groups of two to four people each, and ask each group to write out the procedures for cleaning and sanitizing a specific piece of equipment in the kitchen. Pick a piece of equipment that is common to most people in the class.

The groups should list the specific cleaning and sanitizing agents they would use, the specific tools, and the step-by-step procedures for the process, from start to finish.

After five or ten minutes, have the teams describe their cleaning and sanitizing procedures. Compare the different teams' work, and award a prize to the team whose procedures were the clearest and the most thorough and accurate.

2. Tools of the Trade

Objective: *After completing this activity, class participants should be able to identify different tools for cleaning and sanitizing and describe the specific use or uses of each tool. Note: This activity should help participants evaluate the adequacy of their establishment's current inventory of tools.*

Directions: Break your class into teams of two or three. Each team has the same challenge.

○ List as many different tools for cleaning and sanitizing an establishment as possible—chemicals, buckets, spray bottles, brooms, brushes, scrapers, gloves, mops, sanitizers, for example.

○ List the specific use or uses for each item.

Allow five minutes for this activity, and then ask to see all of the lists or have each team read its list to the class. Give a prize (or a round of applause) to the team that lists the most tools with their specific uses.

On a flipchart or blackboard, build a master inventory of cleaning tools and supplies from the lists. Add additional items that were not listed by the class. Ask the class if their establishments have all of the tools and supplies necessary to conduct effective cleaning and sanitizing programs. If not, which additional items are needed? Remind the class that food safety depends on proper cleaning and sanitizing, and stress that no establishment can maintain proper cleaning and sanitizing without the proper tools and supplies.

Training Tips on the Job

1. Creating a Cleaning Program

Purpose: *To enlist the help of employees from different areas of the establishment to develop a master cleaning schedule.*

Directions: Enlist a "clean team" of employees from different positions in your establishment. Work closely with the team to design and implement an effective, workable, comprehensive master cleaning schedule. Assign team members to work with you on the following tasks.

○ Identifying cleaning needs. Walk through the establishment with team members and identify tools, equipment, and surfaces that need to be cleaned.

○ Create a master cleaning schedule. Identify what should be cleaned, how it should be cleaned, when it should be cleaned, and by whom.

○ Select cleaning tools and supplies.

Once the master cleaning schedule has been created, schedule a kick-off meeting with all staff to introduce the program. Empower "clean team" members to train their coworkers in the new cleaning and sanitizing procedures in their department. Give team members an appropriate reward for their hard work. Schedule periodic meetings with the "clean team" to monitor the progress of the cleaning program. Modify the program if necessary.

2. "How To HAZCOM"

Purpose: *To enlist the help of employees to assess the effectiveness of the establishment's Hazard Communication Program and to modify the program if deficiencies are found.*

Directions: Enlist a "safety team" of employees from different positions in your establishment. Work with the team to assess the establishment's Hazard Communication Program. Focus on the following areas.

○ Hazardous chemical inventory. Is there a written inventory that indicates the chemicals by name and the places they are stored? Is the list current?

○ Labeling procedures. Are chemical containers labeled with the chemical's name, the manufacturer's name and address, and the potential hazards of the chemical?

○ MSDS. Are there MSDSs for each chemical stored at the establishment? Are these sheets accessible to all employees at all times while on the job?

○ Training. Are employees aware of the health hazards of all chemicals, and have they been trained to use them properly? Do they know what PPE is available, and can they use it properly? Do employees know where to find MSDSs, and do they know how to read them?

○ Written HAZCOM plan. Does the establishment have a written plan that describes how it will meet the requirements of the Hazard Communication Standard?

You may assign different "safety team" members to assess the current status in each of the above areas. Each team member should report his or her findings to the group so an action plan can be created to correct any deficiencies. The "safety team" should meet frequently (perhaps monthly) to assess how well the establishment is meeting the requirements of the HAZCOM program. "Safety team" members should receive an appropriate reward for their effort.

DISCUSSION QUESTIONS

1. What is the main difference between cleaning food-contact surfaces as compared to non-food-contact surfaces?

2. Why should foodservice managers not formulate their own detergents?

3. What are the steps that should be taken (in order) when performing manual warewashing?

4. How should clean and sanitized tableware, utensils, and equipment be stored?

5. What are the differences between chlorine, iodine, and quats sanitizers, and in what concentration should they be used?

6. What must be included in a master cleaning schedule?

7. How should stationary equipment be cleaned and sanitized?

8. Where should MSDS be stored?

MULTIPLE-CHOICE STUDY QUESTIONS

1. Which cleaning procedure must be performed daily in an establishment?

   A. Washing and sanitizing milk dispensers
   B. Washing ceilings and light fixtures
   C. Sanitizing storage shelves
   D. Washing dining area windows

2. Your restaurant purchases concentrated sanitizing chemicals. You want to make a spray solution for use in sanitizing food-contact surfaces in the establishment. What should you do to ensure that you have a proper sanitizing solution?

   A. Compare the color of the solution to another solution of known strength.
   B. Try out the solution on a food-contact surface.
   C. Test the solution with a sanitizer test kit.
   D. Use very hot water in making the solution.

3. Your supervisor has asked you to sanitize the food preparation areas of the kitchen. Which of the following is a proper procedure?

   A. Spray with a strong sanitizing solution.
   B. Wash with a detergent, rinse, then wipe with a sanitizing solution.
   C. Wipe with a sanitizing solution, then rinse with clean water and wipe dry.
   D. Scrape and brush off soil, then wipe with sanitizing solution.

4. You have decided to clean out your ice maker's storage bin. Which of the following caused you to make this decision?

   A. You haven't cleaned the bin in two whole days.
   B. You noticed water at the bottom of the bin.
   C. It was close to the end of your work shift and a convenient time to clean the bin.
   D. The establishment's master cleaning schedule indicated that it was time to clean the bin.

5. Your soup-and-sandwich bar serves fresh breads from a wooden
    bakers' table. At the end of the day, what do you need to do to get
    the table ready for the next day?

    A. Clean the table with a stiff brush and a detergent solution. Rinse it
       with clean water, and swab the surface with a sanitizing solution.

    B. Vacuum the crumbs from the table. Spray it with detergent and
       water, and then rinse and towel it dry.

    C. Cover the table surface with cloths that have been soaked in
       sanitizing solution. Allow them to remain on the table
       overnight, rewetting as needed.

    D. Brush the crumbs off the table. Wash it with clean water and
       then dry and sand it lightly.

6. Which of the following is not a proper step in cleaning and sanitizing
    a standing food mixer?

    A. Remove the detachable parts and wash them in the warewashing
       machine.

    B. Dry the detachable parts with a clean cloth and reassemble the
       machine.

    C. Clean the food and dirt from around the base of the mixer.

    D. Wash and rinse nondetachable food-contact surfaces. Wipe
       them with a sanitizing solution.

7. Your manager has given you the task of cleaning the kitchen range
    hood. What kind of cleaning agent will you need to use?

    A. A general purpose detergent   C. An acid cleaner
    B. A solvent cleaner   D. An abrasive cleaner

8. The water in your community is very hard. What does this
    condition cause, and how can you deal with the results?

    A. Hard water leaves a film of microorganisms on surfaces; this
       film must be removed with a chlorine sanitizer.

    B. Hard water causes mineral deposits; they can be removed with
       an acid cleaner.

    C. Hard water makes rust pits in some metals; the rust can be
       removed with an abrasive cleaner.

    D. Hard water scratches stainless steel; it can be repolished with a
       silica cleaner.

9. What should you do to ensure that the sanitizing solution you are using on a food preparation surface will work properly?

   A. In using the solution, rinse it from the surface and then apply it a second time.
   B. After using the solution, test the surface to confirm that no more microorganisms are present.
   C. Use a test kit while preparing the solution to check its concentration.
   D. After preparing the solution, heat it to the temperature recommended by the manufacturer.

10. The posted directions on your high-temperature warewashing machine tell you to keep the final rinse water temperature between 180°F (82°C) and 195°F (91°C). What is the reason for this direction?

    A. At higher temperatures the water will vaporize before sanitizing the items.
    B. Hotter water can cause some metals to be deformed or even melted.
    C. Cooler water can mean that washing will take a longer time.
    D. Cooler water can cause spray nozzles to clog with water minerals.

11. You notice that your warewashing machine is not cleaning your glassware as well it has in the past. How should you respond to the situation?

    A. Try a different kind of detergent.
    B. Increase the wash water temperature above the recommended level.
    C. Check that the wash and rinse nozzles are clear.
    D. Close the restaurant until you can get professional service.

12. Your warewashing machine is not working, and you must wash some tableware by hand. What is the first thing you should do to your three-compartment manual warewashing sink?

    A. Fill the first sink with hot water and detergent.
    B. Clean and sanitize the sinks and drainboards.
    C. Prepare the sanitizing solution in the third sink.
    D. Place clean, fresh towels on the drainboard next to the third sink.

13. Which element below would most likely be included in a master cleaning schedule?

   A. Written procedures for cleaning and sanitizing the storerooms

   B. A description of protective equipment needed when cleaning the floors

   C. A record of past menu changes that have caused changes in the schedule

   D. An estimate of the date for the next health department inspection

14. What is the purpose of a Material Safety Data Sheet (MSDS)?

   A. To protect restaurant customers from potentially dangerous food substances

   B. To provide information regarding the proper use of potentially dangerous tools in the kitchen

   C. To keep foodservice workers informed of the hazards associated with the chemicals they work with

   D. To detail the specifications of proper construction materials for establishments

15. Your establishment purchases liquid warewashing detergent in five-gallon jugs. You want to transfer some to a smaller container. What must you provide on the label you put on the new container?

   A. The expiration date of the detergent

   B. The chemical contents of the detergent

   C. The proper amount of the detergent to use

   D. The potential hazards for the use of the detergent

ADDITIONAL RESOURCES

Books and Periodicals

Hernandez, J. (1998). Keeping it clean. *Restaurant Hospitality, 82*(9), 118.

National Restaurant Association Educational Foundation. (1997). *AWARE™: Employee and customer safety: Handling hazardous materials safely: Manager's module* [Brochure]. Chicago: Author.

National Restaurant Association. (1996). *Sanitation survival kit for restaurant operators.* Washington, DC: Author.

Marriott, N. G. (1994). *Principles of food sanitation.* New York: Chapman & Hall.

Web Sites

1999 FDA Model Food Code

http://vm.cfsan.fda.gov/~dms/fc99-toc.html

Complete outline of the FDA's latest code for regulating operations that provide food directly to consumers. Also includes a quick synopsis of changes from the 1997 Food Code.

Colgate-Palmolive

http://www.nrn.com/operations/cproducts.htm

Information on Colgate-Palmolive cleaning products.

Ecolab

http://www.ecolab.com

International cleaning and sanitation company provides product and service information and news releases on its Web site.

FDA-CFSAN Chemistry Information via the Internet

http://vm.cfsan.fda.gov/~dms/chemist.html

FDA food-safety information with a chemistry focus. Includes research information, publications, and databases.

FDA-CFSAN Pesticides, Metals, and Chemical Contaminants

http://vm.cfsan.fda.gov/~lrd/pestadd.html

Technical information on pesticides, toxic metals, and chemicals that can adversely affect food.

Food Explorer

http://www.foodexplorer.com

Information for food industry professionals about processing and packaging, product development, and business and marketing.

Gojo

http://www.gojo.com

The Web site for the makers of hand sanitizers offers product information and ordering information.

Kimberly-Clark

http://www.kimberly-clark.com

The Web site for this producer of personal care and household products offers an overview of the company and its products.

National Pest Control Association

http://www.pestworld.org

The association representing pest management firms offers resources and information for homeowners, professionals, and the media.

National Restaurant Association

http://www.restaurant.org

The National Restaurant Association site provides information on government agencies affecting the restaurant industry, the latest training and certification updates, and links to state restaurant associations and hospitality schools and universities.

North American Association of Food Equipment Manufacturers

http://www.nafem.org

The North American Association of Food Equipment Manufacturers (NAFEM) represents approximately 700 companies throughout the United States, Canada and Mexico that manufacture commercial foodservice equipment and supplies. This site allows you to access Web sites for many foodservice equipment manufacturers.

OSHA (Occupational Safety and Health Administration)

http://www.osha.gov

OSHA's Web site provides news releases, regulations, and a library to search related topics and past publications.

Proctor & Gamble

http://www.pg.com

The Web site of this manufacturer of personal care and food products offers complete product information, as well as research and development information.

Reckitt & Colman

http://www.reckitt.com

The Web site of this producer of household products and over-the-counter pharmaceutical brands offers product and brand information.

Chapter 12

Integrated Pest Management

TEST YOUR FOOD-SAFETY KNOWLEDGE

1. **True or False:** Once pests enter the building it is very difficult to control them. *(See Introduction, page 12-2.)*

2. **True or False:** Prevention is the most important aspect of integrated pest management. *(See Introduction, page 12-2.)*

3. **True or False:** Glue traps should be used to control roaches. *(See Cockroaches, page 12-7.)*

4. **True or False:** The establishment should have a corresponding MSDS on hand any time pesticides are used. *(See Using and Storing Pesticides, page 12-15.)*

5. **True or False:** Pesticides can be stored in food storage areas if they are properly labeled and closed tightly. *(See Using and Storing Pesticides, page 12-15.)*

Table of Contents

Learning Objectives

After completing this chapter, you should be able to:

○ Develop an integrated pest management (IPM) program.

○ Identify ways to prevent pests from entering the facility.

○ Detect signs of infestation and identify the pests involved.

○ Evaluate and select a pest control operator (PCO).

○ Work with a PCO to choose the best method to control pests in your facility.

Key Terms

Integrated pest management (IPM)

Pest control operator (PCO)

Infestation

Air curtains

Droppings

Residual spray

Contact spray

Pesticide

Glue board

Pests such as insects and rodents can pose serious problems for establishments. Not only are they unsightly to customers, they also damage food, supplies, and facilities. The greatest danger from pests comes from their ability to spread diseases, including foodborne illnesses.

Once pests have infested a facility, it can be very difficult to eliminate them. Developing and implementing an **integrated pest management (IPM)** program will help prevent pests from infesting your establishment. It will require an ongoing and continuous program to get rid of any pests that do invade the premises.

For your IPM program to be successful, you must work closely with a licensed **pest control operator (PCO).** These professionals use safe, up-to-date methods to effectively prevent and control pests. Prevention is critical in pest control. If you wait until there is evidence of pests in your establishment, you may already have a major infestation.

THE IPM PROGRAM

Integrated pest management is a program that uses prevention and control measures to contain or eliminate a pest **infestation** in a foodservice or restaurant establishment.

IPM requires continuous effort and cooperation between management and a PCO. Working together, you can develop and follow a program using both preventive measures and control measures to keep pests out. There are three basic rules of an IPM program.

○ Deny pests access to the facility.

○ Deny pests food, water, and a hiding or nesting place.

○ Work with a licensed and registered PCO to eliminate pests that do enter.

Denying Pests Access

Pests can enter an establishment in one of two ways. They are either brought inside with deliveries, or they enter through openings in the building itself. To prevent pests from entering your establishment with deliveries, start by using reputable suppliers. Check all deliveries before they enter your facility. Refuse any shipment in which you find pests or signs of infestation, such as egg cases and body parts (legs, wings, etc.).

Obviously, holes, cracks, and other openings in the exterior of a building allow pests to get inside and hide, but there are many other ways that allow pests to get inside your establishment. Pay particular attention to the following areas.

Key Point

Check all deliveries before they enter the facility, and refuse any shipment in which you find pests or signs of infestation.

Doors, Windows, and Vents

○ **Screen all windows and vents with at least sixteen mesh per square inch screening.** Anything larger might let in mosquitoes or flies. Check screens regularly and clean and replace as needed.

○ **Install self-closing devices or door sweeps on all doors.** Repair gaps and cracks in doorframes and thresholds. Use weather stripping on the bottoms of doors with no threshold.

○ **Install air curtains (also called air doors or fly fans) that blow a steady stream of air, creating an air shield around doors that are left open.** Insects avoid air curtains, and those that do fly toward the openings are blown back out.

○ **Drive-through windows should remain closed when not in use.**

○ **Keep all exterior openings closed tightly.** Check them for proper fit as part of a regular cleaning schedule.

Pipes

Mice, rats, and insects such as cockroaches use pipes as highways through a facility.

○ **Use concrete to fill holes or sheet metal to cover openings around pipes** *(see Exhibit 12a).*

○ **Install screens over the ventilation pipes and ducts on the roof.**

Floors and Walls

Rodents often burrow into buildings through decaying masonry or cracks in building foundations. They move through floors and walls the same way.

○ **Seal all cracks in floors and walls.** Use a permanent sealant recommended by your PCO, local health department, or building contractor.

○ **Properly seal spaces or cracks where fixed equipment is fitted to the floor.** Use an approved sealant or concrete, depending on the size of the spaces.

○ **Paint a white stripe around the edge of your storeroom that is six inches from the walls.** This stripe will remind employees to store supplies six inches from the walls. It will also help you monitor the area for signs of infestation, since hairs, tracks, and droppings will show up better against the white stripe.

○ **Cover floor drains with hinged grates.** Rats are good swimmers and can enter buildings through the drainpipe.

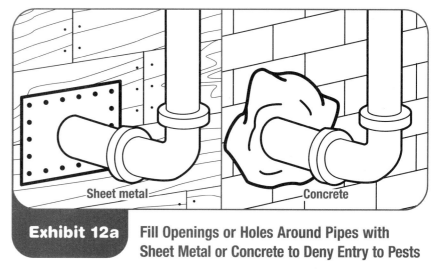

Sheet metal   Concrete

Exhibit 12a Fill Openings or Holes Around Pipes with Sheet Metal or Concrete to Deny Entry to Pests

Garbage should be stored in a clean, tightly covered container that is in good condition. Dispose of garbage quickly so it will not attract pests.

Clean and sanitize a facility thoroughly. Careful cleaning will eliminate a pest's food supply, destroy insect eggs, and reduce the number of places pests can take shelter.

Deny Food and Shelter

Pests are usually attracted to damp, dark, dirty places. A clean and sanitary establishment offers them little in the way of food and shelter. The stray pest that might get in cannot thrive or multiply in a clean kitchen. Stick closely to your master cleaning schedule. Start with these preventive measures to control pests.

○ **Dispose of garbage quickly and correctly.** Garbage attracts pests and provides a breeding ground for them. Keep garbage containers clean, in good condition, and tightly covered in all areas (indoor and outdoor). Clean up spills around garbage containers immediately. Wash, rinse, and sanitize containers regularly.

○ **Store recyclables in clean, pest-proof containers as far away from your building as local regulations allow.** Bottles, cans, paper, and packaging material provide shelter, and even food, for pests.

○ **Properly store all food and supplies as quickly as possible.**

- Keep all food and supplies at least six inches off the floor and six inches away from the walls.
- When possible, keep humidity lower than 50 percent. Low humidity helps prevent roach eggs from hatching. Use dehumidifiers and ventilation that force air outside of the building.
- Refrigerate foods such as powdered milk, cocoa, and nuts after opening. These foods attract insects, but most insects become inactive at temperatures below 41°F (5°C).
- Use FIFO, so pests don't have time to settle into these products and breed.

○ **Clean and sanitize the facility thoroughly.** Careful cleaning eliminates the food supply, destroys insect eggs, and reduces the number of places pests can safely take shelter.

- Clean up food and beverage spills immediately, including crumbs and scraps.
- Clean and sanitize toilets and restrooms as often as necessary.
- Train employees to keep lockers and break areas clean. Food and dirty clothes should not be kept in or under lockers.
- Keep cleaning tools and supplies clean and dry. Roaches love to hide in wet mops.
- Water in buckets will attract rodents.

Grounds and Outdoor Serving Areas

The popularity of *al fresco* (outdoor) dining presents new concerns. Birds, flies, bees, and wasps can be annoying and dangerous to the health of your customers. The key to protecting customers lies in minimizing the presence of these pests by denying food and shelter to them (*see Exhibit 12b*).

○ **Mow the grass, pull weeds, get rid of standing water, and pick up litter.**

○ **Cover all outdoor garbage containers.**

○ **Remove dirty dishes and uneaten food from tables and clean them as quickly as possible.**

○ **Do not allow employees or customers to feed birds or wildlife on the grounds.**

○ **Locate insect electrocutor traps ("zappers"), if used, away from facility doors, serving areas, food, combustible materials, employees, and customers.**

○ **Call your PCO to remove hives and nests.**

IDENTIFYING PESTS

Pests may still get into your establishment even if you take careful preventive measures. Pests are good hitchhikers. They can hide in delivery boxes and may even come in on employees' clothing or personal belongings. Learn how to spot signs of pests and identify what kind they are. If possible, record the time, date, and location of any pest sighting (or evidence of pests) and report this to your PCO. Early detection gives your PCO a chance to start treatment as soon as possible.

Exhibit 12b **Minimize Pests in Outdoor Dining Areas**
Cover garbage containers, remove dirty dishes, and clean up spills to deny food to pests.

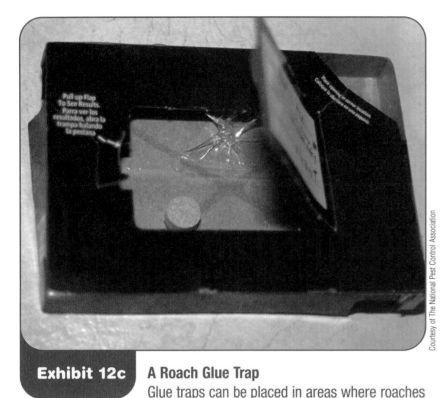

Exhibit 12c   A Roach Glue Trap
Glue traps can be placed in areas where roaches are typically found to monitor their presence.

Cockroaches

Roaches often carry disease-causing microorganisms such as *Salmonella*, fungi, parasite eggs, and viruses. Research shows that many people are allergic to residue left by roaches on food and surfaces. Roaches reproduce quickly and can adapt to some pesticides, making it difficult to control them.

There are several different types of roaches. Most live and breed in dark, warm, moist, hard-to-clean places. Here are some of the places where you will typically find them.

- Behind refrigerators, freezers, and stoves
- In sink and floor drains
- In spaces around hot-water pipes
- Inside equipment near motors and other electrical devices
- Under shelf liners and wallpaper
- Underneath rubber mats

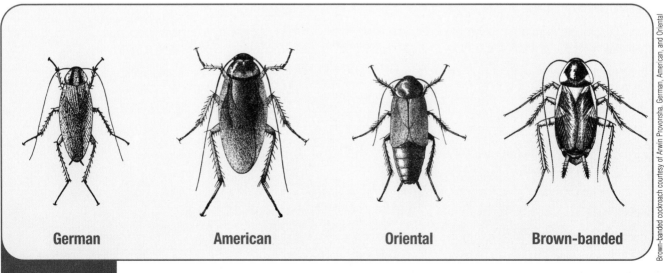

German   American   Oriental   Brown-banded

Exhibit 12d   Common Cockroaches Found in Establishments

○ In delivery bags and boxes

○ Behind unsealed coving (especially rubber-based)

Roaches generally feed in the dark. If you see a cockroach in daylight, you may have a major infestation. Only the weakest roaches come out in daylight. If you suspect you have a roach problem, check for these signs.

○ A strong oily odor

○ **Droppings** (feces), which look like grains of black pepper

○ Egg cases, which are capsule shaped; brown, dark red, or black; and may appear leathery, smooth, or shiny

You may have problems with more than one type of roach. Glue traps should be used to monitor your establishment and find out what type of roaches might be present. These open-ended cardboard containers have sticky glue on the bottom that traps roaches *(see Exhibit 12c)*. Place the traps in places that roaches can typically be found. If possible, place them on the floor in the corner where two walls meet. Check the traps after twenty-four hours and show them to your PCO. The type and stage (nymph or adult) of roaches will determine what type of treatment is needed. Common types of roaches include the Brown-banded, German, American, and Oriental *(see Exhibit 12d)*.

Flies

The common housefly is also a great threat to human health. Because they feed on garbage and animal wastes, flies transmit foodborne illnesses such as typhoid fever and dysentery. They spread microorganisms such as *Streptococcus* and *Staphylococcus* that can stick to their feet, hair, and mouths, and contaminate food with feces and vomitus, which is swarming with bacteria. Flies have no teeth, so they can eat only liquids or dissolved foods. They inject a long spike and vomit into solid food, let it dissolve, and then eat it.

In general, houseflies have the following characteristics.

○ They prefer calm air and the edges of objects, such as rims of garbage cans.

○ They are drawn to odors of decay, garbage, and animal wastes to find food and lay their eggs.

○ In warm weather, flies reproduce rapidly. For their eggs to hatch, they need warm, moist, decayed material that is located out of the sun. Eggs hatch into maggots, which can grow into adult flies in six days.

Other types of flies can be just as bothersome. Fruit flies are somewhat less harmful because they are primarily attracted to spoiled fruit, not animal waste. They can still transmit diseases, however. "Biting" flies, such as deer flies and horse flies, can be an additional nuisance to outdoor diners.

Other Insects

○ **Beetles, weevils, and moths.** Flour-moth larvae and beetles are usually found in dry storage areas. Look for insect bodies, wings, or webs; clumped-together foods; and holes in food and packaging. Cover food tightly, keep storage areas clean and sanitary, and use FIFO methods as preventive measures. Keep storage areas cool and humidity low.

○ **Ants.** Ants often nest in walls and floors near stoves and hot water pipes. They are drawn to grease and sweet foods. Control ants by cleaning up all food scraps and spills.

○ **Termites and carpenter ants.** These insects can cause great structural damage. They bore into wood, weakening walls, floors, and ceilings. They are rarely visible. Carpenter ants nest in wood, but forage for food elsewhere. Look for signs of sawdust that may have fallen from the ceiling. Call your PCO at once if you suspect termites or carpenter ants.

○ **Spiders.** Look for webs in corners and around fixtures. Control spiders by checking for and removing webs daily. The bites of some spiders, such as the black widow and brown recluse, are poisonous and can cause illness, but they are rarely fatal. If you see either type, call your PCO.

○ **Bees and wasps.** Bees, wasps and hornets can sting outdoor diners. Some people are allergic to bee or wasp venom and can go into shock, or even die, from only one or two stings. In general, bees, wasps, and hornets are drawn to sweet foods. Yellow jackets are drawn to proteins in early summer and sweets in late summer.

  ● Wasps and hornets often build nests under eaves. Call your PCO to remove them.

  ● Yellow jackets usually nest in the ground. They aggressively defend these nests, so call your PCO to remove them.

  ● Bees make hives in hollow trees, under eaves, or in other protected places. Bees usually will not bother people unless the hive is threatened. If you have a hive, call your PCO to remove it.

  ● Deer ticks. Ticks can carry Lyme disease and can cause other serious illnesses. Minimize risks by keeping the grass short around dining areas.

○ **Mosquitoes and gnats.** Mosquitoes carry diseases such as malaria, typhoid, yellow fever, and encephalitis. These members of the fly family feed on the blood of animals and humans. Mosquitoes are drawn to heat and are most active at dawn and dusk. Gnats and other small, biting insects are annoying to outdoor diners. To minimize the presence of mosquitoes and gnats, keep all outdoor areas free of stagnant water and move lights away from the building.

Rodents

Rodents are a serious health hazard. They eat and ruin food, damage property, and can spread disease. Most rodents have a simple digestive system. They urinate and defecate as they move around a facility. Their waste can fall or be blown into food and can contaminate surfaces.

Rats and mice are the most common type of rodent found around foodservice facilities *(see Exhibit 12e)*. They hide during the day and search for food at night, but do not travel far. Rats travel only 100 to 150 feet from their nests, while mice travel only ten to thirty feet. Like other pests, they reproduce often. Mice need a hole only the size of a dime to enter a facility; rats need a hole only the size of a quarter.

Rats have very good senses of hearing, touch, and smell. They can stretch to reach as much as eighteen inches, can jump three feet vertically, four feet horizontally, and can even climb straight up brick walls. They are smart enough to avoid poison bait and poorly laid traps. Effective control of rats and mice requires the knowledge and experience of professionals.

A building may be infested with both rats and mice at the same time. Look for these signs.

○ **Droppings.** Fresh droppings are shiny and black. Older droppings are gray.

○ **Signs of gnawing.** Rats and mice gnaw to reach food. Rats' teeth are so strong they can gnaw through pipes and concrete, in addition to wood, cardboard, and cloth sacks *(see Exhibit 12f)*.

○ **Tracks.** Check dusty surfaces by shining a light across them at a low angle.

○ **Nesting materials.** Mice use scraps of paper, cloth, hair, and other soft materials to build nests.

○ **Holes.** Rats nest in burrows, usually in dirt, rock piles, or along foundations.

Exhibit 12f

Signs of a Rodent Infestation
Rats will gnaw through walls to reach food.

Courtesy of the National Pest Control Association

Common house mouse   Norway rat   Roof rat

Courtesy of the National Pest Control Association

Exhibit 12e   **Common Types of Rodents**

Birds

Birds eat animal and vegetable matter, and their droppings carry fungi and bacteria that can make people sick. They also may carry mites and microorganisms that can cause encephalitis and other diseases. Common pigeons, sparrows, and starlings often nest in or near buildings. Seagulls can be a serious problem near the water.

Birds can be drawn to crumbs and food scraps in outdoor dining areas. Keep these areas clean, and remove food from tables quickly. Post signs asking customers not to feed birds. If birds become a problem, call a PCO who specializes in bird control measures.

Other Animals

Though less common, other rodents and animals can infest your building. The danger they pose is mostly from bites and the possibility of spreading rabies. The best method of control is to prevent them from getting inside. If there is evidence that any of these pests have entered your establishment, work with a PCO or your local animal control department to remove them.

○ **Bats.** Bats will nest in high places that are dark, warm, and dry, often in eaves or attics. Bats are beneficial, since they eat insects, and in many places it is against the law to kill them. Bats can spread rabies, however, and will bite if cornered. Established bat colonies may be hard to eliminate. At dark, after bats have left to feed, block all crevices and entrances. Light the area to keep bats away for several days until they find a new place to roost.

○ **Raccoons.** Raccoons have adapted to humans and urban growth. They feed at night, but it's not unusual to see them during the day. They prefer wooded areas, but females will nest in dark places such as attics and chimneys to give birth. Raccoons can contract rabies, and they will bite if cornered.

○ **Squirrels.** Like raccoons, squirrels will nest in attics if given the chance. Keep branches clear of buildings, and fill any holes or cracks around eaves and gutters.

WORKING WITH A PCO

Although you need to take most preventive measures yourself, most control and monitoring measures should be carried out by professionals. Employ a licensed, certified PCO to handle pest control. Working as a team, you and the PCO can prevent and/or eliminate pests and keep them from coming back. Few pest problems are solved simply by spraying **pesticides,** which are chemical agents used to destroy pests. There are many advantages of working with a professional PCO. These include the following.

○ A PCO will work with you to develop an IPM approach to using a combination of sanitation, maintenance, and chemical and nonchemical treatments to solve and prevent problems.

○ He or she specializes in pest control and stays up to date on new equipment and products.

○ A PCO provides prompt service to address problems as they occur. Most contracts should provide regular visits plus immediate service when pests are spotted.

○ A PCO is responsible for keeping records on the use and safe handling of potentially hazardous chemicals and all steps taken to prevent and control pests.

How to Choose A PCO

Hiring a PCO is like choosing any other supplier or service provider. You have to do your homework. Use these guidelines to help make your decision.

○ **Talk to other foodservice managers.** Find out what PCO they use and what their experiences have been. When you call a PCO, ask for references and check them thoroughly. Make sure the PCO has experience working with an establishment.

○ **Make sure the PCO is licensed or certified by your state, as required by federal law.** Certification means that the PCO has passed a test on pesticide use and other control methods.

○ **Ask if the PCO belongs to any professional organizations.** Membership in the National Pest Control Association (NPCA) or state or local groups usually means the PCO has up-to-date information on IPM.

○ **Ask for proof of insurance.** Make sure the PCO has adequate coverage to protect you, your employees and customers, and potential damage to the facility.

○ **Weigh all of the factors, not just price.** Make sure the PCO you choose has the expertise you need and resources to provide the service promised. A low bid may be costly in the long run.

Service Contract

Always require a written contract from your PCO. Service contracts outline the work to be performed and what is to be expected from both you and your PCO. Read contracts carefully. Have your lawyer review them if possible. A contract should include these items.

○ A description of services to be provided, including an initial inspection, regular visits to monitor, follow-up visits, and emergency service

○ A warranty for work to be done

Key Point

Most pest control and monitoring measures should be carried out by professional pest control operators.

Key Point

When choosing PCOs, make sure they are licensed or certified, belong to professional organizations, and are insured.

Always require a contract from your PCO. Service contracts should spell out the work to be performed and what is expected from both you and your PCO.

○ Legal liability of the PCO

○ The period of service

○ Your duties, including preventive measures and facility preparation before and after treatment

○ Records to be kept by the PCO, such as the following.

  ● Pests sighted and trapped, location, species, and actions taken

  ● All chemicals used and MSDS for each

  ● Building and maintenance problems noted and fixed

  ● Maps or photos of the facility noting location of traps, bait, and problem spots

  ● Schedule to check and clean traps, replace bait, change bulbs in electrocutor traps, reapply chemicals, and check for seasonal pests

  ● Regular written summary reports from the PCO

TREATMENT

Effective treatment starts with a thorough inspection. Give the PCO complete access to the building. Walk around the building with the PCO and cooperate fully with the inspection.

○ Prepare employees to answer the PCO's questions.

○ Provide building plans and equipment layouts.

○ Point out possible trouble spots.

After the initial inspection your PCO should outline a treatment plan in writing. In addition to price, the plan should contain the following information.

○ **Exactly what chemicals and procedures will be used for each area or problem.** The PCO should outline the risks involved with each treatment.

○ **The dates and times of each treatment.** The federal government requires a PCO to give you enough advance warning to properly prepare the facility. Employees must be evacuated during the treatment.

○ **Steps you can take to control pests.**

○ **Building defects that may cause problems for both preventive and control measures.**

○ **Timing of follow-up visits.** The PCO should review how well treatment is working and suggest alternate treatments if pests reappear.

CONTROL MEASURES

PCOs can use a variety of pest-control methods that are environmentally sound and safe for establishments. They are trained to know which techniques will work best to control different types of pests in your area. New technologies are being developed all the time. The more you know about each of these methods, the better you can evaluate how well your PCO is doing.

Controlling Insects

There are several ways to control insects, depending on the type of insect and degree of infestation.

- **Repellents.** Repellents are liquids, powders, or mists that keep insects away from an area, but do not kill them. Repellents are often used in hard-to-reach places, such as behind wallboards and plaster.

- **Sprays.** Chemical pesticide sprays are used to control roaches, flies, and ants. Because you can buy them in any supermarket, they are easy to misuse and abuse. Let your PCO, not your employees, use spray to control insects. Letting employees apply chemicals is risky, and improperly applied pesticides may be ineffective. Prepare the area to be sprayed by removing all foods and food-contact utensils. Cover equipment and food-contact surfaces that can't be moved. Wash, rinse, and sanitize food-contact surfaces after the area has been sprayed.

  - **Residual sprays** leave a film of insecticide that insects absorb as they crawl across it. They are used in cracks and crevices like those along baseboards. Sprays can be liquid or a dust, such as boric acid.

  - **Contact sprays** kill insects on contact. They are usually used on groups of insects, such as clusters of roaches or a nest of ants.

- **Bait.** Chemical bait is sometimes used to control roaches or ants. The bait contains an attractant. When insects eat it, the chemical kills them. Advantages of using chemical bait are that kitchen areas don't need to be prepped and people don't need to be evacuated.

- **Traps.** Traps are generally used for flying insects such as flies and mosquitoes. There are several types. Wasp and hornet traps are designed to hold nectar, which attracts them. Once inside they cannot escape. Most fly traps use a light source, usually UV light, to attract insects to the trap.

  - Light-only units simply entice insects to crawl inside where they find it hard to escape.

  - Electrocutors or "zappers" use an electrically charged grid to kill insects attracted to the light.

  - Others use both light and chemical attractant to lure insects onto a glueboard trap.

Key Point

After the initial inspection, your PCO should outline a treatment plan in writing.

Courtesy of The National Pest Control Association.

The placement of traps is important. Never place them above or near food preparation areas or food-contact surfaces. Tests have shown that different insects respond to different types or ranges of UV light. Make sure your PCO is using the most current technology.

Controlling Rodents

Rats and mice tend to use the same routes in your facility. Your PCO will choose the best method to eliminate these pests *(see Exhibit 12g)*.

○ **Traps.** Traps are one safe, effective way to kill rats and mice. If the infestation is large, however, traps will take time. Spring traps use food, such as peanut butter, as bait. Food should be kept fresh. Set traps near or in rodent runways. Check traps often, and remove dead rodents carefully. If a trapped rodent is still alive, have your PCO remove it.

○ **Glue boards.** Glue boards work for killing mice. When these devices are placed in runways, mice get stuck to the board and die in several hours from exhaustion or lack of water or air. Check the boards often. Throw away any with trapped mice. Glue boards are not effective for controlling rats. Rats are strong enough to escape the boards.

○ **Bait.** Chemical bait should be used only by a PCO. Chemical bait should be used outdoors where it cannot contaminate food or food-contact surfaces. It is usually placed in special covered, locked containers near rodent runways and possible entry points. Your PCO may change the baits and their locations often until they begin to work. Rats can easily detect chemical bait and often avoid it.

Controlling Birds

Birds can be a serious problem around outdoor dining areas. While there is no way to eliminate birds, your PCO can use several techniques to keep them from nesting and roosting on your building.

○ **Repellents.** Chemical pastes are sometimes used on gutters and ledges to repel birds. Paste must be used carefully so it doesn't fall into food or on tables.

○ **Netting.** Fine-mesh wire netting is used to keep birds from roosting on statues and bas-relief carvings on buildings.

○ **Wires.** On buildings where birds roost on the roof, a PCO may string wires across the roof to prevent birds from alighting. In some cases, the wires may be electrified to administer a mild shock.

○ **Sound.** In some instances, birds can be frightened away by the sound of bird distress calls. Prerecorded tapes are played through loudspeakers near roosting sites. Birds eventually may ignore the sounds, however, and return to roost.

○ **Balloons.** In some cases, birds can be frightened away successfully with helium-filled mylar balloons. Ask your PCO about this technique.

Glue Board

Multi-use traps

Mouse and Rat traps

Exhibit 12g

Methods for Controlling Rodents
Several devices can be used to control rodents, including glue boards, traps, and chemical bait.
Courtesy of The National Pest Control Association.

USING AND STORING PESTICIDES

While it may seem more cost-effective to purchase pesticides yourself, and apply them in your establishment, there are many reasons not to do so.

○ Pesticides can be dangerous to employees, customers, and food.

○ Pests can develop immunity and resistance to pesticides.

○ Each area of the country has its own pest-control problems, and some control measures are more effective than others.

○ Pesticides are regulated by federal, state, and local laws, and some are not approved for use in establishments.

Rely on your PCO to decide if and when pesticides should be used in your establishment. PCOs are trained to determine the best pesticide for each pest, and how and where to apply it.

Never use over-the-counter pesticides unless they are permitted by local codes and recommended by your PCO. If you do use an approved pesticide, you must follow protective procedures.

○ Clean the area before applying the pesticide.

○ Wear personal protective equipment (PPE) that is appropriate for the pesticide, such as gloves, goggles, and a face mask.

○ Follow label directions carefully.

○ Clean, rinse, and sanitize all food-contact surfaces and utensils after spraying.

○ Change clothes and thoroughly wash hands after using pesticides.

Pesticides are hazardous materials. Any time a pesticide is used or stored on the premises you should have a corresponding MSDS. To minimize the hazard to people, use pesticides only when you are closed for business, and evacuate all employees.

Your PCO should store and dispose of all pesticides used in your facility. If you store any pesticides, follow these guidelines.

○ **Keep pesticides in their original containers.**

○ **Store pesticides in locked cabinets away from food storage and preparation areas.** Store them separately from cleaning supplies.

○ **Store aerosol or pressurized spray cans in a cool place.** Exposure to temperatures higher than 120°F (49°C) could cause them to explode.

○ **Check local regulations before disposing of pesticides.** Many are considered hazardous waste.

○ **Dispose of empty containers according to manufacturer's directions and local regulations.** Rinse bottles and non-aerosol cans three times in a utility sink. Never use a food-prep or warewashing sink. Crush bottles and cans and wrap them in paper. Keep them separate from other trash.

Key Point

Pesticides should be applied only by your PCO. A PCO is trained to determine the best pesticide for each pest, and how and where to apply it.

Courtesy of The National Pest Control Association.

Key Point

Never use over-the-counter pesticides unless they are permitted by local codes and recommended by your PCO. If you do use an approved pesticide, follow protective procedures.

SUMMARY

Pests are clearly a menace to establishments because they can carry and spread a variety of diseases. They are dirty and unsightly, and they can pose a physical danger to employees and customers.

To guard against pests, use an integrated pest management (IPM) program. IPM uses a combination of preventive and control measures to eliminate pests and keep them from infesting your facility. Preventive measures focus on two areas: denying pests access to the facility by closing off openings in the building and inspecting all deliveries, and eliminating sources of shelter and food through good housekeeping and sanitation practices.

If pests are detected, control measures may be necessary. Control measures should be used in combination with preventive measures, never as a substitute. Chemical and non-chemical methods may be used to control pests. Most chemical pesticides are toxic to humans. They should be used only by a licensed or certified pest control operator (PCO). Employees should be informed of the hazards posed by pesticides, and they should be evacuated before pesticides are used.

Working with a PCO to develop and implement a continuous IPM program is the best way to prevent and eliminate pest problems. Choose a PCO who is licensed or certified, reputable, insured, and knowledgeable about all types of pest control. Implementing control measures is best left to a professional PCO, but you should be familiar with the methods and chemicals the PCO plans to use.

A CASE IN POINT

Case Study

The Now We're Cooking Restaurant is located in the middle of town on the first floor of a landmark ninety-year-old building. The building is in a shopping area that includes several other restaurants. Fred, the manager, recently had the interior of the restaurant remodeled with marble floors and new ceilings. He chose materials that are modern and easy to clean. All of the new foodservice equipment is designed for cleanability. The cleaning procedures listed on Fred's master cleaning schedule in food preparation areas are written out in detail and completed by the employees as scheduled, with frequent self-inspections. Spills and grease are cleaned up immediately. Storage procedures include FIFO and metal racks that keep food six inches both off the floor and away from the walls.

During a self-inspection two weeks after the remodeling, Fred finds live cockroaches behind the sinks, in the vegetable storage area, in the public restrooms, and near the garbage storage area.

What factors might Fred have overlooked in his recent remodeling that could have led to this infestation? What should Fred do to eliminate the roaches?

TRAINING TIPS

Training Tips for the Classroom

1. "Pests" versus "People"

Objective: *After completing this activity, class participants will be able to identify preventive measures to keep pests from entering a facility, and to describe control measures that can be taken to treat a pest infestation.*

Directions: Break your class into two groups: the "Pests" and the "People." Divide the "Pests" into four subgroups: roaches, flies, rats, and mice. Divide the "People" into two subgroups: foodservice managers and pest control operators (PCOs). Print up handouts for each group with the following questions noted:

Pests

○ How can you get into a facility?

○ Once inside, where do you typically seek shelter?

○ What do you need to survive and reproduce?

○ What do you prefer to eat? Where and when do you eat?

Foodservice Managers

○ How can you prevent roaches, flies, and rodents from entering the facility?

○ What steps can you take to deny these pests food and shelter once they are inside?

Pest Control Operators (PCOs)

○ What methods can you use to control roaches, flies, and rodents once they are inside a facility?

○ What are the strengths and weaknesses of each method?

Ask each group to appoint a spokesperson to present its answers to the rest of the class. Allow the class to discuss each group's answers and provide additional input. Remind your class that a good Integrated Pest Management program, with vigilant prevention and control measures, is essential in the contest with pests.

2. Help Wanted: PCO!

Objective: *After completing this activity, class participants will be able to identify prevention and control measures for pest infestations.*

Directions: Ask the class for three volunteers. These volunteers are PCOs (pest control operators). The class will assume the role of foodservice manager in charge of hiring a new pest control operator.

While the three PCOs are waiting outside the classroom, ask the class to develop a series of interview questions that focus on testing the PCOs' knowledge of preventing and controlling pest infestations. These may include questions such as the following.

○ How can you determine if an establishment has a cockroach infestation?

○ How would you control a cockroach infestation?

○ How would you prevent a rodent infestation?

After interviewing each of the three candidates separately, have the class evaluate the responses and make a choice. Bring all three PCOs back into the classroom, announce the winner, and let a spokesperson for the class explain why the candidate was chosen.

Training Tips on the Job

1. Pest Patrol

Purpose: *To enlist the help of your employees in the early detection and documentation of a pest infestation in your establishment. This information can then be used by your PCO to quickly isolate specific problems and problem areas.*

Directions: Solicit volunteers from your establishment to become members of the Pest Patrol. Explain the signs of insect and rodent infestations to them. Ask members of the Pest Patrol to watch for and report any signs of pest infestation.

When an employee observes signs of an infestation, the sighting should be recorded in a "Pest Log" posted in or near the manager's office.

The log should record all relevant information including what kind of pest was sighted, when, where, and by whom.

2. Pest Prevention Inspection

Purpose: *To identify areas in your facility where pests may enter, and to determine steps that should be taken to prevent entry.*

Directions: Conduct a thorough inspection of your entire facility. If possible, ask your PCO to perform the inspection with you. Pay particular attention to the following areas:

○ Doors, windows, and vents   ○ Floors and walls

○ Pipes exiting the building   ○ The building's foundation

○ Receiving areas (Note: Review your inspection practices to make sure you are inspecting for signs of infestation in incoming deliveries.)

Once possible entry points have been identified, a plan should be outlined for correcting these problems. While conducting the inspection, look for possible sources of food and shelter for potential pests. Check the way you currently clean the facility, store foods and supplies, and store and dispose of garbage and recyclables. Identify ways in which you might improve your current practices.

DISCUSSION QUESTIONS

1. Identify some ways that pests can be denied access to a building.

2. What pest problems are associated with outdoor dining? How can you prevent them?

3. List three signs that a foodservice facility is infested with cockroaches. List three signs that a foodservice facility is infested with rodents.

4. Why is it important to be able to identify the type of pest problem you have?

5. What factors should be considered when choosing a pest control operator?

MULTIPLE-CHOICE STUDY QUESTIONS ???

1. Which of the following is not a sign of a rodent infestation?
   A. Shiny black droppings
   B. Scraps of paper and cloth gathered in the corner of a drawer
   C. Small holes in a wall
   D. A strong oily odor

2. Your manager has asked you to check how pesticides used at the establishment are being stored. What should you check for?
   A. The container should be labeled with full manufacturers' information
   B. The brand of pesticide should be approved for commercial use by OSHA
   C. The pesticide used must be effective against cockroaches
   D. The containers should be recyclable

3. The health department inspector has reported that the restaurant next door to you has a major problem with cockroaches. She checks your kitchen and finds all of the following situations present. Which one would suggest that you might also have a cockroach problem?

   A. She sees that the screen on the service door needs to be repaired.
   B. She picks up a bit of sawdust from a corner of the dry storeroom.
   C. She sees black grains that look like pepper under the refrigerator.
   D. She finds small holes burrowed into the storeroom.

4. Which of the following is not a measure for controlling pests?

   A. Sealing the space around a pipe that exits the building
   B. Setting up a spring trap in the corner of the room
   C. Placing a chemical bait trap in the dishwashing area
   D. Setting out a multi-catch trap behind the icemaker machine

5. Your county health and safety consultant has recommended a program of Integrated Pest Management for your bed and breakfast establishment. What should you do first?

   A. Create a plan that emphasizes control measures.
   B. Position glue boards around the establishment.
   C. Work with a PCO to create a prevention and treatment plan.
   D. Work with the county government to develop a treatment plan.

6. You have an outdoor dining area at your establishment. Which of the following practices would you want to avoid?

   A. Using lights that do not attract insects
   B. Using citronella candles
   C. Installing a "bug zapper" in a serving area
   D. Mowing the grass prior to opening the patio

7. Your storeroom has a white line painted on the floor around the perimeter of the room. It extends six inches from the wall. What is the purpose of this line?

   A. It seals the floor against moisture seeping from the walls.
   B. It serves as a reminder to employees to keep stored items away from the walls.
   C. It works as a repellent against nesting mice.
   D. It works as a repellent against cockroaches.

8. What is the primary reason that an establishment must act responsibly to control infestations of roaches?

   A. Roaches can disturb the foundation of the establishment.

   B. Roaches can scare away customers.

   C. Roaches eat large quantities of stored foods.

   D. Roaches carry disease-causing microorganisms that can make customers ill.

9. If pesticides are used in an establishment, containers should be

   A. discarded in the Dumpster when empty.

   B. rinsed in a food-prep sink when empty.

   C. rinsed in a warewashing sink when empty.

   D. rinsed in a utility or service sink when empty.

Resources

ADDITIONAL RESOURCES

Books and Periodicals

Buettner, R., Fan, M., & George, T. (1998). Guess who's crawling to dinner…uninvited: Majority of eateries in city are losing the war with vermin. (1993). *New York Daily News,* June 2, p. 7.

Mixon, J. A. (1991). *Guidelines for the storage and care of food products: A technical assistance manual: Vol. V.* (3rd ed.). Washington, DC: Food Industry Services Group, USDA Food and Nutrition Service.

Pest management in restaurants: An integrated approach. (1997). Washington, DC: National Restaurant Association, Technical Services, Public Health and Safety Dept.

Web Sites

1999 FDA Model Food Code

http://vm.cfsan.fda.gov/~dms/fc99-toc.html

Complete outline of the FDA's latest code for regulating operations that provide food directly to consumers. Also includes a quick synopsis of changes from the 1997 Food Code.

FDA-CFSAN Chemistry Information on Internet

http://vm.cfsan.fda.gov/~dms/chemist.html

FDA food-safety information with a chemistry focus. Includes research information, as well as publications and databases.

FDA-CFSAN Pesticides, Metals, and Chemical Contaminants

http://vm.cfsan.fda.gov/~lrd/pestadd.html

Technical information on pesticides, toxic metals and chemicals that can adversely affect food.

National Pest Control Association

http://www.pestworld.org

The association representing pest management firms offers resources and information for homeowners, professionals, and the media.

National Restaurant Association

http://www.restaurant.org

The National Restaurant Association site provides information on government agencies affecting the restaurant industry, the latest training and certification updates, and links to state restaurant associations and hospitality schools and universities.

Occupational Safety and Health Administration (OSHA)

http://www.osha.gov

OSHA's Web site provides news releases, regulations, and a library to search related topics and past publications.

UNIT 4

SANITATION MANAGEMENT

Almost 50 Billion meals are eaten or prepared away from home each year. This makes training managers and employees a critical component to protecting public health. I effectively use the ServSafe training materials for the Texas Agricultural Extension Services Food Protection Management training program. This program has a ninety-percent pass rate.

Peggy Van Laanen, Ed., R.D., LD, CFCS
Associate Professor and Extension Nutrition Specialist
Texas Agricultural Extension Service
Texas A&M University

Chapter 13
Food-Safety Regulations and Standards

TEST YOUR FOOD-SAFETY KNOWLEDGE

1. **True or False:** The FDA writes the food regulations that must be followed by each establishment. *(See The Food Code, page 13-3.)*
2. **True or False:** Foodservice inspectors are generally employees of the Centers for Disease Control and Prevention (CDC). *(See Government Regulatory System, page 13-3.)*
3. **True or False:** You should ask to accompany the inspector during the inspection. *(See Traditional Inspection System, page 13-6.)*
4. **True or False:** Critical violations noted during an inspection should be corrected within one week of the inspection. *(See Traditional Inspection System, page 13-8.)*
5. **True or False:** A HACCP-based inspection focuses on the flow of food in an establishment as opposed to the sanitary appearance of the facility. *(See HACCP-Based Inspections, page 13-10.)*

Key Terms

USDA (U.S. Department of Agriculture)
FDA (U.S. Food and Drug Administration)
USPHS (U.S. Public Health Service)
Model Food Code Regulations
Health inspector
CDC (U.S. Centers for Disease Control and Prevention)
EPA (U.S. Environmental Protection Agency)
NMFS (U.S. National Marine Fisheries Service)

There are several reasons why it is important to have a foodservice inspection program. Most important is the fact that failure to ensure food safety can jeopardize the health of your customers and could cost you your business. All establishments, including quick-service and fine dining restaurants, delis, hospitals, nursing homes, and schools, must follow standard food-safety practices critical to the safety and quality of the food served.An inspection system lets the establishment know how well it is following these practices.

OBJECTIVES OF A FOODSERVICE INSPECTION PROGRAM

All establishments that serve the public must provide safe food and are subject to inspection. It does not matter whether there is a charge for the food, or whether the food is consumed on or off premises.

The purpose of an inspection program is as follows.

○ To evaluate the minimum sanitation and food-safety practices within the establishment.

○ To protect the public's health by requiring establishments to provide food that is safe, uncontaminated, and properly presented.

○ To convey new food-safety information to an establishment.

○ To provide an establishment with a written report, noting deficiencies, so that the establishment can be brought into compliance with safe food practices.

GOVERNMENT REGULATORY SYSTEM

Every country has a history of government involvement in the development of health laws. The first government health laws in this country were adopted more than four hundred years ago. Today, government control in the United States is exercised at three levels: federal, state, and local.

At the federal level, the **U.S. Department of Agriculture (USDA)**, the **Food and Drug Administration (FDA)**, and the **U.S. Public Health Service (USPHS)** are directly involved in the inspection process.

The USDA is responsible for inspection and quality grading of meats, meat products, poultry, dairy products, eggs and egg products, and fruits and vegetables shipped across state boundaries. The USDA provides these services through the Food Safety and Inspection Service (FSIS) agency.

The FDA is the agency that writes recommendations for foodservice regulations (based on input from the Conference for Food

Key Point

The FDA writes the Model Food Code, which provides recommendations for foodservice regulations.

Protection [CFP]). These recommendations are commonly known as the **Model Food Code.** In addition, the FDA inspects foodservice operations that cross state borders (interstate establishments such as those on planes and trains, as well as food manufacturers and processors) because they overlap the jurisdictions of two or more states. The FDA shares responsibility with the USDA for inspecting food processing plants to ensure standards of purity, wholesomeness, and compliance with labeling requirements. The USPHS inspects cruise ships that cross international borders.

In the United States, most food **regulations** that affect restaurant and foodservice operations are written at the state level (except regulations for interstate or international establishments, which are determined at the federal level). Each individual state decides whether to adopt the Model Food Code or some modified form of it.

State regulations may be enforced by local (city or county) or state health departments. In a large city, the city health department will probably be responsible for enforcing health codes. In smaller cities or in rural areas, a county or state health department may be responsible for enforcement. In any case, the manager must be familiar with the local agencies and their enforcement system. City, county, or state **health inspectors** (also called sanitarians, health officials, or environmental health specialists) conduct foodservice inspections in most states. They generally are trained in food safety, sanitation, and in public health principles and methods.

THE FOOD CODE

The Model Food Code is written by the FDA and lists the government's recommendations for foodservice regulations. Currently, these recommendations are updated every two years to reflect developments in the foodservice industry and the field of food safety.

The Model Food Code is intended to assist state health departments in developing regulations for a foodservice inspection program. It is not an actual law. Although the FDA recommends adoption by the states, it cannot require it. Rather, the Model Food Code represents the FDA's best advice for a uniform system of regulation to ensure food safety. Some states use the Food Code as a basis for developing their own codes, instead of adopting it in its entirety. Food codes are written very broadly, and generally cover the following areas.

○ Foodhandling and preparation: sources, receiving, storage, display, service, transportation

○ Personnel: health, personal cleanliness, clothing, practices

Key Point

Recommendations for foodservice regulations are written at the federal level, regulations are made at the state level, and enforcement is carried out at the local level.

Federal Level

Regulations recommended

State Level

Regulations written

Local Level

Regulations enforced

Key Point

Managers must contact local health departments to find out which specific regulations apply to their operations.

○ Equipment and utensils: materials, design, installation, storage

○ Cleaning and sanitizing of the facility and equipment

○ Utilities and services: water, sewage, plumbing, restrooms, waste disposal, integrated pest management (IPM)

○ Construction and maintenance of floors, walls, ceilings, lighting, ventilation, dressing rooms, locker areas, storage areas

○ Mobile and temporary foodservice units

○ Compliance procedures: foodservice inspections and enforcement actions

Adoption and interpretation of food-safety standards may vary widely from one state to another or in some cases one locality to another. Therefore, foodservice managers must consult with local health departments to find out which specific regulations apply to their operations.

Currently, when a state adopts, develops, or amends its food code, it is required to provide a time for public input and comment. This is a time when any interested person or association can comment on how the proposed changes will impact their establishment(s). If you are interested in participating in this process, make sure you keep in contact with your health department or state restaurant association.

Key Point

Establish-ments can provide input on the code requirements in their area.

The lack of uniformity in food codes can frustrate efforts by the restaurant industry to establish uniform food-safety standards. For example, some jurisdictions require refrigerated food temperatures to be 45°F (7°C) or lower, while others require 41°F (5°C) or lower. States may also differ in the recommended inspection frequency. Some states require inspections for foodservice operations at least every six months; others schedule inspections more or less frequently. Some states may even differ as to which establishments should be inspected; for example, some states do not inspect convenience food stores. Some states may have separate regulations, or agencies, for different types of foodservice operations (such as vending machines or delis in grocery stores).

Although all inspectors focus on food-safety practices, the areas emphasized during the inspection can vary among individual inspectors. For example, some inspectors are more concerned with refrigeration temperatures while others may focus on the physical appearance of the facility. One inspector may examine the flow of food while another may focus primarily on the personal hygiene of employees. It is the responsibility of the manager to keep food safe and wholesome throughout the establishment at all times, regardless of the inspector or the inspection process.

FOODSERVICE INSPECTION PROCESS

Well-managed establishments will perform continuous self-inspections to protect food safety, in addition to the regular inspections performed by the health department. Establishments with high standards for sanitation and food safety consider health department inspections only a supplement to their self-inspection programs.

There are many benefits that come from establishing a good self-inspection program. A good program can result in safer food, improved food quality, a clean environment for employees and customers, and higher inspection scores. Strive to exceed the expectations of both your customers and the health department. The higher you set your standards, the more likely you are to do well on your health department inspection. In addition, your customers will notice your commitment to providing them with a safe and sanitary dining experience.

During health department inspections, the local health code serves as the inspector's guide. You should keep a current copy of your local or state sanitation regulations. Regularly compare the code to procedures at your establishment, but remember that code requirements are only minimum standards to keep food safe.

Keep in mind that changes continue to occur in the inspection process. Some areas use traditional scored inspection systems with a number or letter grade, while others use HACCP-based inspections, or a combination of the two.

Key Point

Health department inspections should be a supplement to a self-inspection program.

Traditional Inspection System

Some health departments are required to conduct inspections at least every six months. However, the frequency will vary depending on the area and type of establishment. Many use a risk-based approach to inspection frequency. Determining factors can include the following.

○ **Size and complexity of the operation.** Larger operations, which offer a considerable number of potentially hazardous foods, may be inspected more frequently.

○ **The inspection history of the establishment.** Establishments with a history of low sanitation scores or consecutive violations may be inspected more frequently.

○ **The clientele's susceptibility to foodborne illness.** Nursing homes, schools, daycare centers, and hospitals may receive more frequent inspections.

○ **The thoroughness of the operation's HACCP program.** Establishments with HACCP programs may be inspected less frequently.

○ **The workload of the local health department and the number of inspectors available.**

Key Point

The frequency of regulatory inspections will vary based on several factors, including complexity of the establishment's menu, the size of the operation, and the population it serves.

In most cases, the inspector will arrive for an inspection without prior warning and ask for the manager or the person in charge of the operation. In your absence, make sure your employees know who is in charge. Establishments should not refuse entry. In some jurisdictions inspectors may have the authority to gain access or to revoke the establishment's permit for refusal to allow inspection.

It is best to accompany the inspector during the inspection so that you can answer any questions and identify the actual location of any deficiency. This will allow you the possibility of correcting deficiencies immediately, which demonstrates a sense of urgency. Accompanying the inspector will also give you the opportunity to learn from the inspector's comments and suggestions and gain advice on sanitation. Some companies have specific policies on how to handle an inspection. Make sure that you are aware of those policies.

After the inspection, the inspector will discuss the results and the score (if a score is given) and arrange for any follow-up, if necessary. Managers will be asked to sign the inspection report to acknowledge that they have received it. Follow your company's policy regarding this issue. A copy of the report is then given to the manager or person in charge at the time of the inspection. All deficiencies noted on the report should be acted upon. Critical deficiencies should be corrected immediately or within forty-eight hours of the inspection. All other deficiencies should be corrected as soon as possible. Copies of all reports should be kept on file in the establishment, and referred to when planning improvements and assessing facility goals. Copies of reports are kept on file at the health department. These are considered public documents and may be made available to the public upon request.

The following suggestions will enable managers and operators to get the most out of sanitation inspections.

1. **Ask for identification.** Many inspectors will volunteer their credentials. Do not let anyone enter the back of the facility without proper identification. Ask the purpose of the visit; make sure that you know whether it is a routine inspection, the result of a customer complaint, or for some other purpose.

2. **Cooperate.** Most inspectors have learned to expect some defensiveness and resentment from foodservice operators. This can be interpreted as having something to hide. Answer all of the inspector's questions to the best of your ability. Instruct employees to do the same. Explain to the inspector that you wish to accompany him or her during the inspection. This will encourage open communication and a good working relationship.

3. **Take notes.** As you accompany the inspector, make a note of any problem that is pointed out. Make it clear that you are willing to correct problems.

Key Point

Accompany the sanitarian during the inspection so you can answer any questions and identify the actual location of any deficiency.

If a deficiency can be corrected right away, do so, or tell the inspector when it can be corrected. Taking your own notes will help you remember exactly what was said. If you believe that the inspector is incorrect about something, note what was commented upon. Then ask the inspector's supervisor for a second opinion.

4. **Keep the relationship professional.** Don't offer food or drink before, during, or after an inspection. This could be viewed as bribery.

5. **Prepare to provide records requested by the inspector.** You may ask the inspector why they are needed. If a request appears inappropriate, you can check with the inspector's supervisor or with your lawyer about limits on confidential information. Records that you provide to the inspector will become part of the public record. Inspectors may ask for records of purchases to verify that food is from an approved source, as well as records of IPM treatments, and a list of all chemicals used in the facility. HACCP records may be requested in some cases. HACCP records may be an important part of an inspection as they document the establishment's efforts to ensure food safety. Having a HACCP system in place clearly shows the inspector that you are committed to food safety.

6. **Discuss violations and time frames for correction with the inspector.** The inspection report should be studied closely. Deficiencies and comments should be discussed in detail with the inspector *(see Exhibit 13a)*. If any deficiencies were corrected on the spot, make sure they are noted. In order to make complete and permanent corrections, you will need to know the exact nature of the violation, how it impacts food safety, how to correct it, and whether or not the inspector will do any follow-up. The inspector sees many operations and may offer expert advice on how to correct deficiencies.

7. **Follow up.** Take the inspection report with you through your facility and correct the problems. Determine why each problem occurred by evaluating sanitation procedures, the master cleaning schedule, and employee

Exhibit 13a **The Inspection Process**
Violations and comments should be discussed in detail with the inspector.

(see Exhibit 13b).

foodhandling practices and training. Establish new procedures or revise existing ones to correct the problem permanently *(see Exhibit 13b).*

Many health departments use a traditional inspection system that uses a demerit scoring scale *(see Exhibit 13c).* Usually the highest possible score is one hundred points. For every violation, between one and five points are subtracted from one hundred to get the final score. Minor violations worth one or two points must be corrected by the time of the next routine inspection. Larger violations worth four to five points must be corrected within a time frame given by the inspector. Other health departments use a letter to score the establishment. Whatever scoring system is used, make sure you understand the scoring scale and your options for improving an unsatisfactory score.

Exhibit 13b   Inspection Follow-Up

Determine why each violation occurred by evaluating sanitation procedures, the master cleaning schedule, and employee foodhandling practices and training. Establish new procedures or revise the existing ones.

Establishments are generally given a short amount of time (forty-eight hours or less) to correct major violations and improve their overall score. If a low score is received upon reinspection, the establishment may be fined or even closed.

In some states, if the inspector determines that a facility poses an immediate health hazard, he or she may ask for a voluntary closure, or issue an immediate suspension of a permit to operate. A suspension requires the approval of the local health offices and means that operations at the establishment must cease immediately. A closure is issued when the health department feels that an establishment poses an immediate and substantial health hazard to the public. Examples of hazards that call for closure include:

Key Point

An establishment can be closed when the health department feels it poses an immediate and substantial health hazard to the public.

- A significant lack of refrigeration
- A backup of sewage into the establishment itself or its water supply
- An emergency, such as a building fire or flood
- A serious infestation of insects or rodents
- A long interruption of electrical or water service

FOOD AND SANITATION INSPECTION REPORT

Date	
Address	Inspection Time
Inspector	

Establishment Name

Managers Name

Based on an inspection this day, the items marked below identify the violations in operation or facilities which must be corrected by the next routine inspection or such shorter time as may be specified in writing by the regulatory. Failure to comply with any time limits for corrections specified in this notice may result in cessation of your Food Service operations.

Item #	Wt	
		FOOD
01	5#	From approved, licensed sources, free of spoilage, no home processed foods, no dented or damaged cans.
02	1	Food containers properly labeled.
		FOOD PROTECTION
03	5*	Potentially hazardous foods must be maintained at 40°F or 140°F or above; dairy products, meat, poultry, fish, cooked or broiled potatoes, beans and rice. Cooling: 1) Shallow pans, product 3 inches in depth, refrigerated. 2) Pre-cooled in ice bath with frequent stirring to 45°F. **NO COOLING AT ROOM TEMPERATURE.**
04	4*	Adequate equipment to maintain proper food temperature.
05	1	Accurate thermometers; conspicuous.
06	2	Potentially hazardous foods properly thawed. 1) in a refrigerator, 2) under cold running water, 3) by microwave, 4) as part of cooking.
07	4*	Unwrapped & potentially hazardous foods not re-served.
08	2	Food protected from contamination; ie, covered, off floors, etc.
09	2	Handling of food (ice) minimized, proper utensils provided and used.
10	1	Food dispensing utensils properly stored when in use.
		PERSONNEL
11	5*	Personnel with infectious disease, cuts or burns restricted from handling food, clean utensils/equipment.
12	5*	Hands washed, good hygienic practices. No smoking, eating or drinking in kitchen.
13	1	Clothes clean, hair restrained.
		FOOD EQUIPMENT
14		Food contact surfaces non-toxic, smooth, durable, non-absorbent and easily cleanable.
15	1	Non-food surfaces smooth, durable, non-absorbent and easily cleanable.

Item #	Wt	
16	2	Dishwashing facilities properly designed, constructed, maintained, installed, located, operated.
17	1	Dishwashing facilities provided with accurate thermometers, pressure gauge, chemical test kit.
18	1	Soiled equipment, dishes and utensils pre-flushed, scraped, soaked.
19	2	Wash and rinse water clean, hot.
20	4*	Washing and Sanitizing 3-compartment sink: 1) wash 3) sanitize 2) rinse 4) air dry. Sanitizers: 50 ppm chlorine or 200 ppm quaternary ammonia or 12.5 ppm iodine for 1 minute. Dishmachine: 180°F final rinse temperature or 50 ppm chlorine at dish level.
21	1	Wiping cloths: Stored in 100 ppm chlorine or 25 ppm iodine.
22	2	Food contact surfaces of equipment: cutting boards, meat slicers, can openers, work counters, etc. shall be washed, rinsed and sanitized. Surfaces free of abrasives and detergents.
23	1	Non-food contact surfaces of equipment & utensils clean.
24	1	Clean equipment and utensils properly stored and handled to prevent contamination.
25	1	Single service items properly stored and used to prevent contamination.
26	2	No re-use of single-service articles.
		WATER
27	5*	Safe water source. Hot and cold water provided at all times.
		SEWAGE
28	4*	Sewage and waste (mop) water properly disposed
		PLUMBING
29	1	Plumbing properly installed and maintained
30	5*	No cross-connection between potable and wastewater. Backflow devices present.
		TOILET AND HANDWASHING FACILITIES
31	4*	Handwashing sinks accessible.

Item #	Wt	
32	2	Handsinks provided with soap and single service towels. Toilet rooms clean, in good repair, doors self-closing, trash receptacles.
		GARBAGE AND REFUSE
33	21	Garbage containers sanitarily maintained; covered
34	1	Outside are maintained, clean.
		INSECT AND RODENT CONTROL
35	4*	No insects, rodents or other animals present. Outer openings protected. IPM policies in place.
		FLOORS, WALLS AND CEILINGS
36		Floors clean and constructed to be smooth, durable, nonabsorbent and properly covered.
37	1	Walls, ceilings and attached equipment clean and constructed to be smooth, durable and nonabsorbent.
		LIGHTING
38	1	Lighting adequate shielded.
		VENTILATION
39	1	Rooms and equipment vented as required.
		DRESSING ROOMS
40	1	Employee's personal items stored in locker or separate area
		OTHER OPERATIONS
41	5*	Toxics/Cleaning agents stored away from food preparation and utensil washing areas, toxics labeled.
42	1	Premises clean, maintained. No unnecessary articles. Cleaning maintenance equipment properly stored. Authorized personnel only.
		FIRE EXTINGUISHERS
43	1	Sufficient number, servicing performed, personnel familiar with use.
		FOOD AND SANITATION TRAINING PROGRAM
44	1	

Exhibit 13c **Traditional Establishment Inspection Report Form**

The suspension order may be posted at a public entrance to the establishment; however, this would not be required if the establishment closed voluntarily. Regulations vary from one jurisdiction to another, but to reinstate a permit to operate, the establishment must eliminate the hazard or hazards that caused the suspension, and then pass a reinspection.

The health department may request a hearing if it feels that an establishment must correct a hazard, but the department does not wish to suspend the establishment's permit. The establishment may also request a hearing if it feels that an action by the health department is unjustified. There is often a time limit to request such a hearing (usually within five to ten days after an inspection). Check local regulations to determine these limits.

One limitation of the traditional inspection system centers on the fact that multiple violations of the same type do not cause any more demerits than a single violation. For example, leaving one potentially hazardous food at room temperature incurs the same number of demerits as leaving all foods at room temperature. Clearly, leaving all foods at room temperature causes a greater health concern.

As a result, a new inspection system is being implemented in some areas. In this type of inspection system, a narrative format is used to report violations rather than a demerit system. When the inspector sees problems with the facility, he or she writes up the problems and recommends corrections in paragraph form. Reference numbers to the code and repeat violations are noted. Multiple violations of the same type receive greater emphasis. In the old scoring system violations were considered *critical* (those worth four or five demerits each) or *noncritical* (one or two demerits each). The new system allows inspectors to use their professional judgment regarding some of the violations. It also provides establishments with more feedback. See *Exhibit 13d* for an example of the new inspection report form.

HACCP-Based Inspections

Some health departments use HACCP-based inspections, which focus on the flow of food rather than on the sanitary appearance of the facility. Inspectors may observe the way an establishment receives, stores, prepares, and serves food, and the Critical Control Points for each step. Since this type of inspection might be viewed as complex and time-consuming, it may only be performed under special circumstances. A health department may perform a HACCP-based inspection to do the following.

○ Trace the source of contamination following a report of foodborne illness.

○ Evaluate an establishment with significant hazards.

○ Assist an establishment in converting to a HACCP system.

DEPARTMENT OF HEALTH AND HUMAN SERVICES
PUBLIC HEALTH SERVICE
FOOD AND DRUG ADMINISTRATION

FDA

FOOD ESTABLISHMENT INSPECTION REPORT

Violations cited in this report shall be corrected within the time frames specified below, but within a period not to exceed 10 calendar days for critical items
(§ 8–405.11) or 90 days for noncritical items (§ 8–406.11)

VIOLATIONS: CRITICAL _____ NONCRITICAL _____

PERMIT NUMBER: _____ DATE: _____

ESTABLISHMENT: _____ CITY: _____ STATE: _____ ZIP: _____

ADDRESS: _____ TELEPHONE: _____

PERSON IN CHARGE / TITLE: _____

INSPECTOR / TITLE: _____ OTHER: _____ TIME: _____

INSPECTION TYPE: ROUTINE FOLLOW-UP COMPLAINT

Critical (X)	Repeat (X)	Code Reference	Violation Description / Remarks / Corrections

Food Establishment Inspection Report Page _____ of _____

Exhibit 13d New Establishment Inspection Report Form Recommended by the FDA

Other health agencies are using the HACCP-based approach to decide how often different establishments should be inspected, or to determine how the inspections should be conducted. Scoring is done differently. Instead of using a point system with demerits, HACCP-based inspections determine critical violations. Many of these critical violations are similar to the four- and five-point items identified in the traditional inspection system. *Exhibit 13e* shows one type of form used during a HACCP-based inspection.

FEDERAL REGULATORY AGENCIES

Several federal government agencies work to protect the sanitary quality of foods purchased by an establishment. Earlier in this chapter, the roles of the FDA, USDA, and USPHS were discussed. A few other agencies should be mentioned.

The **U.S. Centers for Disease Control and Prevention (CDC)**, located in Atlanta, Georgia, are agencies of the U.S. Public Health Service. The Centers provide the following services.

○ Investigate outbreaks of foodborne illness.

○ Study the causes and control of disease.

○ Publish statistical data and case studies in the *Morbidity and Mortality Weekly Report (MMWR)*.

○ Provide educational services in the field of sanitation.

○ Conduct the Vessel Sanitation Program, an inspection program for cruise ships.

The **Environmental Protection Agency (EPA)** sets standards for air and water quality, and regulates the use of pesticides (including sanitizers) and the handling of wastes.

The **National Marine Fisheries Service (NMFS)** of the U.S. Department of Commerce implements a voluntary inspection program that includes product standards and sanitary requirements for fish-processing operations.

VOLUNTARY CONTROLS WITHIN THE INDUSTRY

Few industries have devoted as much effort to regulating themselves as the restaurant and food processing industries. Scientific and trade associations, manufacturing firms, and foodservice corporations have vigorously pursued programs to raise the standards of the industry through research, education, and cooperation with government. Although participation in these programs is voluntary, these organizations have actively promoted professional standards, recommended legislative policy, sponsored uniform enforcement procedures,

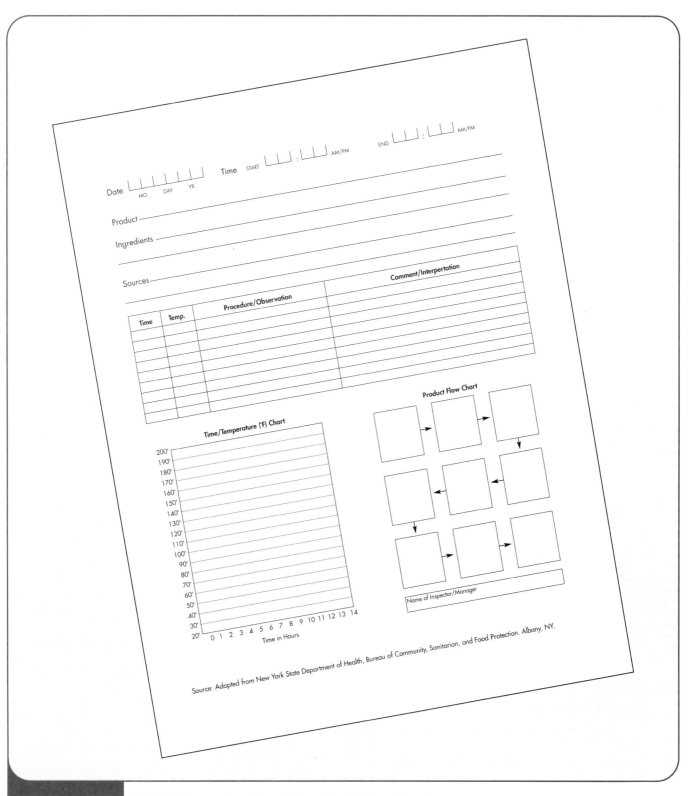

Exhibit 13e HACCP Inspection Form Using Food Flow Charting

and provided educational opportunities. The overall results in food safety have included the following.

○ An increased understanding of foodborne illness and its prevention

○ Improvements in the design of equipment and facilities which aid in cleaning

○ Industry-wide efforts to maintain the sanitary quality of foods during processing, shipment, storage, and service

○ Efforts to make foodservice laws science based, and more uniform and practical

While many organizations have contributed to these endeavors, those having the most relevance for the foodservice operator and manager include the following.

○ The National Restaurant Association, founded in 1919, is the leading business association for the restaurant and foodservice industry. The Association's mission is to represent, educate, and promote the rapidly growing industry. The Association is a strong voice that advocates proper food-safety practices and science-based regulations. The Technical Services and Public Health and Safety Department of the Association provides information to the industry on topics such as sanitation, pest control, nutrition, and the Americans with Disabilities Act. In addition, it works closely with government agencies to develop and implement proper guidelines and regulations regarding the food supply. Through its excellent Research Department, the Association generates both operator and consumer research on a variety of issues and trends, including the annual Restaurant Industry Operations Report and the Restaurant Industry Forecast. The Association's successful lobbying arm represents the interests of the industry by being in constant contact with legislators. Each May, the National Restaurant Association holds the Restaurant, Hotel-Motel Show, one of the largest trade shows in the United States. More than one hundred thousand people come to Chicago during the show to see everything from tablecloths and plates to industrial warewashers and food-safety systems. For more information about the National Restaurant Association and its services and member benefits, call 800-424-5156 or 202-331-5900.

○ The National Restaurant Association Educational Foundation is a not-for-profit organization dedicated to developing, promoting, and providing educational and training solutions for the restaurant and hospitality industry. The Foundation has developed training programs and educational curricula regarding issues such as workplace safety, responsible beverage alcohol service, prevention of sexual harassment, and food safety and sanitation. In 1993, the Foundation formed the International Food Safety Council to focus industry attention on the importance of food-safety

Key Point

The National Restaurant Association works closely with government agencies to develop and implement proper guidelines and regulations regarding the food supply.

Key Point

The National Restaurant Association Educational Foundation is dedicated to developing, promoting, and providing educational and training solutions for the restaurant and hospitality industry.

training and to promote the industry's food-safety efforts to the public. Restaurant and foodservice professionals certified in ServSafe automatically become members of the Council. Members display the Council seal to communicate their food-safety commitment to their employees and to the public. The Council also provides information and resources to help members manage and continually enhance food safety.

As the educational resource for the industry, the Foundation administers the largest scholarship program of its kind for the restaurant and hospitality sector. Whether students are beginning their restaurant and hospitality education or continuing a successful career, the Foundation offers scholarships to help them. Scholarships are available for senior high school students, undergraduate and graduate students, industry professionals, and educator/administrators.

For more information about the National Restaurant Association Educational Foundation and its products and services, call 800-765-2122 ext. 701 (outside the Chicago area) or 312-715-1010 ext. 701 (within the Chicago area). For information or applications for the Scholarships Program, call 312-715-1010 ext. 733.

○ The National Environmental Health Association (NEHA) is an organization of environmental specialists and professionals, including those responsible for food inspection services and environmental health programs. NEHA's Registered Sanitarian (RS) program has established national standards for education, experience, and testing for health inspectors.

○ The International Association of Milk, Food, and Environmental Sanitarians (IAMFES) was a pioneer in the highly successful United States milk sanitation program. It publishes information on milk and food safety, and on the generally accepted standard procedures for investigating a foodborne illness.

○ The Institute of Food Technologists (IFT) is a multidisciplinary scientific society with expertise in technology, research, education, manufacturing, and the safety of foods in the foodservice industries. IFT sponsors food and restaurant industry research in a variety of areas such as food quality, safety, and process controls.

○ The National Society of Professional Sanitarians (NSPS) and the American Academy of Sanitarians are composed of food industry professionals and sanitarians. They are concerned with environmental health protection policies at the national, state, and local levels.

○ The Association of Food and Drug Officials (AFDO) develops and publishes food sanitation codes and encourages food protection through the adoption of uniform legislation and enforcement procedures.

○ The Council of Hotel and Restaurant Trainers (CHART) is an association of foodservice trainers and human resource professionals. CHART was formed in 1971 to provide members with a forum in which to grow professionally and to increase their effectiveness as trainers.

○ The Frozen Food Industry Coordinating Committee has developed the Code of Recommended Practices for the Handling of Frozen Food, which describes procedures to be used from the processing stage to retail food service.

○ The National Pest Control Association (NPCA) consists of licensed and certified Pest Control Operators (PCOs) throughout the United States. NPCA provides guidelines and training materials for integrated pest management (IPM) treatment programs, for hazard communications, and for other topics relating to safety and sanitation.

○ NSF International (formerly the National Sanitation Foundation) will evaluate and list foodservice equipment if it meets their standards, and give it the NSF International mark. NSF International develops and publishes widely accepted standards for equipment design, construction, and installation, which are updated every five years. Listing of equipment that meets these standards is updated every six months. NSF's approach demonstrates the kind of progress that the foodservice industry can achieve through voluntary programs.

○ Underwriters Laboratories, Inc. (UL) performs a similar service, by listing equipment that meets NSF International standards. In addition, UL lists electrical equipment that passes its own safety requirements.

SUMMARY

Public and private organizations and agencies can offer valuable assistance to foodservice managers in meeting their commitment to food safety. It is up to the manager to use available help and to maintain a sanitary establishment.

Federal governmental agencies create standards that affect the establishment directly or indirectly. The Model Food Code, developed by the Food and Drug Administration (FDA), serves as a guideline for a large number of state and local regulations. In addition, the FDA regulates the purity and safety of foods in interstate commerce.

While state and local health codes vary throughout the nation, virtually all of them contain provisions governing food safety; personal hygiene; sanitary facilities, equipment, and utensils; safe operating practices; training; and enforcement procedures. The inspector, a representative from the state or local health department, is a professional in sanitation and public health.

One highly effective way to produce safe food is to use the HACCP food-safety system and to follow a daily program of self-inspection. The effectiveness of this system will be reflected in inspection reports. HACCP-based inspections, a departure from the traditional emphasis on sanitary facilities, are being adopted by some regulatory agencies. Inspectors focus on causes of foodborne illness by observing the way an establishment receives, stores, prepares, and serves food, and may designate Critical Control Points for each step.

Professional and trade organizations in food service and public health recommend guidelines for foodservice sanitation, investigate sanitation problems, and promote best practices. These associations also develop educational programs and conduct research into the causes and prevention of foodborne illness.

A CASE IN POINT I

Case Study

Carolyn, the inspector from the city's public health department, was standing at the door of Jerry's Diner. Jerry greeted her and they both walked to the kitchen. Carolyn pulled out a HACCP worksheet. While Jerry explained that the cook was preparing a special of stir-fried chicken and vegetables, Carolyn noted the ingredients and their sources. Then she observed the preparation procedures and noted times and temperatures, which she plotted on a time-temperature graph. She also filled in a product flowchart and indicated the Critical Control Points for the chicken special. Jerry told her what corrective actions his employees were trained to carry out if the standards of the Critical Control Points were not met.

Next, Carolyn checked the concentration of the sanitizing solution in the three-compartment sink that the dishwasher used for manual cleaning, rinsing, and sanitizing of equipment. She also checked the handwashing station for the foodhandlers.

Jerry and Carolyn went to Jerry's office where they discussed the report. Jerry compared his flowchart for the stir-fried chicken and vegetables with the one Carolyn had done, and determined where changes could be made to improve the monitoring system.

Did Jerry handle the inspection correctly? What does Jerry need to do following this inspection?

TRAINING TIPS

Training Tips for the Classroom

1. Who Regulates What?

Objective: *After completing this activity, class participants should be able to identify the purpose of different regulatory and food-safety agencies.*

Directions: Create a handout to review the different regulatory and food-safety agencies discussed in Chapter 13.

Divide the sheet into two columns:

List of Agencies

○ Food and Drug Administration

○ United States Department of Agriculture

○ Centers for Disease Control and Prevention

○ Food Safety and Inspection Services

○ National Marine Fisheries Services

○ Environmental Protection Agency

○ Local Health Department

Purpose of the Agency

Ask class participants to identify the purpose of each agency on the handout. Allow ten minutes for the class to complete the activity, and then discuss.

2. "If you're from the health department, where is your badge?"

Objective: *After completing this activity, class participants should be able to identify proper procedures for guiding a health inspector through an establishment.*

Directions: Conduct two role-plays.

Role-play 1: The wrong way

Select two class participants who have a flair for the dramatic, and ask one to play a health department inspector conducting a health inspection, and the other a restaurant manager. Ask the participants to refer to the proper procedures for guiding an inspector through an establishment as presented in Chapter 13, and do the opposite of what is suggested. Ask them to develop a dialogue that is adversarial, somewhat exaggerated, humorous, but realistic. Have them limit their role-play to five to eight minutes. After the role-play, discuss with the class what went wrong. Discuss the correct procedure for guiding an inspector through the establishment, as presented in chapter 13.

Role-play 2: The right way

In this role-play you will demonstrate the proper way to guide an inspector through the establishment. Choose one class participant to act out this role-play with you. Depending on the class, you can choose to be the manager or the inspector in this activity. After the role-play discuss with the class what went "right".

Training Tips on the Job

1. Inspection Report Review and Preview

Purpose: *To evaluate the effectiveness of your food-safety system using the inspection requirements of your health department.*

Directions: Conduct an inspection of your establishment using the inspection form used by your health department. Don't be easy on yourself, don't make excuses, and don't make exceptions. Pretend that you are conducting the inspection of a competitor's establishment!

After the inspection, determine your food-safety score. You might solicit help from your health department in determining how to use and score the form. What corrective actions must be taken immediately, and what corrective actions can be taken later?

Before you actually make any corrections, you might ask your chef or assistant manager (or both) to conduct a similar inspection. How do the inspection reports compare? Do each of you see food safety from the same perspective? Chances are, your inspections will differ in some areas. Discuss these differences, and take corrective action.

DISCUSSION QUESTIONS

1. What is the function of the FDA regarding food protection?

2. Which governmental agency is responsible for inspecting and grading meat and poultry shipped across state lines?

3. What is the role of federal, state, and local agencies regarding the regulation of establishments?

4. What should a manager do during an inspection?

5. What should a manager do following an inspection?

6. How would a health agency typically determine the frequency of inspection?

7. Name some of the significant industry organizations that help managers deal with sanitation regulations and standards and employee training.

MULTIPLE-CHOICE STUDY QUESTIONS

1. An establishment can be closed for all of the following reasons except

   A. a significant lack of refrigeration in the establishment.
   B. a backup of sewage in the establishment.
   C. a serious infestation of insects or rodents in the establishment.
   D. a minor violation in the establishment that was not corrected within twenty-four hours.

2. Which of the following is a goal of the food-safety inspection process?

   A. To evaluate the sanitation and food-safety practices within the establishment
   B. To protect the public's health
   C. To convey new food-safety information to establishments
   D. All of the above

3. Which operation would most likely be subject to a federal food-safety inspection?

   A. A hospital

   B. A cruise ship that crosses international waters

   C. A local ice cream store with a history of safety violations

   D. A food kitchen run by church volunteers

4. A person shows up at a restaurant claiming to be a health inspector. What should the manager do?

   A. Ask to see identification.

   B. Ask to see an inspection warrant.

   C. Ask for a hearing to determine if the inspection is necessary.

   D. Ask for a twenty-four-hour postponement to prepare for the inspection.

5. How does a HACCP-based food-safety inspection differ from an ordinary inspection?

   A. It focuses more on how food flows through the establishment.

   B. It needs to be done more frequently.

   C. It focuses more on the sanitary appearance of the facility.

   D. It uses a point system with demerits.

6. Which of the following agencies enforce food safety in a restaurant?

   A. The FDA

   B. The Centers for Disease Control (CDC)

   C. State or local health departments

   D. The USDA

7. Violations noted on the inspection report should be

   A. discussed in detail with the inspector.

   B. corrected within forty-eight hours or less if they are critical.

   C. explored to determine why they occurred.

   D. All of the above.

8. The responsibility for the sanitary operation of an establishment rests with

   A. the state health department.   C. the health inspector.

   B. the manager/operator.   D. the FDA.

9. Food codes developed by state agencies are

    A. minimum standards necessary to ensure food safety.

    B. maximum standards necessary to ensure food safety.

    C. voluntary guidelines for establishments to follow.

    D. inspection practices for grading meats and meat products.

ADDITIONAL RESOURCES

Books and Periodicals

Frable, F. (1997). Industry needs central resource for code, regulation information. *Nation's Restaurant News, 31*(20), 44.

Hertneky, P. B. (1996). You and your health inspector. *Restaurant Hospitality, 80*(6), 57.

Marriott, N. G. (1994). *Principles of food sanitation.* New York: Chapman & Hall.

Murray, J. (1994). The model food code. *FoodService Director, 7*(4), 82.

Penner, K. (1992). *Exploring public policy options: Food safety in food service.* Manhattan, KS: Kansas State University Cooperative Extension Service.

Solis, O. C. (1997). Partner or adversary? *Food & Service, 58*(3), 16.

Web Sites

1999 FDA Model Food Code

http://vm.cfsan.fda.gov/~dms/fc99-toc.html

Complete outline of the FDA's latest code for regulating operations that provide food directly to consumers. Also includes a quick synopsis of changes from the 1997 Food Code.

American Public Health Association (APHA)

http://www.apha.org

Information on disease prevention and health promotion as well as other resources for public health professionals.

Archives of FDA Publications

http://www.fda.gov/opacom/archives.html

A complete database of FDA materials from the last six years, including press releases, speeches, and consumer publications.

FDA Center for Food Safety and Applied Nutrition (CFSAN)

http://vm.cfsan.fda.gov/list.html

Comprehensive site from CFSAN offers a wealth of food-safety information, from foodborne illness to food labeling. CFSAN strives to be a leader in food safety, and to protect consumers from economic fraud, promote sound nutrition, and encourage innovation.

FDA Office of Regulatory Affairs

http://www.fda.gov/ora/inspect_ref/iom

This page includes information provided to FDA investigators and inspectors to assist them in their daily activities, such as an investigations operations manual.

FDA Plan Review Guide

http://vm.cfsan.fda.gov/~dms/prev-toc.html

The document on this site was developed to serve as a guide to facilitate greater uniformity and ease in conducting plan review, whether your position is a regulator or an owner/operator wishing to build or expand.

Food and Drug Administration (FDA)

http://www.fda.gov

Web site for the Food and Drug Administration features link to the FDA's Center for Food Safety and Applied Nutrition.

Food and Drug Law Institute

http://www.fdli.org

The Web site for this nonpartisan, non-profit, educational organization provides a neutral forum for the examination of the laws, regulations, and policies related to drugs, medical devices, other health care technologies, and foods.

National Food Safety Database

http://www.foodsafety.org

A compilation of food-safety database information from government, consumer, and public health organizations, this is a one-stop Web site for food-safety information on the Internet.

National Restaurant Association

http://www.restaurant.org

The National Restaurant Association site provides information on government agencies affecting the restaurant industry, the latest training and certification updates, and links to state restaurant associations and hospitality schools and universities.

United States Department of Agriculture (USDA)

http://www.usda.gov

The United States Department of Agriculture's Web site features information, publications, and other educational materials about the nation's agriculture.

United States Department of Agriculture's Food Safety and Inspection Service

http://www.fsis.usda.gov

The USDA's Food Safety and Inspection Web site offers the latest food-safety news, educational materials, and HACCP implementation materials.

Chapter 14
Employee Food-Safety Training

TEST YOUR FOOD-SAFETY KNOWLEDGE

1. **True or False:** A cook's helper with two years of job experience no longer needs training. *(See Assess Training Needs, page 14-5.)*

2. **True or False:** The "magic apron" is an accepted training method for new employees. *(See One-on-One Training, page 14-8.)*

3. **True or False:** Newly trained employees should have a chance to practice what they have learned. *(See Key Elements of Effective Training, page 14-3.)*

4. **True or False:** A written test is the only objective way to evaluate the success of a training session. *(See Evaluate, page 14-16.)*

5. **True or False:** On-the-job observation will tell managers how employees are applying their training to their jobs. *(See Evaluate, page 14-16.)*

Table of Contents

Learning Objectives

After completing this chapter, you should be able to:

○ Assess the training needs of employees.

○ Design, execute, and reinforce a training program.

○ Evaluate the success of a training program.

○ Identify when and where food-safety training should take place.

○ Recognize the importance of food-safety certification.

Key Terms

Presentation
Feedback
Application
Training program
Training need
Objective
Demonstration

Lecture
Role-plays
Job aids
"Magic apron"
Technology-based training
Inservices
Evaluation

Training means teaching employees how to do a job properly. This sounds simple, but training is actually an immense task. In your establishment, it is up to you to train all employees in a wide range of areas and to include food-safety practices as a key element in the training process.

Training programs must be established for both new employees and current employees. For new employees, food-safety training must be mandatory. It is dangerous to assume that new employees know the safe and sanitary procedures used in your establishment. Current employees may know the correct procedures, but they may not always follow them because they hurry, forget, or lack motivation. For current employees, it may be important to schedule short retraining sessions, updates on new procedures, or motivational sessions that reinforce methods and practices.

In this chapter, we will discuss the following aspects of food-safety training for employees.

○ Assessing and analyzing training needs for food safety.

○ Planning, executing, and reinforcing the training program.

○ Evaluating the success of the training effort.

PURPOSE OF FOOD-SAFETY TRAINING

Food-safety training provides employees with the knowledge and skills needed to handle food safely in your establishment. The final responsibility for food safety rests with the manager.

Employee training is an important factor in every operation's financial statement. At first glance, training appears costly. The manager, of course, must evaluate every facet of the business for cost and might ask, "How does training contribute to my profits?" It is true that training in food-safety practices may require time away from regular tasks for both employees and managers. It may also require the services of professional trainers, and selection and use of training materials such as videos, slides, books, and posters to reinforce safe practices. However, training of staff will have a positive return on investment in the long run. Benefits from food-safety training include:

○ Avoiding the costs associated with an outbreak of foodborne illness. These costs may include legal fees and medical bills.

○ Preventing the loss of reputation and revenue when an establishment is forced to close because food-safety standards are not met.

○ Improvement in employee morale and a reduction in turnover. Most employees want to do their jobs right, and expect to receive training. Training helps instill employee confidence.

○ Increased customer satisfaction. Customer satisfaction will be higher when they see that an establishment is committed to serving safe food.

A successful training program has these essential elements:

○ Clearly defined and measurable objectives

○ Training that supports the objectives

○ Evaluation to ensure the objectives have been achieved

○ A work climate that reinforces training

○ Management support

In order for the training program to be effective, employees must see that the commitment to food safety comes from the top down. Management should lead by example. If managers show a commitment to food safety by their behavior and attitude, employees are likely to follow.

KEY ELEMENTS OF EFFECTIVE TRAINING

Successful training involves three key elements: presentation, application, and feedback. **Presentation** is the delivery of content to the learner, which can be accomplished through a variety of methods. Once the content is presented, the learner must have the opportunity to practice, apply, or respond to the content in order to retain it. While learners are practicing or applying the content, they must receive **feedback** (positive or negative reinforcement) on their performance. The feedback received must be specific and immediate. **Application** or practice without feedback will be ineffective.

As a general rule, one-third of the time spent training should be devoted to the presentation of content, while the remaining two-thirds should be devoted to activities that allow class participants to apply what they have learned, with feedback.

An effective trainer will have good presentation skills and be knowledgeable in the science of food-safety principles and procedures. In addition, a good trainer recognizes that each trainee has unique learning skills and backgrounds. People learn at different rates and through different media.

DEVELOPING THE TRAINING PROGRAM

A **training program** is a structured sequence of events that leads to learning. To be effective, the training program must be well organized. Some large establishments set up a training department and have a training professional on staff to develop and deliver their program. Other establishments rely on the manager to develop and deliver training. A foodservice

Key Point

Food-safety training is a positive investment.

Key Point

As a general rule, one-third of the time spent training should be presentation of content, while the remaining two-thirds should be spent applying information learned, with feedback.

manager can provide an effective training program through careful planning and the use of supportive tools.

There are several steps involved in the development and delivery of an effective training program. Each of these steps appears below and will be explained in more detail in the following sections.

1. Assess training needs.
2. Establish learning objectives.
3. Choose training delivery methods.
4. Select instructor.
5. Choose training materials.
6. Schedule training sessions.
7. Select training area.
8. Prepare trainer.

Assess Training Needs

The first task in developing a training program is to assess the training needs in your establishment. A **training need** is a gap between what your employees are required to know to perform their job and what they actually know. For new hires, the need may be apparent. For current employees, the need is not always as obvious. Determining the gap in performance may require work. To identify food-safety training gaps, a manager or trainer can do the following.

○ Observe employee job performance.
○ Question or survey employees to identify areas of weakness.
○ Review past health inspection reports for violations relating to employee performance.
○ Test employees' food-safety knowledge.

Your staff needs the correct knowledge and skills regarding food safety. Beginning on their first day on the job, new foodhandlers will require training in the following areas.

○ The importance of food safety.
○ Personal hygiene: health, personal cleanliness, clothing, and hygienic practices
○ Food preparation: potentially hazardous foods, time-temperature control, use of thermometers, proper techniques for preparation, cooking, holding, cooling, and reheating
○ Cleaning and sanitizing procedures for the facility and equipment
○ Proper methods of handling hazardous materials
○ Pests: harborage, pest identification, and preventive control measures

Key Point

A training need is a gap between what your employees are required to know to perform their job and what they actually know.

Training needs will differ for each job within the establishment. All employees will need certain information. Other information will be unique to specific jobs. For example, everyone needs to know the proper way to wash hands, however only cooks need to know the minimum required cooking temperature for chicken.

Regardless of their positions, all employees need to be periodically retrained on the food-safety practices listed above. It is the manager's responsibility to keep employees informed about changes in the science of food safety and of industry best practices.

Key Point

Food-safety training should be job specific.

Establish Learning Objectives

Once training needs have been identified, the next step is to define the objectives of the training program. An **objective** states what the learner will be able to do after instruction is finished. Objectives need to be clearly stated in measurable terms. Use action verbs such as *operate, demonstrate,* or *practice* rather than *be aware, observe, understand,* or *notice.* Some examples of clearly stated objectives are listed below.

- Demonstrate the proper procedure for calibrating a bi-metallic stemmed thermometer using the ice-point method
- Given a variety of food items and storage space, diagram the proper placement of each food item
- List the internal cooking temperature for meat, seafood, and poultry

Key Point

An objective states what the learner will be able to do after instruction is finished.

Select Training Delivery Methods

There are many methods for delivering content to the learner. Some delivery methods are better suited to one-on-one learning and others to group learning. No single delivery method is best for training at all levels. In general, however, using several methods of delivery will result in more effective learning. Some methods of delivering training include the following:

- Demonstration
- Lecture
- Role-play
- Job aids
- One-on-one training (on-the-job training)
- Technology-based training
- Group training
- Inservices

Remember, the delivery method chosen should allow trainees to apply what they have learned and receive feedback on their performance. However, the choice will also depend upon the number of people that need to be taught, the cost, and—most importantly—how the trainees will learn best.

Every participant learns differently. When training adults, you should consider several principles of adult learning when choosing a method of delivery. Ask yourself if the methods you have chosen allow learners to do the following:

○ Talk to each other (interpersonal principle)
○ Reflect on the content and determine how it applies to their job (introspective principle)
○ See the task being performed (spatial principle)
○ Be physically active and perform the task (physical principle)
○ Hear the instructions spoken (auditory principle)
○ Read the materials and take notes (verbal principle)
○ Reason through real-life situations (logical principle)

Depending on the content of the training, you may use one or more of these principles.

The choice of delivery method must also be based on the specific training objective being taught. For example, if the manager wants to train an employee to properly wash, rinse, and sanitize the salad-preparation counter, the best method may be to demonstrate the process and then have the employee perform the task (using the physical principle of adult learning). Demonstration allows for immediate feedback. The manager can explain the key points of a task while performing it, then observe the progress as the employee practices the same task. In this example, using a demonstration is the best method for delivering the desired content. More importantly, demonstration also adheres to most of the adult-learning principles mentioned previously.

Demonstration

Demonstration is the process of illustrating a skill or task before another person or group. When conducting a demonstration, follow these guidelines.

○ Preface the demonstration with an explanation of what participants should look for when the skill or task is demonstrated.
○ Emphasize the key points as you demonstrate the task or skill.
○ Explain how each step fits into the task sequence. Participants need to know where they are, step by step, in the process of performing the task.

○ Demonstrate the skill or task slowly, so that the participants can see what is happening. Then repeat the skill or task at normal speed.

○ Before the participants are given an opportunity to perform the task, ask them to explain each of the steps in sequence.

○ Ask the participants to demonstrate the skill or task. Provide appropriate feedback correcting any errors as they occur.

Lecture

A **lecture** is a prepared oral presentation used to deliver content to a group of participants. Lectures are most effective when mixed with other presentation methods and media. Lectures are better received and accepted when the following techniques are used.

○ Start with an interesting statement, observation, quotation, or question.

○ Use relevant humor where appropriate.

○ Use interesting and relevant examples, anecdotes, analogies, and statistics.

○ Ask frequent questions to solicit audience participation.

○ Use frequent small group discussions and activities.

○ Build in review.

As with other presentation methods, use only one-third of the time for lecturing, and allow two-thirds of the time for applying information learned and providing feedback.

Key Point

Lectures are most effective when mixed with other presentation methods and media.

Role-Play

In **role-plays,** participants enact a situation in order to try out new skills or apply what has been learned. Different types of role-plays are suitable for different types of learning situations. These situations include the following:

○ **Confrontation.** This involves a situation where a participant is confronted by another participant and must answer questions, handle problems, provide satisfaction, solve a complaint, etc. For example, a role-play can be used to teach a manager how to deal with a health inspector during inspection.

○ **Consultation.** This involves a situation where a participant tries to help a client (another participant) solve a problem. For example, a role-play can be used to teach a waitress how to provide a customer with information about a menu item in the ordering process.

○ **Court techniques.** This involves a situation where a person is "tried" in a mock courtroom, as participants work out the consequences of a mishandled task. For example, one participant could role-play a manager who is being held responsible for a foodborne illness outbreak, while other participants cross-examine him or her about activities at the establishment.

Job Aids

Job aids can be used to deliver content to employees. In the classroom, job aids may include materials such as worksheets, checklists, samples, flowcharts, procedural guides, glossaries, diagrams, decision tables, etc. Other job aids may include visual reminders such as posters that illustrate the proper steps in handwashing, manual warewashing, or other tasks. Employees can then use these job aids once they return to their jobs. Job aids are particularly useful when:

○ Tasks are performed infrequently.

○ Tasks are complex.

○ The sequence of performance is critical.

○ The consequences of making a mistake are severe.

○ Safety is a concern.

One-on-One Training

One-on-one training is effective when only a few people need training. Delivering this type of training has the following advantages:

○ It allows for the special needs of individual employees.

○ It can take place on the job, so a separate training location is not needed.

○ It enables the manager to monitor employee progress.

○ It allows for immediate feedback.

○ It allows for the opportunity to apply information that has been learned.

One-on-one training does have some disadvantages. Most importantly, its effectiveness depends upon the ability of the person who will deliver the training. The trainer must be selected very carefully. Many establishments certify trainers or validate that they have the appropriate skills before they are allowed to deliver training.

Expecting employees to learn job responsibilities by themselves while they work on the job is sometimes referred to as the **"magic apron"** method. The assumption is made that employees will somehow "magically" know how to do the job once they put on the apron. This assumption is, of course, false. The employee must actually be learning the job from the assigned employee, not trying to learn the job with the help of his "magic apron."

Technology-Based Training

Technology-based training programs such as CD-ROM and CD-Interactive (CD-i) offer another method of one-on-one training, where the trainer is replaced by the computer or other technology. The advantages of this method include:

○ **Standardized delivery.** The computer delivers the training the same way every time.

○ **Standardized feedback.** Each time a participant responds to a situation, the computer can provide standardized feedback.

○ **Customizable instruction.** The participants can choose their own learning path.

○ **Increased performance practice.** The participants are allowed to practice a skill until they are proficient.

Group Training

If several employees require food-safety training, group sessions may be more practical than one-on-one training. Group training can be delivered through a number of methods, including those discussed earlier in this chapter. Group training has several advantages (see Exhibit 14a).

○ It is more cost effective for larger groups.

○ Training is more uniform.

○ You know precisely what employees have been taught.

One disadvantage of group training is that it is designed to meet the overall needs of a group of employees rather than the needs of each individual. Slower learners, less-skilled employees, or those with limited English proficiency may not understand all the material that is presented. A trainer must remember that it is not enough to just cover the material. Class participants must be involved in the training process. If they are involved, they will retain the information. Studies on training effectiveness show that participants retain the following:

○ 10 percent of what they read

○ 20 percent of what they hear

○ 30 percent of what they see

○ 50 percent of what they hear and see

○ 70 percent of what they say

○ 90 percent of what they say and do

Exhibit 14a **Group Training**
Group training can be cost effective, and may offer a more uniform delivery of content.

Key Point

Class participants retain 90 percent of what they say and do.

Inservices

Ongoing training may be provided in short sessions called inservice training or **inservices.** Typically, inservice training uses the group method. In some settings, such as healthcare foodservice operations, inservices are conducted on a monthly basis, and may be required by law. These short training sessions may be used to reinforce previously learned procedures, introduce new procedures, or motivate employees to follow existing procedures.

Select an Instructor

The manager is often the most likely person to train staff in food safety. He or she knows the operation and is responsible for food-safety practices within the establishment. If the manager chooses not to personally conduct the training, an instructor should be selected, based on these criteria.

○ Knowledge of food-safety practices

○ Understanding of the operation's food-safety challenges

○ Demonstrated skill in human relations

○ Proven ability to help others learn

Individuals who may be able to conduct food-safety training include the following.

○ **The immediate supervisor of the trainee or trainees.** Many times this person is a logical choice because of his or her working relationship with the employee.

○ **Staff trainer.** In large establishments, a training professional is often on staff to provide food-safety instruction and support the training needs of the operation.

○ **Representative of the health department.** The local sanitarian or a public health educator may agree to teach a food-safety training session. This fosters a cooperative working relationship with this agency.

○ **Representative from a professional or educational organization.** These organizations often provide trainer-preparation courses for operators and also have staff instructors who conduct courses on site or at other locations. Check with suppliers and your state restaurant association.

Very often, you will get successful results when you use a combination of these resource people.

Choose Training Materials

Proper use of training materials saves time, adds interest, helps participants to learn and retain information, and makes the trainer's job easier.

Key Point

Contact your local health department for help in teaching food safety in your establishment.

The manager or trainer should be guided by the "Three A's" when choosing materials: To be useful, training materials must be *accurate, appropriate,* and *attractive.*

Accurate. Materials must be factual, up to date, and complete. To ensure that training materials are accurate, they should be purchased from professional or educational organizations, industry suppliers, or from an authority in foodservice training.

Appropriate. Materials must be suitable for the purpose they are to serve. Written materials must be matched with the reading comprehension levels of the participants. For employees with limited English proficiency, training materials in other languages are available. In addition, materials should suit the abilities of the trainer. Limitations imposed by the training location will influence your choice of training materials. Some rooms do not have chalkboards or flipcharts. Reliable equipment must also be available in order to use videotapes, slides, CDs and CD-is, overhead transparencies, and other audiovisual or technological aids.

Attractive. In order to teach people, you must first gain their attention. When teaching subjects that do not generally have a wide appeal, make information exciting and memorable with eye-catching audiovisual aids, whenever possible.

Materials Used in a Typical Lesson Plan

The manager or trainer should start each training session by clearly stating the objectives of the session. It is a good idea to provide students with handouts with these objectives clearly stated or to write them on flipchart, easel, or chalkboard.

Sometimes a pre-test or quiz may be given, to assess the prior knowledge of the participants before the information is presented. The class material is then presented, supplementing the oral presentation with appropriate printed or audiovisual materials, after which a brief written, oral, or practical exam is given. This exam can be one that has been prepared by an educational or training organization or one that is written by the trainer. The advantage of using prepared exams is that they are usually validated for content and evaluated for level of reading comprehension.

Any mistakes should be discussed and corrected immediately after the exam is scored. To reinforce the training, signs, posters, bulletin boards, reminders in pay envelopes, and other materials can be used.

Schedule Training Sessions

Developing and implementing a master training schedule at your establishment can be a useful method to determine training priorities and

Key Point

To be useful, training materials must be accurate, appropriate, and attractive.

show your commitment to training. Both orientation and ongoing training should be included in this schedule. Special situations, such as the training of a large group of new employees, the opening of a new restaurant, a reopening after remodeling, or preparations for a convention banquet may require changes in the master training schedule. For legal reasons, it is important to record that training took place, even if the training session was brief.

Training sessions should not be too long. The ideal length is probably about twenty to thirty minutes per segment. Successful training sessions, however, may be as long as an hour and a half, if well planned. The best length depends on the type of learning activity to be used. Training sessions may need to be given more than one time during the day in order to cover all shifts of employees.

Training sessions should be scheduled during slow times for the establishment: before opening, after closing, or on a day when the establishment is normally closed, for example. If there is a separate area in the facility for training purposes, scheduling can be more flexible.

Select the Training Area

Training should be done in a comfortable location. An on-site location allows for demonstrations and provides the opportunity to relate instruction to your establishment. Employee work stations, for example, can be used as training areas, particularly for one-on-one sessions. In most establishments, an employee lounge, an executive office, or a section of the facility not in use can also serve for training sessions.

Consider the following suggestions when choosing an area for group training.

○ The area should be an appropriate size so no one feels crowded.

○ There should be tables or desks so students can take notes.

○ Seating should be adequate and comfortable.

○ Seating should be arranged to encourage open discussion.

○ A blackboard or flip chart is often needed.

○ If a projector, sound equipment, videotape player, computer, or other electronics are used, the room must have enough electrical outlets.

○ If audiovisual materials are to be used, it must be possible to darken the room so trainees can see better.

○ The area should be free of distractions.

Prepare the Trainer

Training employees requires excellent communication and organization skills. The following are suggestions for the trainer on how to develop these skills and conduct a successful training session.

○ **Make sure you are knowledgeable in all areas of food safety, and familiar with the details of all food-safety practices.** Feeling comfortable with the subject is important in case trainees ask questions. If you are uncertain of your subject, the participants will know.

○ **Prepare for the presentation.** If you will be lecturing, make an outline that you can refer to. Practice the presentation until it feels natural. If you are using an overhead projector, you might want to have the transparencies framed so that they will not stick together. Make sure they are numbered in case they get out of order. Familiarize yourself with the room, and check all audiovisual equipment to make sure it is in working order and that you know how to use it.

○ **During your presentation, maintain eye contact with the participants.** Ask questions to ensure that the participants can apply the material presented.

○ **Keep your delivery conversational and informal.** This makes the atmosphere more comfortable. Vary the tone of your voice for emphasis and do not speak too quickly. A moderate rate will allow trainees to take notes and ask questions.

○ **Use simple language.** Using unnecessarily technical terms will cause participants to lose interest or become lost. Pause often and ask questions to make sure your audience understands the material.

○ **Treat seriously all questions and comments made by the participants.** Answer them in a straightforward way. You can give participants positive feedback by saying, "That's a good question."

○ **Look for cues that employees are not picking up the information or that they are bored and losing interest.** These cues could include fidgeting, looking at watches, doodling, and not asking or answering questions. Use as many varied visuals as possible to reinforce content.

○ **Keep the sessions short.**

○ **Keep the training as practical as possible and relate information to their jobs.** If the subject is too general, people tend to lose interest.

Key Point

Give specific examples of how the training content relates to the employee's job.

CONDUCTING THE TRAINING SESSION

For effective training, you must tune your training into the participants' favorite radio station WII-FM, What's In It For Me? Participants want to know how the training will enable them to do their job faster, easier, better. In addition, they want to know what benefits they are going to gain and what losses they are going to avoid.

Participants have to want to learn what you are teaching. It is the job of the trainer to capture the trainees' attention and to help them remember the material. Here are some guidelines to follow in order to achieve this.

Keep It Simple and Short

In any training, it is important to take a simple, straightforward approach. Because the time for training sessions is often limited, focus on the most important points. Present no more than five to nine major points in a short training session. Some establishments use regular pre-shift meetings to present brief training tips.

Individualize Training

Make trainees feel that you are talking to them individually and that you have a real interest in them. Encourage employee participation during sessions, and show them that you are interested in what they have to say. Letting employees know that training is a company priority will increase your program's chances of success.

Recognize Employee Achievement in the Training Program

Wall charts that track individual progress through the training course and certificates of completion can serve as incentives.

Be Creative

You can put questions and answers in a game show format. Create teams and have them compete to give correct answers for points or small prizes *(see Exhibit 14b on the next page).* You can have one trainee draw sanitation procedures on a chalkboard or flip chart while the rest of the group guesses what is drawn. If the group is not interested in drawing, they can act out the words. You might tell the trainees to write a rhyme that describes a foodborne illness and have them read their creations for the rest of the group to guess.

At the end of each chapter in this text there are specific training tips that can be used in the classroom and back in the establishment. Also, a variety of unit-level training tools and games are available through the National Restaurant Association Educational Foundation.

Provide Feedback

Praise and positive feedback are part of all good learning experiences and should be a part of the training process. Praise indicates to employees that they are performing well and have been successful in the training program. Generous praise reinforces employees' positive attitudes about their jobs. When they see that management values their best effort, they will continue their good practices.

Evaluate

The training program is not complete until it has been evaluated. The process of **evaluation** is important because it will tell you if training has provided the participants with the knowledge and skills needed to do their job, or do it better. To evaluate training, the manager must carefully judge the performance of the participants against the learning objectives.

Exhibit 14b **Training Games**
Games can be used by trainers to hold the attention of participants, and help them retain information.

When evaluating the training program's effectiveness, ask the following questions: Did the training produce results on the job? If the intended results were not produced, why not?

If training was ineffective, several factors may be involved. For example, the trainee may have acquired the knowledge, but is simply not applying it. Perhaps the equipment used for training is different from the equipment used on the job. Perhaps there are negative consequences, such as peer pressure, for doing the job the way the employee was trained to do it.

In some cases, an employee has learned the material, but is simply applying it incorrectly. There may be several reasons for this. Perhaps the employee has been improperly trained and his performance is consistent with the bad practices he was taught. Perhaps the employee has learned the skill but it is not being reinforced on the job. This requires a management intervention to ensure that the proper skills are being taught, practiced, and reinforced.

Key Point

When evaluating the training program's effectiveness ask the following questions: Did the training produce results on the job? If the intended results were not produced, why not?

Key Point

Evaluation should always be based on the objectives of the training.

Evaluation should always be based on the training objectives. There are several ways in which objectives can be measured. Most commonly, objectives are measured through written or oral tests. Test results can help a manager determine if some areas of content need to be reviewed. Objectives can also be measured by evaluating an employee while he or she performs a task or skill required by the objective. Evaluation works best when a combination of written and performance-based tests are used.

FOOD-SAFETY CERTIFICATION

The National Restaurant Association and federal and state regulatory officials recommend food-safety training and certification, particularly for managers and supervisors. Certification demonstrates that a person comprehends the basic food-safety principles and recommended food-safety practices that will prevent foodborne illness.

Manager certification is already a requirement in some states. In other states, some cities and counties may require certification although the state as a whole may not require it. The National Restaurant Association Educational Foundation Web site provides a jurisdictional summary of manager certification requirements for the entire country.

Regardless of state regulations, many proactive establishments and managers have made a strong commitment to food safety by making food-safety training and education an integral part of their establishments. You, too, should make a commitment to training, as well as certification when possible, even if your state or city does not require it. As a conscientious foodservice manager, you can train your employees, monitor their practices, and make food safety a part of everyone's job description. In this way, you will be able to ensure that the food you serve will be safe.

On the following page you will find a press release template that can be used to communicate to the public your commitment to food-safety training and to ServSafe® training in particular *(see Exhibit 14c)*.

National Food Safety Education Month

Food safety is vital all year, but each September the International Food Safety Council sponsors National Food Safety Education Month to focus attention on the importance of food-safety education and training, and to build public awareness of the restaurant industry's commitment to serving safe food. To get involved and demonstrate your dedication, log on to www.foodsafetycouncil.org or call your state restaurant association.

[Print on your company's letterhead]

FOR IMMEDIATE RELEASE
[Insert month day, year]

CONTACT:
[Insert name]
[Telephone Number]

[Company's Name] DEMONSTRATES COMMITMENT TO FOOD SAFETY BY TRAINING STAFF

[Insert City, Date] – [Name of Company] is demonstrating a strong commitment to serving safe food by [certifying [#] of employees; or planning to certify [#] of employees][give time frame] in the National Restaurant Association Educational Foundation's ServSafe® food-safety training program.

Accepted by more than 95 percent of state and local jurisdictions as satisfying certification requirements, the ServSafe program is the industry standard in food-safety training. Through the program, participants learn important food-safety procedures, such as sanitation, time and temperature constraints, and HACCP (Hazard Analysis Critical Control Point) principles, which are essential in providing safe food to consumers. "[Insert Restaurant]'s food-safety training efforts show its customers and employees how seriously the company takes food safety, and that serving safe food is a priority," said A. Reed Hayes, President and COO of the Educational Foundation.

"As experts in food preparation and service, it is natural for us as restaurant [owners or operators] to take a leadership role in addressing the critical issue of food safety," said [Company Executive or Owner]. "We support the restaurant industry's position that the best way to help prevent foodborne illness is through food-safety training. By training our staff in the ServSafe program, we demonstrate our concern for our guests and our commitment to food safety."

In addition, each employee certified in the ServSafe food-safety training program automatically becomes a member of the International Food Safety Council, a coalition of all segments of the restaurant and foodservice industry that is dedicated to encouraging food-safety training within the industry. The Council also sponsors National Food Safety Education Month each September.

[You may want to insert your company name, along with a brief description of your operation, the number of units you operate, where they are located and how many employees you have, etc.]

The National Restaurant Association Educational Foundation develops, promotes, and provides educational and training solutions for the restaurant and hospitality industry. To find out more about the ServSafe program or any other Foundation training program, visit the Educational Foundation Web site at http://www.edfound.org.

Exhibit 14c **Standard Press Release Form**

How To Use The Press Release:
○ Retype the sample language to fit on your business or personal stationery or news-release letterhead, if available.
○ Insert the appropriate information requested in the brackets.
○ Add additional professional and personal information where indicated.
○ Press release should be double-spaced when printed.
○ To further promote food safety, send the press release to local news media outlets when employee(s) become ServSafe® certified or when your company commits to putting employees through food-safety training. Photocopy as many copies as you need to mail, E-mail, or fax to the local outlets.

SUMMARY

One of the foodservice manager's most important responsibilities is to train employees in the principles and practices of food safety. Effective training helps protect the public from foodborne illness.

The benefits from an effective training program appear in improved food safety, higher employee morale, and the financial stability of the establishment. Food-safety training needs to be a priority in all establishments. Continuous training programs on an individual or group basis for all employees are essential for running a successful, quality establishment. Setting a good example is a primary role for the manager before, during, and after the training program.

Training is a process that includes acknowledgment of each individual's learning skills and level. It should be comprised of presentation, application, and feedback. Feedback is praise or constructive criticism based on the employee's performance. Follow-up by the manager through monitoring and supervising is a necessary step in the training process.

Assessing your food-safety training needs is the first step in the process. First, you must determine the gap between what your employees are required to know and what they actually know. The next step is to determine the specific subject matter to be covered. Training objectives are written to define the expected results. These objectives form the standards of achievement for the program and can later be used to evaluate its effectiveness.

Next, the training method and materials are selected. Training can be conducted one-on-one or in groups, depending on the number of participants, the costs, and the priority of the need. The length of time and location for training sessions will depend on the audience and the subject being taught. A combination of training techniques and materials is the best approach for food-safety training. The instructor must be well prepared. He or she must know the subject well and be able to answer any questions from participants.

The final step in a training program is evaluating the outcome. The process of evaluation is important because it will tell you if the training has provided the participants with the knowledge and skills needed to do their jobs, or do them better. To evaluate the training, the manager must carefully judge the performance of each participant based on the learning objectives of the training.

Praise is an important aspect of learning and should follow training to reinforce good food-safety practices. Wall charts that show daily or weekly progress, or individual certificates or pins, can serve as incentives.

The National Restaurant Association and federal and state regulatory officials recommend food-safety training and certification, particularly for managers and supervisors. Certification shows that a person comprehends the basic food-safety principles and recommended food-safety practices that will prevent foodborne illness.

A CASE IN POINT

Case Study

Paul has two employees who must receive food-safety training. One is an assistant cook, Albert, who used to be a buser and needs further training. The other, Maria, is an inexperienced server. Albert received general training in the fundamentals of food safety about three months ago, but Maria has had no training. Both Albert and Maria require specialized training. Paul maintains an ongoing food-safety training program, so he has already trained the rest of his employees and evaluated the effectiveness of his original program.

Paul decides to have the kitchen supervisor train Albert using the one-on-one method, so the supervisor can demonstrate procedures and give immediate feedback to Albert as he practices them. By relying on the supervisor to train the cook, Paul can have both the cook and server trained at the same time.

Paul writes learning objectives, using his standardized job description for servers. He gives Maria a copy of the job description and learning objectives and sets up a food-safety videotape in his office for her to watch. While the videotape is playing, Paul goes to the back of the facility to talk with a clerk about an expected shipment. After half an hour Paul returns to his office to give Maria a written test. While Maria is completing the test, Paul goes to the kitchen to observe the supervisor who is training Albert, the cook. Noting that everything looks good in the kitchen, Paul goes to the dining room.

Do you think Paul's training program for the cook and server will be effective?

What else does Paul need to be doing?

TRAINING TIPS

Training Tips for the Classroom

1.   *The Key Ingredients of a Successful Trainer: A Recipe for Success!*

Objective: *After completing this activity, class participants will be able to identify the traits, characteristics, and qualities of a successful trainer.*

Directions: After discussing Chapter 14, ask your class to make a list of the traits, characteristics, or qualities a trainer should possess in order to be effective.

Allow a period of three to five minutes for participants to write their lists. Then allow students to share their lists. A good way to do this is as follows.

Ask a student to explain one ingredient on his or her list. Then solicit thoughts, comments, or reactions from the rest of the class. Some questions to stimulate conversation might include, "Is this ingredient on anyone else's list?" or "Do you think it's a critical success factor for a trainer?"

Suggest that the class add the ingredient to their lists, if they agree it is valid. Then have the first student select another student to discuss one ingredient on his or her list, and continue on until everyone in the class has had a chance to contribute at least one ingredient.

The combined list of ingredients will no doubt add up to a real recipe for success in training!

2.   *Training Tools That Work*

Objective: *After completing this activity, class participants will be able to identify the advantages and disadvantages to using various training tools in the classroom.*

Directions: Many different training tools can be used in the process of food-safety training. All have advantages and disadvantages. Training tools can include the following:

- A coursebook
- A training guide or manual
- Training videos
- Slides
- Overheads
- Handouts
- Flipcharts
- Posters
- Computer-based presentations
- Role-plays
- Dry-erase board or chalkboard
- CDs and CD-is
- Demonstration materials (equipment, kitchen tools, thermometers, etc.)

Make a list of training tools, such as those listed above. You can print up the list as a handout, or ask the students to generate the list while you write the items on a chalkboard or flipchart. Then divide the class into two groups. Ask one group to list the advantages of each of the training tools on the list. Ask the other group to list the disadvantages of the training tools.

After a short time (about ten minutes), review with the entire class the advantages and disadvantages of each training tool. Write these on the board, if possible. Lead the class to see that the effectiveness of specific tools may depend on a number of different factors, including size of class, type of audience, layout of room, time allotted, quality of training tool, and personal preferences of the trainer.

Point out that no training tool works all the time, that every tool has some advantages and disadvantages, that every trainer has favorite tools, and that all training tools must be thoughtfully selected and properly used.

Training Tips on the Job

1. Conduct a Food-Safety Training Materials Audit

Purpose: *To audit existing food-safety training materials to determine if they are appropriate and accurate.*

Directions: The training materials your operation uses may or may not include valid and up-to-date food-safety information. An audit should help you assess of your current materials as they relate to current food-safety information, and help you determine how they can be improved.

First, collect all training materials currently in use, and highlight any parts related to food safety. Then assess the materials, by asking the following questions.

- Is it accurate and up to date with company policy?
- Does it meet or exceed local health department codes?
- Is it job specific?
- Is it based on clearly stated and measurable objectives?
- Is it user friendly?
- Is it thorough?
- Is it necessary?
- Does it effectively convey the information?
- Are materials sufficient in size and scope?

Based on your answers, the training materials may need to be modified or enhanced so that all of the above questions are satisfied.

DISCUSSION QUESTIONS

1. What are the benefits for an establishment that invests in continuous food-safety training?

2. New employees need training in what general food-safety topics?

3. What can be used to help you identify the training needs of your employees?

4. List at least two training methods mentioned in the text and describe the advantages and disadvantages of each.

5. How are training objectives related to training evaluation?

MULTIPLE-CHOICE STUDY QUESTIONS

1. Mary has hired eight college students to run her lakeside hot-dog stand for the summer. She would like to develop a brief training program to prepare them for the job. What should Mary do first?

   A. Write objectives   C. Assess needs
   B. Choose delivery methods   D. Evaluate

2. Which restaurant employee will most likely need some food-safety training?

   A. A hostess who becomes the assistant salad chef
   B. A meat chef who purchases some new knives
   C. A manager who is promoted to a larger branch restaurant
   D. A waiter who becomes a bartender

3. The ideal length for a training session is

   A. 5 to 10 minutes.   C. 1 to 2 hours.
   B. 20 to 30 minutes.   D. 2 to 3 hours.

4. Which of the following is not a good way for a restaurant manager to identify employee training needs?

   A. Reading the comments on a health department inspection report
   B. Reading a textbook about training restaurant employees
   C. Asking employees to tell what they are not certain about
   D. Observing the practices of employees on the job

5. What do participants want to know about training?

   A. What benefits they will gain from the training
   B. What losses they will avoid
   C. How the training will enable them to do their job faster, easier, better
   D. All of the above

6. Which of the following is a good example of using positive reinforcement?

   A. Praising employees who exhibit good food-safety practices on the job
   B. Hiring a full-time instructor/training manager
   C. Demonstrating key sanitation practices personally
   D. Firing employees who develop bad habits over time

7. Which delivery method is recommended for doing on-the-job training?

   A. Lecture   C. Job aids
   B. Role-play   D. Demonstration

8. Which is a major disadvantage of using a written test to evaluate the effectiveness of training?

   A. Having knowledge about something doesn't prove that you can apply it.
   B. Test items are hard to create.
   C. Writing about something doesn't necessarily prove knowledge.
   D. A written test takes time away from other important tasks.

9. Which of the following is not true of training objectives?

A. Objectives should describe the end result that the instruction is intended to produce.

B. Objectives need to be clearly stated in measurable terms.

C. The objective states what the learner will be able to do prior to instruction.

D. The words used in a objective should be action verbs.

ADDITIONAL RESOURCES

Books and Periodicals

Bax, B. (1997). *Handbook for safe food service management*. Upper Saddle River, NJ: Prentice-Hall.

Craig, R. L. (Ed.). (1987). *Training and development handbook: A guide to human resource development* (3rd ed.). New York: McGraw-Hill. (Sponsored by the American Society for Training and Development.)

Pike, R. W. (1994). *Creative Training Techniques Handbook* (2nd ed.). Minneapolis MN: Lakewood Books.

Rothwell, W. J., & Kazanas, H. C. (1998). *Mastering the instructional design process: A systematic approach* (2nd ed.). San Francisco: Jossey-Bass.

Web Sites

1997 FDA Model Food Code

http://vm.cfsan.fda.gov/~dms/fc-toc.html

Complete outline of the USDA's latest code for regulating operations that provide food directly to consumers. Also includes a quick synopsis of changes from the 1995 Food Code.

Archives of FDA Publications

http://www.fda.gov/opacom/archives.html

A complete database of FDA materials from the last six years, including press releases, speeches, and consumer publications.

Council of Hotel and Restaurant Trainers

http://www.chart.org

The Web site for this learning forum for hospitality-industry professionals involved in human resource training, developmental management, and management. Includes resources, links, and publications.

FightBac!

http://www.fightbac.org

The Web site for the Partnership for Food Safety Education, a coalition of government, industry, and consumer groups.

International Food Safety Council

http://www.foodsafetycouncil.org

The Web site offers information on food safety, National Food Safety Education Month, and other ways to become involved in the Council.

National Association of College and University Food Services

http://www.nacufs.org

NACUFS is a trade association for foodservice professionals at institutions of higher learning in the United States and Canada. Their Web site includes educational materials as well as tours of members' foodservice operations.

National Restaurant Association

http://www.restaurant.org

The National Restaurant Association site provides information on government agencies affecting the restaurant industry, the latest training and certification updates, and links to state restaurant associations and hospitality schools and universities.

National Restaurant Association Educational Foundation

http://www.edfound.org

This site offers information on education and training options; it also includes a vehicle for locating training classes.

Appendixes

APPENDIX A
STORAGE TEMPERATURES FOR FRESH FRUITS

Source: Produce Marketing Association

Fruit	Storage Temperatures (°F/°C)	Relative Humidity (%)	Comments
Apples	32°F to 35°F (0°C to 2°C)	90–95%	Will ripen at room temperature.
Avocados	CA: 40°F to 42°F (4°C to 6°C), FL: 55°F (13°C)	85–90% 85–95%	Will ripen at 65°F to 70°F (18°C to 21°C). FL: 55°F (13°C). Will ripen at room temp at 85 to 95%RH.
Bananas	60°F	85–90%	Will ripen at 62°F to 64°F (17°C to 18°C) at 85–95%RH.
Blueberries	31°F to 34°F (-1°C to 1°C)	90–95%	Should not be washed before storage.
Cantaloupe	38°F to 40°F (3°C to 4°C)	85–90%	Will ripen at room temperature.
Cherries	32°F to 34°F (0°C to 1°C)	90–95%	Will not ripen after picking; keep dry; store away from other foods with strong odors.
Cranberries	38°F to 40°F (3°C to 4°C)	85–90%	
Grapefruit	CA/AZ: 55°F to 58°F (13°C to 14°C), FL/TX: 50°F to 55°F (10°C to 13°C)	85–90%	Will not ripen after picking; chill damage occurs at temps below 40°F (4°C).
Grapes	32°F to 35°F (0°C to 2°C)	90–95%	Will not ripen in storage; do not wash before storage; do not handle excessively.
Kiwifruit	32°F to 35°F (0°C to 2°C)	90–95%	Can be stored up to 12 weeks at 32°F (0°C); will ripen quickly if stored near ethylene-producing fruits.
Lemons	50°F to 55°F (10°C to 13°C)	85–90%	Long refrigerated shelf life; store away from foods with strong odors.
Limes	50°F to 55°F (10°C to 13°C)	85–90%	Do not hold as long as lemons.
Mangos	50°F to 55°F (10°C to 13°C)	90–95%	Will ripen at room temperature; refrigerate to retard ripening.
Melons	45°F to 50°F (7°C to 10°C)	85–90%	Will ripen at room temperature; refrigerate to retard ripening.
Nectarines	32°F (0°C)	95%	Immature fruit will not ripen further.
Oranges	CA: 45°F to 48°F (7°C to 9°C), FL: 34°F to 40°F (1°C to 4°C), TX: 32°F to 35°F (0°C to 2°C)	85–90% 85–90% 85–95%	
Peaches	32°F (0°C)	95%	Will not ripen properly after picking.
Pears	32°F to 35°F (0°C to 2°C)	90–95%	
Pineapple	45°F (7°C)	85–90%	Prone to chill damage; store as close to 45°F (7°C) as possible.
Plums	32°F (0°C)	95%	Does not have a long shelf life, even under refrigeration.
Raspberries	32°F to 35°F (0°C to 2°C)	90–95%	Do not sprinkle before storage; refrigerate immediately after receiving; use quickly.
Strawberries	32°F to 35°F (0°C to 2°C)	90–95%	Will not ripen after being picked; do not wash until just before use; very perishable.
Tangerines	40°F (4°C)	85–90%	
Tomatoes	55°F to 60°F (13°C to 16°C)	85–95%	Do not refrigerate.
Watermelon	55°F to 70°F (13°C to 21°C)	85–90%	Holding at room temperature improves flavor; quality deteriorates at temperatures of 50°F (10°C) or below.

This table is a general guideline for best product quality and overall safety. Where applicable, always use any product by its use-by date marked on package. If purchase date is unknown, or if quality or safety is compromised in any way, discard product.

APPENDIX B
STORAGE TEMPERATURES FOR FRESH VEGETABLES

Source: Produce Marketing Association

Vegetables	Storage Temperatures (°F/°C)	Relative Humidity (%)	Comments
Artichokes	32°F to 34°F (0° to 1°C)	90–95%	Do not dampen when unrefrigerated, may cause mold.
Asparagus	32°F to 35°F (0° to 2°C)	90–95%	Shelf life is improved when stood on end in an inch of water.
Beans	45°F to 50°F (7°C to 10°C)	90–95%	Do not rinse until before use; allow adequate air flow around containers.
Broccoli	32°F to 35°F (0° to 2°C)	85–90%	
Cabbage	32°F (0°C)	90–95%	Will lose moisture at room temperatures; refrigerate whole heads with leaves intact.
Carrots	32°F to 34°F (0° to 1°C)	90–95%	Extremely long shelf life; remove tops to prevent moisture loss.
Cauliflower	32°F to 35°F (0° to 2°C)	85–90%	Susceptible to damage in handling and storage; store boxes upside down; dry atmosphere will cause curd to dry out.
Celery	32°F to 35°F (0° to 2°C)	90–95%	Highly perishable; wilts in high humidity.
Corn	32°F to 34°F (0° to 1°C)	85–90%	Refrigerate to slow conversion of sugar to starch.
Cucumbers	45°F to 50°F (7°C to 10°C)	85–95%	Most cucumbers are shipped waxed to prevent moisture loss.
Eggplant	45°F to 50°F (7°C to 10°C)	85–90%	Easily damaged; will decay if bruised.
Lettuce	32°F to 35°F (0° to 2°C)	90–95%	Rotate FIFO; revive wilted lettuce by plunging in cold water; store away from ethylene-producing produce.
Mushrooms	32°F to 35°F (0° to 2°C)	85–90%	Do not wash or cut until just before using.
Onions (white)	45°F to 50°F (7°C to 10°C)	65–70%	Can withstand long-term storage; air circulation is beneficial to prevent other produce from absorbing odors.
Onions (green)	32°F (0°C)	90–95%	Very perishable; trimming end and soaking in water will revive wilting.
Peas	32°F to 35°F (0° to 2°C)	50%	Very perishable; refrigerate and rotate supplies.
Peppers	45°F to 50°F (7°C to 10°C)	85–90%	Very perishable at room temperature.
Potatoes	45°F to 50°F (7°C to 10°C)	85–90%	Store in cool, dark area; should not be exposed to light or freezing temperatures.
Spinach	32°F (0°C)	90–95%	Wilts quickly at room temperature; adding ice to containers prolongs life.
Sprouts	36°F to 40°F (2°C to 4°C)	90–95%	Cover during storage to prevent moisture loss.

This table is a general guideline for best product quality and overall safety. Where applicable, always use any product by its use-by date marked on package. If purchase date is unknown, or if quality or safety is compromised in any way, discard product.

Thermophile

APPENDIX C
REFRIGERATED STORAGE OF FOODS

Sources: Tyson, Egg Board, *Safe Food Storage Times and Temperatures* by Marl L. Tamplin, Ph.D.

Food	Recommended Product Temperatures (°F/°C)	Maximum Storage Periods
Meat		
Roasts, steaks, chops	35°F to 41°F (2°C to 5°C)	2 to 5 days
Steaks	35°F to 41°F (2°C to 5°C)	2 to 5 days
Chops	35°F to 41°F (2°C to 5°C)	3 to 4 days
Ground and stewing	35°F to 41°F (2°C to 5°C)	1 to 2 days
Variety meats	35°F to 41°F (2°C to 5°C)	1 to 2 days
Whole ham	35°F to 41°F (2°C to 5°C)	7 days
Half ham	35°F to 41°F (2°C to 5°C)	3 to 5 days
Ham slices	35°F to 41°F (2°C to 5°C)	3 to 5 days
Canned ham	35°F to 41°F (2°C to 5°C)	9 months to 1 year
Frankfurters	35°F to 41°F (2°C to 5°C)	1 week
Bacon	35°F to 41°F (2°C to 5°C)	5 to 7 days unopened
Luncheon meats	35°F to 41°F (2°C to 5°C)	3 to 5 days
Leftover cooked meats	35°F to 41°F (2°C to 5°C)	1 to 2 days
Gravy, broth	35°F to 41°F (2°C to 5°C)	1 to 2 days
Poultry		
Whole chicken, turkey, duck, goose	32°F to 36°F (0°C to 2°C)	1 to 2 days
Giblets	32°F to 36°F (0°C to 2°C)	1 to 2 days
Stuffing	32°F to 36°F (0°C to 2°C)	1 day
Cut-up cooked poultry	32°F to 36°F (0°C to 2°C)	1 to 2 days
Fish		
Fresh fish	32°F to 36°F (0°C to 2°C)	1 to 2 days
Fish (smoked)	30°F to 41°F (-1°C to 5°C)	1 to 2 days
Clams, crab, lobster (in shell)	30°F to 41°F (-1°C to 5°C)	2 days
Scallops, oysters, shrimp	30°F to 41°F (-1°C to 5°C)	1 day
Eggs		
Eggs in shell	45°F (7°C)	*4 to 5 weeks beyond pack date
Leftover yolks	40°F to 45°F (4°C to 7°C)	1 to 2 days
Leftover whites	40°F to 45°F (4°C to 7°C)	4 days
Dried eggs (whole eggs and yolks)	40°F to 45°F (4°C to 7°C)	Up to 1 year (unreconstituted)
Reconstituted dried eggs		Use immediately
Cooked Dishes with eggs, meat, milk, fish, poultry	32°F to 36°F (0°C to 2°C)	Serve day prepared
Dairy Products		
Fluid milk	35°F to 41°F (2°C to 5°C)	5 to 7 days after date on container
Butter	35°F to 41°F (2°C to 5°C)	2 weeks
Hard cheese (cheddar, parmesan, romano)	35°F to 41°F (2°C to 5°C)	1 month
Soft cheese	35°F to 41°F (2°C to 5°C)	1 week
Dry milk (nonfat)	35°F to 41°F (2°C to 5°C)	1 year unopened
Reconstituted dry milk	35°F to 41°F (2°C to 5°C)	1 week

This table is a general guideline for best product quality and overall safety. Where applicable, always use any product by its use-by date marked on package. If purchase date is unknown, or if quality or safety is compromised in any way, discard product.
*Recommended by the American Egg Board. Most eggs arrive at a distribution site within a few days of being packed.

APPENDIX D
STORAGE OF FROZEN FOODS

Sources: Tyson and *Safe Food Storage Times and Temperatures* by Marl L. Tamplin, Ph.D.

Food	Maximum Storage Period at 0°F to 10°F (-12°C to -18°C)
Meat	
Beef, roasts and steaks	6 to 9 months
Beef, ground and stewing	3 to 4 months
Pork, roasts and chops	4 to 8 months
Pork, ground	2 months
Lamb, roasts and chops	6 to 9 months
Lamb, ground	3 to 5 months
Veal	8 to 12 months
Variety meats	3 to 4 months
Ham, frankfurters, bacon, luncheon meats	2 weeks
Leftover cooked meats	2 to 3 months
Gravy, broth	2 to 3 months
Sandwiches with meat filling	1 to 2 months
Poultry	
Whole chicken, turkey, duck, goose	12 months
Giblets	3 months
Cut-up cooked poultry	4 to 6 months
Fish	
Fresh fish	2 to 3 months
Frozen fish	3 to 6 months
Clams, lobster	3 months
Scallops, shrimp	3 months
Ice Cream	3 months; original container; quality maintained better at 10°F (-12°C)

This table is a general guideline for best product quality and overall safety. Where applicable, always use any product by use-by date marked on package. If purchase date is unknown, or if quality or safety is compromised in any way, discard product.

APPENDIX E
SHELF LIFE OF DRIED GOODS

Source: *Food Storage Times and Temperatures* by Marl L. Tamplin, Ph.D.

Food	Recommended Maximum Storage Period if Unopened
Baking Materials	
Baking powder	8 to 12 months
Baking soda	2 years
Chocolate, baking	6 to 12 months
Chocolate, sweetened	2 years
Cornstarch	2 to 3 years
Dried bread crumbs	6 months
Flour	6 to 8 months
Honey	12 months
Rice, white	2 years
Yeast, dry	18 months
Beverages	
Coffee, cans	2 years
Coffee, ground, not vacuum packed	2 weeks
Coffee, instant	8 to 12 months
Tea, bags	18 months
Tea, loose	2 years
Tea, instant	3 years
Canned Goods	
Fruits (in general)	1 year
Fruits, acidic (citrus, berries, sour cherries)	6 to 12 months
Fruit juices	9 months
Seafood (in general)	1 year
Pickled fish	4 months
Soups	1 year
Vegetables (in general)	1 year
Vegetables, acidic (tomatoes, sauerkraut)	7 to 12 months
Dairy Foods	
Cheese, parmesan (grated)	10 months
Milk, condensed	1 year
Milk, evaporated	1 year
Non-dairy creamer	9 months
Fats and Oils	
Mayonnaise	2 months
Salad dressings	10 to 12 months
Salad oil	6 to 9 months
Shortening, solid	8 months

Food	Recommended Maximum Storage Period if Unopened
Grains and Grain Products	
Cereal grains for cooked cereal	6 months
Cereals, ready-to-eat	6 to 12 months
Flour, bleached	9 to 8 months
Macaroni, spaghetti, and other dry pasta	2 years
Rice, white	2 years
Rice, flavored or herb	6 months
Seasonings	
Flavoring extracts	2 years
Monosodium glutamate	Indefinite
Mustard, prepared	2 to 6 months
Salt	Indefinite
Sauces (steak, soy, etc.)	2 years
Spices and herbs (whole)	2 years to indefinite
Paprika, chili powder, cayenne	1 year
Seasoning salts	1 year
Vinegar	2 years
Sweeteners	
Sugar, granulated	2 years
Sugar, confectioners	18 months
Sugar, brown	4 months
Syrups, corn, honey, molasses, sugar	1 year
Miscellaneous	
Dried beans	1 to 2 years
Cookies, crackers	1 to 6 months
Dried fruits	6 to 8 months
Dried prunes	6 months
Gelatin	2 to 3 years
Ketchup	1 month
Jams, jellies	1 year
Nuts (whole or packaged meats)	6 months
Pickles, relishes	1 year
Potato chips	1 month

This table is a general guideline for best product quality and overall safety. Where applicable, always use any product by use-by date marked on package. If purchase date is unknown, or if quality or safety is compromised in any way, discard product.

APPENDIX F: RESPONDING TO AN OUTBREAK OF FOODBORNE ILLNESS

This book is designed to help you take steps to ensure that the food you serve is safe. Despite your best efforts, however, an outbreak of foodborne illness can occur in your establishment at any time. How you respond when that happens can determine whether or not you end up in the middle of a crisis.

Some of the most respected companies in the industry have suffered from crises involving foodborne illness. Most companies survive, but not without a tremendous loss of both credibility and business. In some cases, crises have been severe enough to shut down businesses altogether.

Public relations experts say there are two ways to determine if you're in a crisis. First, an event occurs that could potentially threaten your business and your reputation. Next, the media hear about it.

The media's job is to look for and report news. A problem or incident in your establishment can escalate into a crisis with the fuel that media attention provides. Often, the size and duration of the crisis can be judged by the number of reporters who respond to the event.

Dealing with the media may be the most difficult facet of any crisis. At the same time you are trying to determine what went wrong and how to fix the problem, the media want to know what happened, who was responsible and what you're going to do about it.

In many cases, operators don't have answers to these questions right away, which is why crises often play out in phases that include:

○ Surprise

○ Lack of reliable information

○ An escalating flow of events

○ Loss of control

○ Intense outside scrutiny from customers and the media

○ Short-term focus

○ Panic

Averting a Crisis

You may be able to avert a crisis by responding quickly when you do receive customer complaints. Take all customer complaints seriously. Express your concern and be sincere, but do not admit responsibility or accept liability. Listen carefully and promise to investigate and respond.

Write down all facts about the incident. Record the exact food and beverages in question and the time when the customer became ill. These facts can help you identify the illness and determine whether your food is the cause. Consider developing an incident report to guide you through the process. Questions to ask include:

○ What did you eat and drink at our establishment and when?

○ When did you become ill? What were the symptoms and how long did you experience them?

○ Did you eat anything else before or after eating at our establishment? What and where? Who else ate the same foods, and did they become ill?

○ Did you seek medical attention? Where and how soon after becoming ill? What diagnosis and treatment did you receive?

Evaluate the complaint. If it is isolated, use it as an opportunity to review food-safety procedures with employees. Remind them of the importance of personal hygiene. If you have a HACCP system in place, review all documentation on the date in question and verify that the system is working. Pay close attention to temperature and equipment issues.

If more than one person has complained, you have the potential for a crisis on your hands. Take steps to control the situation and reassure customers that you are doing everything you can to identify and fix the problem. If you have a crisis management plan in place, call your crisis team together.

First, contact your local health department. Health department officials often can help you determine the cause and source of the foodborne illness. They also can act as an information resource for the media. It's important that local health officials be partners, not adversaries, while you investigate the possible cause of a foodborne illness. Therefore, make sure you develop a working relationship with them. Don't wait for a crisis to call them.

You may also want to call in outside experts. You may need help isolating the cause of the illness or dealing with the situation. Resources you can call on include food-testing labs, public relations firms or issues experts, and even your own management personnel.

Managing a Crisis

If you have received two or more complaints, it's usually only a matter of time before the media hear about it and you do have a crisis on your hands. Every crisis evolves differently, but in general, you can minimize potential damage by taking certain steps.

○ Put a team together to gather information, plan courses of action, and manage events as they unfold. In large multi-unit operations, the team may comprise the president and senior managers from finance, operations, marketing, franchising, human resources, public relations, and training departments. Actual crisis management teams are usually much smaller and may vary in makeup depending on the establishment and the situation. An independent establishment's team might include the owner, general manager, and chef.

○ Appoint a single spokesperson to handle all media queries and communications. Designating a point person usually results in more consistent messages and allows you to control media access to your staff. Train the spokesperson in media relations and interview skills so he or she knows what to expect and how to respond. Crisis situations can be very stressful. Training can help your spokesperson handle these situations more easily. Make sure all of your staff knows who the spokesperson is, and instruct them to direct any and all questions to that person.

○ Work with, not against, the media. Be as proactive as you can, as early as you can. Contact the media and arrange a press conference to communicate what you know before they contact you. That way you can control what the media reports. Stick to the facts, and be as honest as possible. If you don't have all the facts, say so, and let the media know that you'll communicate them as soon as you do know. Keep a cool head and don't be defensive. The easiest way to magnify or prolong a crisis is to deny, lie, or change your story.

○ Show concern and be sincere. If health officials have confirmed that your establishment is the source of the illness, accept responsibility. Accepting responsibility is not the same as admitting liability. While customers may have become ill from eating food in your operation, the cause may have been beyond your control and not your fault. If you don't express your concern, and mean it, you'll lose credibility with the public, not just customers.

○ Develop a communication system to get information directly to all of your key audiences. Don't depend on the media to relay all the facts. Tell your side of the story to employees, customers, stockholders, and the community. Use any means possible such as a newsletter, a Web site, flyers, and newspaper or radio advertising.

○ Fix the problem and communicate what you've done both to the media and to your customers. Each time you take a step to resolve the problem, let the media know. Hold news briefings when you have news, and go into each briefing or press conference with an agenda. Take control. Don't simply respond to questions.

Crisis Preparedness

The best way to manage a crisis is to avoid having one in the first place. Just as a HACCP plan lets you monitor and correct potential problems to prevent foodborne illnesses, a crisis management plan can avert much of the negative publicity associated with an outbreak of foodborne illness.

The process of preparing for a crisis usually decreases the chance of a crisis actually occurring. There are a number of steps you can take to prepare for the possibility of a crisis. First, assemble a team from different areas of your operation. Use this appendix as a guide to some of the issues the team should address, and assign tasks and responsibilities. The National Restaurant Association Educational Foundation also has a Crisis Management Reference Book that describes in detail the process of developing a plan.

○ Certify all your managers in food safety. Thoroughly train employees in good food-safety practices. Implement a HACCP plan to give you greater control over food safety and create a paper trail in case something ever goes wrong.

○ Cultivate a good relationship with your local health department. Don't wait for local health officials to inspect your establishment as required by law. Work with them to develop a food-safety program and invite them to monitor your progress. Hire an outside firm to inspect your establishment, using higher standards than your health department requires. If something does go wrong, you want the health department to be able to come to your defense.

○ Identify and assess all potential risks. While the greatest threat to customers may be from foodborne illnesses, don't forget that other crises can include robberies, severe weather, fire, or some other trauma.

○ Develop simple instructions on what to do in each type of crisis. In an outbreak of foodborne illness, steps include isolating the suspect food, obtaining samples of the suspect food, preventing further sale of the food, excluding suspect employees from handling food, contacting the local health department, and so forth. In each case, decide:

  ● Who will manage the situation

  ● Who will act as spokesperson

  ● Who you want to communicate to

  ● What should be said

○ Assemble a contact list of names and numbers and post it by the phones. The list should include all crisis-management team members and outside resources such as police, fire and health departments, testing labs, issues experts, and management or headquarters personnel.

○ Train your spokesperson to be professional, truthful, understanding, and concerned when dealing with the media and others. Public relations firms or companies specializing in media training can help. The National Restaurant Association also offers "Twelve Minutes That Can Save Your Business—Effective Media Training," a video and manual that help operators deal with media situations.

○ Develop a list of media responses or a Q&A sheet suggesting what to say in the event of each type of crisis, including foodborne illness, robbery, fire, or other disaster you may have identified. Create sample press releases that can be tailored quickly to each incident.

○ Put together a list of media contacts to call for press conferences or news briefings. Include a media relations plan with do's and don'ts of dealing with the media.

○ Include instructions on how to communicate with employees, whether through shift meetings, E-mail, a telephone tree, or some other vehicle.

○ Assemble everything in a crisis kit for the establishment. The kit can be in the form of a three-ring notebook or binder enclosing these materials. Keep the kit in an accessible place, such as the manager's or chef's office.

○ Test the plan by running a simulation. Hire a public relations or consulting firm with crisis management experience to enact a crisis and test your team's readiness. In most cases, the firm will design a simulated crisis that will be as close as possible to what could happen in a real-life situation.

An outbreak of foodborne illness has the potential to damage your business beyond repair. Investing time and resources in a crisis management plan can insure against that. Remember these three key rules of crisis management.

1. Take steps to prevent a crisis from occurring by practicing good food-safety habits.

2. Prepare for the possibility of a crisis by developing contingency plans.

3. If a crisis does occur, take control of the situation. Use your plan to manage the crisis thoughtfully, honestly, and as quickly as possible.

ADDITIONAL RESOURCES

APPENDIX A

Books and Periodicals

Beck, B. (1984). *Produce: A fruit and vegetable lovers' guide.* New York: Friendly Press.

The import issue: How safe is U.S. produce? (1997). *FoodService Director,10*(6), 72.

National Restaurant Association. (1984). *Buying, handling, and using fresh fruits.* Washington, DC: Author.

Pijpers, D. (1986). *The complete book of fruit.* New York: Gallery Books.

The Produce Marketing Association. (1995). *The food service guide to fresh produce.* Newark, DE: Author.

Schneider, E. (1986). *Uncommon fruits and vegetables: A commonsense guide.* New York: Harper & Row.

Web Sites

Produce Marketing Association

http://www.pma.com

The members of this not-for-profit trade association market fresh fruits and vegetables to the foodservice industry. Purchasing guidelines are available.

USDA Agricultural Marketing Service

http://www.ams.usda.gov/standards/frutmrkt.htm

Comprehensive site featuring a list of standards for fresh fruit from the USDA's Agricultural Marketing Service.

Other Useful Addresses

International Fresh-Cut Produce Association

www.fresh-cuts.org

The Web site for the association representing the fresh-cut produce industry. Features information, resources, and publications.

Produce Marketing Association

1500 Casho Mill Road, PO Box 6036, Newark, DE 19714-6036
Phone (302)738-7100 Fax (302)731-2409
E-mail webmaster@mail.pma.com

North American Produce Safety Kit: A Complete Media Guide to Food Safety Experts and Information

http://www.pma.com/news/issues/fsftykit.html

United Fresh Fruit & Vegetable Association

717 North Washington Street, Alexandria, VA 22314-1977
Phone (703)836-3410 Fax (703)836-2049

APPENDIX B

Books and Periodicals

The Produce Marketing Association. (1995). *The food service guide to fresh produce.* Newark, DE: Author.

Web Sites

International Fresh-Cut Produce Association

www.fresh-cuts.org

The Web site for the association representing the fresh-cut produce industry. Features information, resources, and publications.

Produce Marketing Association

http://www.pma.com

The members of this not-for-profit trade association market fresh fruits and vegetables to the foodservice industry. Purchasing guidelines are available.

USDA Agricultural Marketing Service

www.ams.usda.gov/standards

Comprehensive site featuring the USDA Agricultural Marketing Service's list of standards for various fresh foods, including poultry and vegetables.

APPENDIX C

Books and Periodicals

Bertagnoli, L. (1995). Enter the new ice age. *Restaurants & Institutions, 105*(8), 124.

Bertagnoli, L. (1995). Rethinking refrigeration. *Restaurants & Institutions, 105*(23), 120.

Castagna, N. G. (1997). Know your cook-chill. *Restaurants & Institutions, 107*(16), 84.

Durocher, J. (1997). Cool breeze: Specialty refrigeration for restaurants. *Restaurant Business, 96*(21), 161.

Durocher, J. (1998). The big chill. *Restaurant Business, 97*(8), 125.

Hunt-Ashby, B. (1995). *Protecting perishable foods during transport by truck* (Handbook 669).Washington, DC: USDA Agricultural Marketing Service Transportation and Marketing Division.

Sanitation survival kit: Don't serve illness to your customers. (1998). Washington, DC: National Restaurant Association.

Slomon, E. (1997). Refrigeration: The cold, hard facts. *Pizza Today, 15*(8), 60.

Tamplin, M. L. (1994). *Safe food storage times and temperatures* (Fact Sheet HE 8490). Gainesville: University of Florida. A series of the Home Economics Department, Florida Cooperative Extension Service, Institute of Food and Agricultural Sciences.

Web Sites

American Egg Board

http://www.aeb.org

The American Egg Board's Web site features product, safety, and industry information for foodservice professionals as well as recipes and other information for the general public.

National Food Safety Database

http://www.foodsafety.org

A compilation of food-safety database information from government, consumer, and public health organizations, this is a one-stop Web site for food-safety information on the Internet.

Tyson

http://www.tyson.com

Tyson's Web site offers a consumer kitchen page with recipes and instruction on preparing chicken products. An entire page is also devoted to foodservice professionals and offers information on the industry, Tyson products, and menu planning.

APPENDIX D

Books and Periodicals

Tamplin, M. L. (1994). *Safe food storage times and temperatures* (Fact Sheet HE 8490). Gainesville: University of Florida. A series of the Home Economics Department, Florida Cooperative Extension Service, Institute of Food and Agricultural Sciences.

Web Sites

National Food Safety Database

http://www.foodsafety.org

A compilation of food-safety database information from government, consumer, and public health organizations, this is a one-stop Web site for food-safety information on the Internet.

Tyson

http://www.tyson.com

Tyson's Web site offers a consumer kitchen page with recipes and instruction on preparing chicken products. An entire page is also devoted to foodservice professionals and offers information on the industry, Tyson products, and menu planning.

APPENDIX E

Books and Periodicals

Bendall, D. (1998). Making the most of shelving. *Restaurant Hospitality, 82*(11), 146.

Food Marketing Institute. (1996). *The food keeper.* Washington, DC: Author.

Lorenzini, B. (1994). Storing supplies safe and sound. *Restaurants & Institutions, 104*(14), 106.

Mixon, J. A. (1991). *Guidelines for the storage and care of food products: A technical assistance manual: Vol. V.* (3rd ed.). Washington, DC: Food Industry Services Group, USDA Food and Nutrition Service.

National Restaurant Association. (1996). *Sanitation Survival Kit for Restaurant Operators.* Washington, DC: Author.

Tamplin, M. L. (1994). *Safe food storage times and temperatures* (Fact Sheet HE 8490). Gainesville: University of Florida. A series of the Home Economics Department, Florida Cooperative Extension Service, Institute of Food and Agricultural Sciences.

Web Sites

National Food Safety Database

http://www.foodsafety.org

A compilation of food-safety database information from government, consumer, and public health organizations, this is a one-stop Web site for food-safety information on the Internet.

United States Department of Agriculture (USDA)

http://www.usda.gov

The United States Department of Agriculture's Web site features information, publications, and other educational materials about the nation's agriculture.

APPENDIX F

Books and Periodicals

Augustine, N. R. (1995). *Managing the crisis you tried to prevent.* Harvard Business Review, 73 (6), 147.

Barton, L. (1994). Crisis management: Preparing for and managing disasters. *Cornell Quarterly, 35* (2), 59.

Confronting a crisis. (1998). *Best Practices. (2)* 3, 6-10.

Doeg, C. (1995). *Crisis management in the food and drinks industry: A practical approach.* In *Practical Approaches to Food Control and Food Quality: Vol. 2.* New York: Chapman & Hall.

Fitzpatrick, K. R. (1995). Ten guidelines for reducing legal risks in crisis management. *Public Relations Quarterly, 40* (2), 33.

In the event of a crisis, the crisis management team would be notified, day or night. (1990). *NCA/Club Director, 1* (9), 10.

National Restaurant Association. (1998). *Effective media training.* Washington, DC: Author.

Glossary

Note: The number(s) in parentheses at the end of each entry refers to the chapter in which this term is used.

Abrasive cleaner
Cleaners that contain a scouring agent that helps scrub off hard-to-remove soils. They may scratch some surfaces. *(11)*

Acid cleaner
Acid cleaners (pH below 7.0) are used on mineral deposits and other soils that alkaline cleaners can't remove, such as scale, rust, and tarnish. *(11)*

Acidity
An acidic substance has a pH below 7.0. Foods with a pH range from 4.6 to 7.0 (slightly acidic) are called potentially hazardous foods. *See high acid foods. (2)*

Acrylic wood
Floor covering made of wood impregnated with plastic. *(10)*

ADA (Americans with Disabilities Act)
Federal law which requires reasonable accommodation for access to a facility by both patrons and employees with disabilities. *(10)*

Aerobic
Microorganisms that grow only when oxygen is present. *(2)*

Air curtains
A ventilation unit that blows a steady stream of air across an open door (or outward) to discourage flying insects from entering. Also called air doors or fly fans. *(12)*

Air gap
A device to prevent water backflow to a potable water supply. It is an unobstructed open space that separates an outlet of the potable water supply (for example, a faucet) from any potentially contaminated source, such water in a sink or a drain pipe. *(10)*

Al fresco dining
Dining areas set up outside or in areas where there are no screens on outdoor openings. *(12)*

Alkaline cleaner
All detergents are mildly alkaline (pH above 7.0) and are used to clean fresh soil from floors, walls, ceilings, prep surfaces, and most equipment and utensils. *(11)*

Alkalinity
An alkaline substance has a pH above 7.0. Most foods are not alkaline. *(11)*

Americans with Disabilities Act
See ADA. (10)

Anaerobic
Microorganisms that do not require oxygen to grow. *(2)*

Application
A key element of training; practice by the learner. *(14)*

Approved suppliers
See suppliers. (5)

Backflow

A type of cross-connection that can occur in a potable water system. Backflow is the unwanted reverse flow of contaminated water through a cross-connection into a potable water system. It can occur whenever the pressure in the potable water supply drops below the pressure of the contaminated supply. *See air gap. (10)*

Bacteria

The most common foodborne microbial contaminants. Bacteria are living, single-celled microorganisms that can cause food spoilage and illness. Some form spores, which can survive freezing and very high temperatures. Bacteria that cause disease are pathogenic. Bacteria that produce toxins are toxigenic. *See aerobic, anaerobic, facultative. (2)*

Bacterial growth

Bacteria reproduce (or grow) by splitting into two. *(See vegetative microorganisms.)* When conditions are favorable, bacteria can grow and multiply very rapidly, doubling the population as often as every twenty minutes, in some cases. Three phases of bacterial growth are the lag phase (slow growth), log phase (rapid growth), stationary phase (equilibrium between growth and die-off), and death phase (decrease in population). *See the acronym FAT-TOM,* which lists the conditions conducive to bacterial growth. *(2)*

Behavioral objectives

Specific actions a learner will be able to perform upon completion of training. *(14)*

Bi-metallic stemmed thermometer

The most common and versatile type of thermometer, it measures temperature through a metal probe with a sensor in the end. Easily recalibrated. Most can measure temperatures from 0°F to 220°F (-18°C to 104°C); accurate to within +/-2°F or +/-1°C. *(5)*

Biological contamination

Presence of microorganisms in food. Biological contamination from microorganisms, or microbial contamination, is the leading cause of foodborne illness throughout the world. *See contaminants. (3)*

Biological toxins

One type of biological contamination that causes foodborne illness. Many biological poisons, or toxins, occur naturally (in some fish, plants, mushrooms); others are caused by the diets of certain animals. Most toxins cannot be destroyed by heating or freezing. *(3)*

Biological, chemical, and physical hazards

See hazards, biological; hazards, chemical; hazards physical. (1, 3)

Blast chiller

Blast chillers can move food through the temperature danger zone quickly. Most cool foods from 140°F to 37°F within 90 minutes. *(10)*

Bodily fluids

Fluid secretions of the human body, such as mucus, saliva, feces, perspiration, and oily secretions in skin and hair. Microorganisms in those fluids can be transmitted to customers via food if employees do not practice good personal hygiene. *(4)*

Booster heater

An extra water heater attached to hot-water lines leading to warewashing machines or sinks; it raises water to temperatures required for heat sanitizing of tableware and utensils (180°F [82°C]). Many warewashing machines have internal booster heaters. *(10, 11)*

Cantilever mounted

Also wall-mounted. Equipment attached to the wall with a bracket to allow for easier cleaning behind and underneath it. *(10)*

Carrier

An animal, insect, or human that carries a pathogenic microorganism, often without showing symptoms of the disease. These carriers may infect others with the disease, yet may never become ill themselves. The carrier state cannot be cured with antibiotic treatment. Carriers may not work as foodhandlers or as child-care attendants. *(2, 4)*

CCP

See Critical Control Point, HACCP.

CDC (Centers for Disease Control & Prevention)

An agency of the U.S. Public Health Service that investigates outbreaks of foodborne illness, studies the causes and control of disease, and publishes statistical data. *(13)*

Ceramic tile

A hard, nonresilient, nonporous, porcelain-type tile, installed with grout, commonly used for floors and walls in establishments. *(10)*

Chemical agent

Chemical agents often used in establishments include cleaning products, polishes, lubricants, sanitizing chemicals, and pesticides. All can be dangerous to customers and employees if not properly used or stored. *(3)*

Chemical contaminant

One type of contamination that causes foodborne illness. Food can become contaminated by chemical substances normally found in establishments, including cleaning chemicals; utensils and equipment that leach toxic metals into foods; pesticides; and food additives and food preservatives, even those generally regarded as safe (GRAS), if not used correctly. *(3)*

Chemical hazard

See hazard, chemical.

Chemical sanitizing

A method of reducing the number of microorganisms on a surface by exposing an object to a sanitizing solution for a specific period of time. Common sanitizing chemicals are chlorine, iodine, and quaternary ammonia. *See heat sanitizing. (11)*

Chlorine

Most commonly used chemical sanitizer. *(11)*

Ciguatera

A foodborne illness caused by a biological toxin. Ciguatera occurs when a person eats certain predatory fish that eat smaller fish that have eaten a certain species of toxic algae. The toxins in the algae accumulate in the tissues of the larger fish. *(3)*

Clean

Free of visible soil; refers only to outward appearance of a surface. *See soil. (1)*

Cleaning

The process of removing food and other types of soil from a surface. Surfaces must first be cleaned and rinsed before being sanitized. *(11)*

Cleaning agent

Chemical compounds that remove food, soil, rust, stains, minerals, or other deposits from surfaces. *(11)*

Code

A collection of regulations. *(13)*

Cold paddles

Utensils that can be filled with water and frozen. Stirring food products with them chills food very quickly. *(7)*

Contact spray

A liquid or powder insecticide that must come into direct contact with insects to kill them. *(12)*

Contaminants, contamination

Presence of harmful substances not originally present in the food. Three types of contaminants are biological hazards, chemical hazards, and physical hazards. Although most food-safety hazards are contaminants introduced by humans, some food hazards occur naturally, such as toxins in certain fish, mushrooms, or plants. *(1)*

Control measures (versus preventive measures)

Steps taken, usually by a PCO, to control or eliminate pests that have infested a building. These include chemical and non-chemical treatment methods. Preventive measures include steps that can be taken to prevent pests from entering a building, such as keeping the facility clean and sanitary and maintaining the building properly. *(12)*

Control point

Any step in the flow of food where a physical, chemical, or biological hazard can be controlled. *See HACCP. (9)*

Corrective action

A predetermined action taken when food doesn't meet a critical limit. For example, when the temperature of a hot food falls below 140°F (60°C), the proper corrective action is reheat the food to 165°F (74°C) for fifteen seconds within two hours. *See critical limit, HACCP. (9)*

Coving

A curved, sealed edge between the floor and wall that makes cleaning easier and eliminates hiding places for insects. *(10)*

Critical Control Point (CCP)

The last step where you can intervene to prevent, control, or eliminate the growth of microorganisms in food. *See HACCP. (9)*

Critical limit

At a CCP, a maximum or minimum value that must be met (usually time, temperature, water activity, pH, or oxygen) to prevent, eliminate, or reduce a hazard to an acceptable limit. *See HACCP. (9)*

Cross-connection

Any physical link through which contaminants from drains, sewers, or waste pipes can enter a potable water supply. *(10)*

Cross-contamination

Transfer of harmful substances or disease-causing microorganisms from one food product to another through direct contact, or contact with utensils, equipment, work surfaces, or employees' hands or clothing. *(1)*

Death phase

The period of bacterial growth when the number of microorganisms dying exceeds the number of microorganisms being produced and the number declines. *(2)*

Deep-chill storage

Deep-chill units hold foods at near-freezing temperatures, between 26°F and 32°F (-3°C to 0°C), for short periods of time. Poultry, meat, seafood, and *sous vide* products stored at these temperatures have a longer shelf life, but deep-chill temperatures may freeze and damage other foods. *(6)*

Demonstration

The process of illustrating a skill or task before another person or a group. *(14)*

Detergent

A water-soluble preparation, chemically different from soap, used in cleaning to break down oils, hold dirt in suspension, and act as a wetting agent. General purpose detergents are mildly alkaline. *(10, 11)*

Droppings

The feces of insects, birds or animals. *(12)*

Dry storage

The holding of nonperishable food items, such as rice, flour, crackers, and canned goods, at 50 to 60 percent humidity and between 50°F and 70°F (10°C and 21°C). *(6)*

Electrocutor trap ("zapper")

A mechanical device that attracts flying insects to a light. Insects flying inside are then killed by electricity. *(12)*

EPA (Environmental Protection Agency)

A government agency that sets standards for environmental quality, to include air and water quality, and regulates the use of pesticides and the handling of wastes. *(10, 13)*

Evaluation

Judging the performance of training participants against the learning objectives. *(14)*

Facultative

Facultative microorganisms grow in the presence or absence of free oxygen. Most bacteria that cause foodborne illness are facultative. *See aerobic, anaerobic. (2)*

FAT-TOM

An acronym that lists the factors necessary for the growth of microorganisms (particularly bacteria), and related food-safety controls: food, acidity, time, temperature, oxygen, moisture. *(2)*

FDA (Food and Drug Administration)

An agency of the U.S. Department of Health and Human Services that regulates food and drug safety; also responsible for developing the Model Food Code. *(13)*

Feedback

Evaluation given to employees about their performance. This may be constructive criticism given to an individual to correct a mistake, or praise given to reinforce proper performance of a skill or procedure. (14)

FIFO (first in, first out)

A method of stock rotation in which supplies with earliest expiration date are shelved in front of supplies with later dates, so the old are used first. All inventory is marked with the expiration date, when it was received, or when it was stored after preparation. *(6)*

Finger cot

A protective covering, similar to a glove, for one finger. *(4)*

Flood rim

Spillover point on a sink. An air gap exists between the top of the flood rim and the sink faucet. *(10)*

Flow of food

The path of food through an establishment, from receiving through storing, preparing, cooking, holding, serving, cooling, and reheating. Food-safety hazards can occur at every stage in the flow of food. *(1)*

Food additive, food preservative

Common food additives and preservatives include sulfites, nitrites, and monosodium glutamate (MSG). Most are generally regarded as safe (GRAS) but excessive amounts of certain food preservatives have caused illness or allergic reactions. *(3)*

Food allergy

The body's negative reaction to a particular food or foods. *(3)*

Food and Drug Administration (FDA)

See FDA.

Food bar

Also self-service bar. Self-service buffet where patrons can choose what they want to eat and serve themselves. Self-serve areas should be monitored by employees trained in food-safety procedures to prevent contamination of food by customers. *(8)*

Food Code, The

(Also referred to as the FDA Model Food Code.) The set of science-based guidelines for food safety for restaurants and establishments. Local, state, and federal regulators use the Food Code as a model to help develop or update their own food-safety rules. *(1)*

Food contaminant

See contamination and chemical contaminant. (1)

Food quality

Proper appearance, flavor, texture, consistency, and nutritional value in food. Food that is stored, prepared, and served properly is more likely have high quality. *(1)*

Food safety

Unsafe food usually results from contamination due to biological hazards, chemical hazards, or physical hazards. To ensure food safety, establish standards that focus on controlling time and temperature, practicing good personal hygiene, maintaining a sanitary facility, preventing cross-contamination, and purchasing food supplies from approved suppliers. *(1)*

Foodborne illness

A disease that is carried or transmitted to people by food. Foodborne diseases are classified as infections, intoxications, or toxin-mediated infections. *(1, 2, 3)*

Foodborne infection

A foodborne illness that results when live, pathogenic microorganisms in ingested food grow in the intestines. Symptoms typically do not appear immediately. *(2)*

Foodborne intoxication

An illness caused by eating food containing toxins produced by pathogens. A person does not need to ingest live microorganisms to become ill, just the toxins, many of which are not destroyed by cooking. Symptoms of a foodborne intoxication typically appear within a few hours. *(2)*

Foodborne-illness outbreak

The Centers for Disease Control and Prevention (CDC) defines a foodborne-illness outbreak as an incident in which two or more people experience the same illness after eating the same food. *(1)*

Food-contact surface

Any surface or utensil that normally touches food. *See non-food-contact surface. (1, 10, 11)*

Food-grade sealant

A lubricant or oil rated safe for use on kitchen equipment or utensils. *(10)*

Foot-candle

A unit of lighting equal to the illumination one foot from a uniform light source. *(10)*

Four-hour rule

Potentially hazardous foods may not be exposed to the temperature danger zone for more than four hours. *(7)*

Frozen storage

The holding of frozen perishable food items at freezing temperatures, typically 0°F(-18°C) or lower. Frozen storage extends the shelf life of food. Freezers should not be used to freeze refrigerated or room-temperature foods. Also, freezing does not kill most microorganisms. *(6)*

FSIS (Food Safety and Inspection Service)

An agency of the USDA that deals with all food-safety issues concerning meat and poultry, their products, and produce shipped across state boundaries. *(13)*

Fungi

Fungi range in size from microscopic, single-celled microorganisms to very large, multicellular organisms. Molds, yeasts, and mushrooms are examples of fungi. Some microscopic molds and yeasts can damage food quality and cause foodborne illnesses. *(2)*

Garbage

Wet waste matter, usually containing food, that cannot be recycled. *(10)*

Gastrointestinal illness

An illness relating to the stomach or intestine. *(4)*

Glue board

Mice are trapped by the glue on these pest control devices, and then die from exhaustion or lack of water or air. *(12)*

GRAS (Generally Regarded as Safe)

The federal government publishes a list of substances GRAS (generally regarded as safe) that are safe to use in food, such as MSG. *(3)*

HACCP (Hazard Analysis Critical Control Point)

A dynamic system that uses a combination of proper foodhandling procedures, monitoring techniques, and record keeping to help ensure the consistent safety of food. *(1, 9)*

HACCP plan

A written document based on HACCP principles, which describes the procedures a particular establishment will follow to ensure the safety of the food served. *(9)*

Hair restraint

A cap, net, hat, or other device used to cover the hair and/or beard. *(4)*

Hand sanitizer

A liquid used to lower the number of microorganisms on the surface of the skin. *(4)*

Handwashing station

A sink set aside for handwashing only, never used for mixing cleaning chemicals or for washing food or utensils. Handwashing stations must be located in restrooms, and must be located in other convenient locations throughout the food-preparation and warewashing areas as well. *(4, 10)*

Hard water

Water that contains minerals such as calcium and iron in concentrations of more than 120 parts per million (ppm). Concentrations of 61 to 120 ppm are considered moderately hard. *(11)*

Hazard Analysis Critical Control Point

See HACCP. (1, 9)

Hazard analysis

Determining where hazards may occur in the flow of food if care is not taken to prevent or control them. *(9)*

Hazards

Biological, chemical, or physical agents that may cause illness or injury if reduced, controlled, or prevented. *(9)*

Hazards, biological

Pathogenic microorganisms that contaminate food, such as certain bacteria, viruses, parasites, and fungi. Biological hazards also exist in certain plants, mushrooms, and fish in the form of harmful toxins. *See contaminants. (1, 3)*

Hazards, chemical

Chemical substances that may contaminate food, such as pesticides, food additives, preservatives, cleaning supplies, and toxic metals that leach from worn cookware and equipment. *See chemical contamination. (1, 3)*

Hazards, physical

Foreign objects that accidentally get into food and contaminate it, such as dirt, metal staples, and broken glass. *(1, 3)*

HCS (Hazard Communication Standard) of OSHA Also known as Right-to Know or HAZCOM. Standard that requires employers to inform employees of chemical hazards that they may be exposed to in the workplace. *See OSHA. (11)*

Health inspector

City, county, or state employees who conduct inspections in most states. They generally are trained in food safety, sanitation, and in public health principles and methods. Also called sanitarians, health officials, or environmental health specialists. *(13)*

Hearing

An official meeting held between the local health department and an establishment to discuss sanitation inspections and possible enforcement activities. The hearing may be requested by either party. *(13)*

Heat sanitizing

Raising the temperature of a food-contact surface to 165°F (74°C) or above to kill microorganisms. The most common way to heat sanitize tableware, utensils, or equipment is to submerge or spray items with hot water. *(11)*

Heat-treated

Plant foods that have been cooked, partially cooked, or warmed, and have thereby become potentially hazardous foods. *(1)*

Hepatitis A

A disease that causes inflammation of the liver and is transmitted to food by poor personal hygiene or contact with contaminated water. *(4)*

High-acid food

Acidic foods have a ph below 7; high-acid foods have ph values below 4.6; examples include sauerkraut, tomatoes, and citrus products. Such foods, if stored improperly, can cause chemical contamination. *(3)*

Highly alkaline cleaner

Heavy-duty detergents that may also contain a grease-dissolving agent. Highly alkaline detergents are used to remove wax, aged or dried soils, and baked-on grease. Warewashing detergents and floor-scrubbing compounds are also are highly alkaline. *(11)*

Highly susceptible population

(Also high-risk population, immunosuppressed persons.) Groups of people at high risk for foodborne illness due to age or health status, such as very young children, pregnant women, older people, people taking certain medications, and those with certain diseases or weakened immune systems. Establishments with a highly susceptible population may include hospitals, nursing homes, daycare centers. *(1, 7)*

High-risk population

See highly susceptible population. (1)

Histamine

A biological toxin typically formed in temperature-abused scombroid fish that causes a foodborne illness, scombroid poisoning (intoxication). *(3)*

Host

A person, animal, or plant on or in which another organism lives and takes nourishment. *(2)*

Hot-holding equipment

Equipment designed to hold hot foods for service at 140°F (60°C) or higher. Hot-holding equipment includes steam tables, *bains maries,* chafing dishes, double boilers, and heated cabinets. Hot-holding equipment should not be used to reheat foods. *(8)*

Hygrometer

An instrument that measures relative humidity in the air. *(6)*

Ice-water bath

A cooling method: food is divided into small quantities in pans, then the pans are put in ice water in a sink or large pot. *(7)*

Immune system

The bodily system that protects the body from illness. Persons with compromised immune systems are more susceptible to foodborne illness. *See highly susceptible population. (1)*

Infected lesion

An infected wound or injury. *(4)*

Infection

See foodborne infection. (2)

Infestation

The situation when pests overrun or inhabit an establishment in large numbers. *(12)*

Inservice

A short training session held for current employees (those already "in service"). May be used to update employees on a new procedure or practice, to reinforce appropriate behaviors, or to motivate. *(14)*

Integrated pest management

See IPM.

Interstate establishments

Establishments that operate across state borders. Examples include foodservice operations on trains, planes, and ships. *(13)*

Intoxication

See foodborne intoxication.

Iodine sanitizer

A chemical sanitize considered to be one of the most effective. Less corrosive and irritating to the skin than chlorine. *(11)*

IPM (Integrated Pest Management)

A system of preventive measures and control measures to prevent pests from entering your facility, and to eliminate existing pest infestations. *(12)*

Irradiation

Ionizing radiation is used to destroy microorganisms and/or inhibit their growth to prevent foodborne illness. Also known as cold pasteurization. *(2)*

Jaundice
Yellow skin and eyes. A common symptom of Hepatitis A. *(4)*

Job aids
Materials or visual reminders that are used to deliver training content to employees. *(14)*

Lag phase
A phase in bacterial growth. When bacteria are first introduced to a new environment, they may go through an adjustment period where their numbers are stable and they are preparing for growth. To control the growth of bacteria, it is important to prolong the lag phase as long as possible. *(2)*

Leach
Under some circumstances, potentially toxic metals can be leached, or drawn out of equipment, storage containers, and certain tableware, cookware, and glassware, and mixed into foods, causing chemical contamination and foodborne illness. *See high-acid food. (3)*

Lecture
A prepared oral presentation used to deliver content to a group of participants. *(14)*

Log phase
A phase in bacterial growth. When conditions are favorable, bacteria can multiply very rapidly. This type of population growth is called exponential growth, or the log phase of the bacterial growth curve. Foods rapidly become unsafe during the log phase. *(2)*

Magic apron
An ineffective type of training where employees are expected to learn proper procedures by themselves while they work on the job. A false assumption is made that employees will somehow "magically" know the job once they put on the apron. *(14)*

MAP (modified atmosphere packaging)
A sealed package in which the oxygen has been removed or replaced with other gases. This type of packaging extends the food's shelf life and reduces or prevents the growth of aerobic microorganisms. Vacuum packaging and *sous vide* foods are forms of MAP. *(5)*

Master cleaning schedule
A detailed schedule that lists all cleaning tasks in an establishment, when and how they are to be cleaned, and who will do the cleaning. *(11)*

Microbial contamination
See biological contamination. (2)

Microorganisms
Also microbes. Small, living organisms that can be seen only with the aid of a microscope. While not all cause disease, some do. These are called pathogens. Four kinds of microorganisms have the potential to contaminate food and cause foodborne illness: bacteria, viruses, parasites, and fungi. *(2)*

Mobile unit
A portable foodservice facility. Mobile units range from simple vending carts that hold and display prepackaged foods to full field kitchens capable of preparing and cooking elaborate meals. *(8)*

Modified atmosphere packaging
See MAP.

Mold

A type of fungus that causes food spoilage. Some molds produce toxins that can cause foodborne illness. Moldy food should be discarded, unless the mold is a natural part of the food (e.g., some cheeses.) *(2)*

Monitoring

A consistent procedure to observe or measure a CCP to make sure critical limits are being met and to produce a record useful for verification. *See HACCP. (9)*

Monosodium glutamate

See MSG.

MSDS

Material safety data sheets that must be provided with all hazardous materials by manufacturers. These sheets list the chemical and common name of the material, potential physical and health hazards, and directions for safe handling and use. *(11)*

MSG (monosodium glutamate)

MSG is used as a flavor enhancer in many prepackaged foods, and is included on the federal government's GRAS (generally regarded as safe) list of chemicals that are safe to use in food. However, in some people MSG can cause illness. Only the recommended amount of MSG should be used in recipes. *(3)*

Mushroom toxins

Many fungi, such as certain varieties of mushrooms, have naturally occurring toxins (poisons) that may make people ill. *(3)*

National Marine Fisheries Service

Provides a voluntary inspection service for processed fishery products. *(13)*

Nitrites

Chemicals used as flavor enhancers, curing agents, and for the prevention of certain microbial growth in meats. Nitrite-treated meats, especially those that have been over-browned or burned, may produce cancer-causing substances. *(3)*

Non-food-contact surface

Surfaces that do not ordinarily come in contact with food, such as tables, walls, floors, legs of equipment, shelves, drains. *See food-contact surface. (11)*

NSF International

Agency that develops and publishes standards of equipment design. Manufacturers can request an evaluation of their equipment which, if approved, will be listed by NSF International as meeting their standards. *(10)*

Objective

States what a learner will be able to do after training or instruction is finished. *(14)*

Off-site service

Food that is delivered to people away from where it was prepared. Examples are home delivery, carry-outs, or deli trays from restaurants or grocery stores; caterers; home-delivery of meals to elderly or ill persons; vending machines; temporary or mobile units. *(8)*

OSHA (Occupational Safety and Health Administration)
The federal agency that regulates and monitors workplace safety. *(11)*

Outbreak of foodborne illness
An incident in which two or more people experience the same illness after eating the same food. *(1)*

Parasite
An organism that needs to live in or on a host organism to survive. Many types live in animals that humans use for food, such as cows, chickens, hogs, and fish. Foodborne parasites range from microscopic, single-celled organisms to larger, multicellular organisms. Several parasites pose significant hazards to food and water. *(2)*

Paralytic shellfish poisoning (PSP)
Shellfish may contain toxins that occur because of algae upon which they feed. Generally associated with mussels, clams, cockles, and scallops. *(3)*

Pathogenic bacteria
Disease-causing bacteria. *(2)*

Pathogens
Microorganisms that can cause disease in living organisms. *(1, 2, 4)*

PCO (Pest Control Officer)
A licensed or certified technician who implements and monitors pest-control programs for companies that contract for services. *(3, 12)*

Personal hygiene
Sanitary health habits that include keeping body, hair, and teeth clean, maintaining good health, wearing clean clothes, and washing hands regularly, especially when handling food and beverages. *(1, 4, 8)*

Pest control operator
See PCO.

Pesticide
Chemical used to control pests, usually insects. *See contact spray, residual spray. (3, 12)*

pH
A measure of the acidity or alkalinity of a substance. Used to control the growth of microorganisms. The pH scale ranges from 0 to 14.0. A pH of 7.0 is neutral. A pH above 7.0 is alkaline. A pH below 7.0 is acidic. Most potentially hazardous foods have pH values from 4.6 to 7.0, that is, they are slightly acidic. *(2)*

Physical contamination
The accidental introduction of foreign particles such as dirt, hair, and glass or metal particles into food. Some physical hazards occur naturally, such as bones in fish or chicken. *See contamination. (3)*

Plant toxins
Some plants have natural toxins that may make some people ill. People may ingest plant toxins by eating toxic plants themselves, or by eating products from animals that have ingested plant toxins. *(3)*

Pooled eggs
Eggs that have been cracked open and combined in a container for quantity cooking. *(7)*

Porosity
The extent to which water and other liquids are absorbed by a substance. Term used in relation to flooring material. *(10)*

Potable water
Water that is safe to drink; an approved water supply. *(10)*

Potentially hazardous foods (PHF)
Foods in which microorganisms can rapidly grow. Potentially hazardous foods often have a history of being involved in foodborne illness outbreaks, have potential for contamination due to methods used to produce and process them, and have

characteristics that generally allow microorganisms to thrive. They are often warm, high in protein, and chemically neutral or slightly acidic. *(1, 2)*

Prerequisite programs

Procedures that protect food from contamination, minimize microbial growth, and ensure the proper functioning of equipment. Also called standard operating procedures (SOPs). *(9)*

Presentation

A key element of training, the delivery of content to the learner through a variety of methods. *(14)*

Preventive measures (versus control methods)

Steps that can be taken to prevent pests from entering a building, such as keeping the facility clean and sanitary and maintaining the building properly. Control measures focus on controlling and minimizing the problems from those that do enter. *(12)*

Pulper

Pulpers grind food and other waste into small parts that are flushed with water. The water is then removed so that the processed solid wastes weigh less and are more compact for easier disposal. *(10)*

Quarry tile

A type of stone tile, generally reddish-brown in color, often used in public restrooms or high-soil areas. *(10)*

Quaternary ammonium (quats)

A group of sanitizers all having the same basic chemical structure. Quats are generally nontoxic, noncorrosive, and stable when exposed to heat but may not kill certain types of microorganisms. *See chlorine, iodine. (11)*

Ready-to-eat foods

Properly cooked foods, and raw, washed, cut, and whole fruits and vegetables (including those that have had their rinds, peels, husks, or shells removed). The FDA identifies most ready-to-eat foods as potentially hazardous. *(1)*

Reasonable care defense

A defense against a food-related lawsuit; if you can prove that your establishment did everything that could be reasonably expected to ensure that the food served was safe. *(1)*

Refrigerated storage

Short-term holding of fresh, perishable, and potentially hazardous food items at internal temperatures of 41°F (5°C) or lower, but above freezing, to slow the growth of microorganisms. *(6)*

Regulation

Details how a law will be put into effect by the government agency responsible for that regulation. *(13)*

Residual spray

A type of pesticide spray that leaves a film behind, which eventually kills insects that walk across it. Residue sprays are typically sprayed by a PCO around the perimeter of a facility on a monthly basis to provide continuous protection against insects that might enter. *See contact spray. (12)*

Resiliency

The ability of something to react to a shock without breaking or cracking. Term usually used in relation to flooring materials. *(10)*

Role-plays

A method in which training participants enact a situation in order to try out new skills or apply what has been learned. *(14)*

Sanitary, sanitation

A sanitary or sanitized object or surface is free from harmful levels of disease-causing organisms and other harmful contaminants. To be effective, sanitation must follow effective cleaning. *(1, 11)*

Sanitizer

A compound used in the cleaning process that lowers the level of microorganisms on a surface to levels that do not cause illness. *See chlorine, iodine, quaternary ammonium. (10, 11)*

Sanitizing hand lotion

A germicidal lotion or hand dip to be used after proper handwashing. *(4)*

Sanitizing

The process of reducing the number of microorganisms on a surface to safe levels. *(11)*

Scale

The buildup of mineral deposits in pipes or equipment from water hardness. Scale is a problem for equipment that uses hot water, boiling water, or steam, for example warewashing machines and steam tables. *(11)*

Scombroid poisoning (intoxication)

A foodborne illness caused by histamine, a biological toxin. When scombroid fish, such as tuna, mackerel, bluefish, skipjack, and bonito, are time-temperature abused, the bacteria associated with them produce the toxin histamine, which causes scombroid intoxication. Scombroid intoxication has also been associated with some nonscombroid fish. *(3)*

Service sink

Sink used exclusively for cleaning mops and disposing of waste water. At least one service sink or one curbed drain area is required in an establishment. *(10)*

Shelf life

A recommended period of time that a food may be stored and remain suitable for use. *(6)*

Shellstock identification tags

The FDA requires that shipments of live molluscan shellfish carry shellstock identification tags to identify the shipper, the date of shipment, date of receipt, and other information. The tags should remain attached to the containers the shellfish came in until the container is empty. Operators then must keep the tags on file for ninety days after the shellfish has been used. Never mix shellfish from one shipment with another. *(5)*

Single-use items

Another name for disposable eating utensils and tableware. These may be used once only. They are generally made from paper, plastic, wood, or aluminum foil. *(8, 10)*

Single-use paper towel

Paper towels designed to be used once, then discarded. Use minimizes risk of cross-contamination. *(4)*

Slacking

The process of gradually thawing frozen food in preparation for deep-fat frying or to allow even heating during cooking. *(7)*

Sneeze guard

A food shield used to prevent contamination of food in salad bars and other self-serve food bars. It is placed fourteen to forty-eight inches above the food, in a direct line between food on display and the mouth and nose of a person of average height. *(8)*

Soap

A cleaning agent used in restaurants and establishments primarily for handwashing. *(4)*

Soil

A general term for foreign material present on a surface that requires removal. In establishments, soil might consist of food debris on tableware; or wax, aged or dried substances; stains and baked-on grease; or scale, rust, and tarnish on other surfaces. *See clean, cleaning, cleaning agent. (11)*

Solid waste

Dry bulky trash that can be recycled, including glass, plastic, paper, and cardboard. *(10)*

Solvents

Solvent cleaners, often called degreasers, are alkaline detergents that contain a grease-dissolving agent. These cleaners work well in areas where grease has been burned on. Solvents are usually effective only at full strength. *(11)*

Source

The source of a microorganism is an item or organism (often a human being) where the microorganism usually resides. Sources may be referred to as vehicles, carriers, or hosts. *(2)*

Sous vide

Food processed by this method is vacuum-packed in individual pouches, partially or fully cooked, and then chilled. These foods are often heated for service in the establishment. *(5)*

Spoilage microorganism

Foodborne microorganisms that cause food to spoil. These microorganisms typically do not cause foodborne illness. *(2)*

Spore

Some bacteria form spores, thick-walled formations within the bacterial cell that serve as a means of protection against unfavorable environmental conditions (high or low temperatures, low moisture, high acidity, and exposure to sanitizing solutions). The spore itself does not reproduce, but it is capable of becoming a vegetative organism when conditions again become favorable. Also the reproductive cell of a fungus. *(2)*

Standard

A measure used as a comparison for quality to determine the degree or level that a requirement should meet. *(13)*

Stationary phase

A phase of bacterial growth where just as many bacteria are growing as dying. Follows log phase of bacterial growth. *(2)*

Studs, joists, rafters

Internal support beams and framework for a building. *(10)*

Sulfites

Preservatives added to some foods to preserve freshness and color. Some people are allergic to sulfites, and some states forbid restaurants to add them to foods. *(3)*

Suppliers, approved (certified)

Reputable and reliable suppliers whose products and practices meet federal and local standards. Reputable suppliers deliver correctly packaged food in adequately refrigerated delivery trucks, train their employees in food-safety practices, adjust delivery schedules to meet your needs, and allow you to inspect their delivery vehicles and production facilities. *(1)*

Surfactants

Agents in all detergents that reduce surface tension between soil and the item being cleaned, making it easier to remove soil. *(11)*

Suspension

A temporary withdrawal of a health department permit for an establishment to operate because the health inspector determined the facility posed an immediate health hazard. Correction of all violations and a satisfactory re-inspection may be required for reinstatement of the permit. *(13)*

Technology-based training

Training programs that are delivered through a computer or other technology. *(14)*

Temperature abuse

Any time potentially hazardous food is exposed to the temperature danger zone of 41°F (5°C) to 140°F (60°C). Foods being prepared or cooked should pass through the temperature danger zone as quickly as possible. Temperature abuse of potentially hazardous foods can cause the rapid growth of microorganisms, potentially causing foodborne illness. *(5)*

Temperature danger zone

The temperature range between 41°F and 140°F (5°C to 60°C) within which most foodborne microorganisms rapidly grow and reproduce. Potentially hazardous foods must not be left in the temperature danger zone for more than four hours total. *(2, 5, 7)*

Temporary operation (unit)

An establishment that operates in one location, for no more than fourteen consecutive days, in conjunction with a special event or celebration. Temporary units usually serve prepackaged foods or foods that require only limited preparation. *(8, 13)*

Terrazzo

A mixture of marble chips and portland cement used for flooring. An attractive, nonporous, nonresilient surface good for use in the back of the house, dining rooms, and restrooms. *(10)*

Time-temperature indicator (TTI)

Time-temperature indicators are used to determine if a product has been time-temperature abused. TTIs are most often placed on temperature-sensitive foods by packers or processors. *(5)*

Time-temperature monitoring

To maximize food safety, and minimize growth of microorganisms in food, time and temperature must be controlled and monitored throughout the flow of food, including receiving, storage, preparation, cooking, holding, cooling, and reheating. *(1)*

Toxigenic bacteria

Bacteria that produce toxins. *(2)*

Toxin-mediated infection

A foodborne illness that results when a microorganism from ingested food grows in the intestinal tract and then produces toxins that cause illness. *(2)*

Toxins

Some microorganisms produce toxins, or poisons, in food that is then ingested. This is called a foodborne intoxication. Thus, a person does not need to ingest live organisms to be become ill, just the toxins, many of which are not destroyed by cooking. *(2)*

Training need

A gap between what your employees are required to know to perform their job and what they actually know. *(14)*

Training program

A structured sequence of events that leads to learning. *(14)*

TTI

See time-temperature indicator. (5)

Tumble chiller

Equipment designed to cool food quickly. Prepackaged food is placed into a drum which rotates inside a reservoir of chilled water. The tumbling action increases the time the food is chilled. *(10)*

Two-stage cooling method

Cooked foods must be cooled from 140°F (60°C) to 70°F (21°C) within two hours and from 70°F (21°C) to below 41°F (5°C) in an additional four hours for a total cooling time of six hours. *(7)*

UL (Underwriters Laboratories)
An agency that lists equipment that meets NSF International's standards. UL's particular emphasis is on electrical safety requirements; it also has a sanitation mark. *(See also NSF International). (10)*

USDA (U.S. Department of Agriculture)
A government agency responsible for the inspection and grading of meats, meat products, poultry, dairy products, eggs and egg products, and fruits and vegetables shipped across state lines. *(13)*

USPHS (U.S. Public Health Service)
A federal government agency that inspects cruise ships that cross international borders. *(13)*

Vacuum breaker
A backflow prevention device. *See backflow. (10)*

Vacuum packaging
Food packaging in which all of the air is removed around the product. *(5)*

Vegetative microorganisms
Bacteria reproduce by splitting in two. Vegetative microorganisms are those bacteria actively splitting in two (growing). *(2)*

Vehicle
Food, plant, person, or object by which pathogenic microorganisms are transferred to another food, person, plant, or object. *(2)*

Vending machine
A type of off-site delivery of food. Vending operators must protect foods from contamination and temperature abuse during transport, delivery, and service. Vending machines that dispense chilled or hot food must have automatic cutoff controls that prevent foods from being dispensed if the temperature stays in the danger zone for a certain amount of time. *(8)*

Verification
Activities that determine whether a HACCP system is working as intended. *(9)*

Virus
The smallest of the microbial contaminants. Viruses are not complete cells and are not considered to be living organisms. They rely on a living host and do not reproduce in food. Some survive freezing and boiling. They contaminate food via poor personal hygiene, and contaminated food and water supplies. *See bacteria, fungi, parasites. (2)*

Warm-air hand dryer
Also hot-air, forced-air. An acceptable method of drying hands at a handwashing station. Some jurisdictions require the use of these machines instead of towels. *(4)*

Warranty of sale
The rules stating how food must be handled in an establishment. *(1)*

Water activity (a_w)
The amount of moisture in food available for the growth of microorganisms. Potentially hazardous foods have water-activity values of 0.85 or above. *(2)*

Index

G